AF255290

THEORY OF EQUATIONS.

DUBLIN UNIVERSITY PRESS SERIES.

THE
THEORY OF EQUATIONS:

WITH

AN INTRODUCTION TO THE THEORY OF BINARY ALGEBRAIC FORMS.

BY

WILLIAM SNOW BURNSIDE, M.A.,

FELLOW OF TRINITY COLLEGE, DUBLIN;
ERASMUS SMITH'S PROFESSOR OF MATHEMATICS IN
THE UNIVERSITY OF DUBLIN:

AND

ARTHUR WILLIAM PANTON, M.A.,

FELLOW AND TUTOR, TRINITY COLLEGE, DUBLIN.

DUBLIN: HODGES, FIGGIS, & CO., GRAFTON-STREET.
LONDON: LONGMANS, GREEN, & CO., PATERNOSTER-ROW.
1881.

DUBLIN:
PRINTED AT THE UNIVERSITY PRESS.

PREFACE.

WE have endeavoured in the present work to combine some of the modern developments of Higher Algebra with the subjects usually included in works on the Theory of Equations. The first ten Chapters contain all the propositions ordinarily found in elementary treatises on the subject. In these Chapters we have not hesitated to employ the more modern notation wherever it appeared that greater simplicity or comprehensiveness could be thereby obtained.

Regarding the algebraical and the numerical solution of equations as essentially distinct problems, we have purposely omitted in Chap. VI. numerical examples in illustration of the modes of solution there given of the cubic and biquadratic equations. Such examples do not render clearer the conception of an algebraical solution ; and, for practical purposes, the algebraical formula may be regarded as almost useless in the case of equations of a degree higher than the second.

In the treatment of Elimination and Linear Transformation, as well as in the more advanced treatment of Symmetric Functions, a knowledge of Determinants is indispensable. We have found it necessary, therefore, to give a Chapter on this subject. It has been our aim to make this Chapter as simple and intelligible as possible to the beginner; and at the same time to omit no proposition which might be found useful in the application of this calculus. . For many of the examples in this Chapter, as well as in other parts of the work, we are indebted to the kindness of Mr. Cathcart, Fellow of Trinity College.

We have approached the consideration of Covariants and Invariants through the medium of the functions of the differences of the roots of equations—this appearing to us the simplest mode of presenting the subject to beginners. We have attempted at the same time to show how this mode of treatment may be brought into harmony with the more general problem of the linear transformation of algebraic forms. In the chapters on this subject we have confined our attention to the quadratic, cubic, and quartic; regarding any complete discussion of the covariants and invariants of higher binary forms as too difficult for a work like the present.

Of the works which have afforded us assistance in the more elementary part of the subject, we wish to mention particularly the *Traité d'Algèbre* of M. Bertrand, and the writings of the late Professor Young* of Belfast, which have contributed so much to extend and simplify the analysis and solution of numerical equations.

In the more advanced portions of the subject we are indebted mainly, among published works, to the *Lessons Introductory to the Modern Higher Algebra* of Dr. Salmon, and the *Theorie der binären algebraischen Formen* of Clebsch; and in some degree to the *Théorie des Formes binaires* of the

---

* *Theory and Solution of Algebraical Equations*, London, 1835; *Analysis and Solution of Cubic and Biquadratic Equations*, London, 1842; and *Theory and Solution of Algebraical Equations of the Higher Orders*, London, 1843.

Chev. F. Faà De Bruno. We must record also our obligations in this department of the subject to Mr. Michael Roberts, from whose Papers in the *Quarterly Journal* and other periodicals, and from whose professorial lectures in the University of Dublin, very great assistance has been derived. Many of the examples also are taken from Papers set by him at the University Examinations.

In the chapter on the Complex Variable we have followed closely the treatment of imaginary quantities given by M. Briot in his *Leçons d'Algèbre*.

In connexion with various parts of the subject several other works have been consulted, among which may be mentioned the treatises on Algebra by Serret, Meyer Hirsch, and Rubini, and papers in the mathematical journals by Boole, Cayley, Sylvester, Hermite, and others.

We have, in the last place, to express our thanks to Mr. Robert Graham, of Trinity College, Dublin, who has read the proof sheets, and verified most of the examples. His thorough acquaintance with the subject has been invaluable to us, and many improvements throughout the work are owing to suggestions made by him.

TRINITY COLLEGE,
 September, 1881.

Note.—The first ten Chapters of this work may be regarded as forming an elementary course. In reading these Chapters for the first time, Students are recommended to omit Art. 53 of Chap. V., and to confine their attention in Chap. VI. to Arts 55, 56, 57, 61, 62, and 63.

TABLE OF CONTENTS.

INTRODUCTION.

CHAPTER I.

GENERAL PROPERTIES OF POLYNOMIALS.

CHAPTER II.

GENERAL PROPERTIES OF EQUATIONS.

CHAPTER III.

RELATIONS BETWEEN THE ROOTS AND COEFFICIENTS OF EQUATIONS, WITH APPLICATIONS TO SYMMETRIC FUNCTIONS OF THE ROOTS.

CHAPTER IV.

TRANSFORMATION OF EQUATIONS.

CHAPTER V.

SOLUTION OF RECIPROCAL AND BINOMIAL EQUATIONS.

CHAPTER VI.

ALGEBRAIC SOLUTION OF THE CUBIC AND BIQUADRATIC.

CHAPTER VII.

PROPERTIES OF THE DERIVED FUNCTIONS.

CHAPTER VIII.

LIMITS OF THE ROOTS OF EQUATIONS.

CHAPTER IX.

SEPARATION OF THE ROOTS OF EQUATIONS.

CHAPTER X.

SOLUTION OF NUMERICAL EQUATIONS.

CHAPTER XI.

DETERMINANTS.

CHAPTER XII.

SYMMETRIC FUNCTIONS OF THE ROOTS.

CHAPTER XIII.

ELIMINATION.

CHAPTER XIV.

COVARIANTS AND INVARIANTS.

CHAPTER XV.

COVARIANTS AND INVARIANTS OF THE QUADRATIC, CUBIC, AND QUARTIC.

CHAPTER XVI.

TRANSFORMATIONS.

CHAPTER XVII.

THE COMPLEX VARIABLE.

NOTES.

ERRATA.

Page 46, line 6, supply θ_1 before $\sqrt[3]{\overline{P}}$.

,, 171, second last line, for $-h$, read $1 - h$.

,, 255, line 3, supply a to the suffix 1.

,, 334, line 4, supply the coefficient 4 before $(a_1 b_2)(b_1 c_2)$, and make corresponding corrections in the subsequent work.

,, 351, line 20, for Ω, read 4Ω.

,, 352, lines 10 and 12, for U, read $-U$.

,, 341, line 2, for $-p_4 p_3$, read $-p_1 p_3$.

THEORY OF EQUATIONS.

INTRODUCTION.

1. **Definitions.**—Any mathematical expression involving a quantity is called a *function* of that quantity.

We shall be employed mainly with such algebraical functions as are *rational* and *integral*. By a *rational* function of a quantity is meant one which contains that quantity in a rational form only ; that is, a form free from fractional indices or radical signs. By an *integral* function of a quantity is meant one in which the quantity enters in an integral form only; that is, never in the denominator of a fraction. The following expression, for example, in which n is a positive integer, is a *rational* and *integral algebraical function* of x :—

$$ax^n + bx^{n-1} + cx^{n-2} + \ldots\ldots + kx + l.$$

Here it is to be observed that our definition has reference to the quantity x only, of which the expression is a function. The several coefficients a, b, c, &c., may be irrational or fractional, and the function still remain rational and integral in x.

A function of x is represented for brevity by $F(x), f(x), \phi(x)$, or some similar symbol.

The name *polynomial* is given to the algebraical function to express the fact that it is constituted of a number of terms containing different powers of x connected by the signs plus or minus.   For certain values of the variable quantity x, one given polynomial may become equal to another differently constituted.   The algebraical expression of such a relation is called an *equation;* and any value of the quantity x which satisfies this equation is called a *root* of the equation.   The determination of all possible roots constitutes the complete *solution of the equation.*

It is obvious that, by bringing all the terms to one side, we may arrange any equation according to descending powers of x in the following manner :—

$$a_0 x^n + a_1 x^{n-1} + a_2 x^{n-2} + \ldots + a_{n-1} x + a_n = 0.$$

The highest power of x in this equation being n, it is said to be an equation of the nth *degree* in x.   For an equation of the nth degree we shall, in general, employ the form here written. The suffix attached to the letter a indicates the power of x which each coefficient accompanies, the sum of the exponent of x and the suffix of a being equal to n for each term.   An equation is not altered if all its terms be divided by any quantity.   We may thus, if we please, dividing by a_0, make the coefficient of x^n in the above equation equal to unity.   We shall find it often convenient to make this supposition ; and in such cases we shall write the equation in the form

$$x^n + p_1 x^{n-1} + p_2 x^{n-2} + \ldots + p_{n-1} x + p_n = 0.$$

An equation is said to be *complete* when it contains terms involving x in all its powers from n to 0, and *incomplete* when some of the terms are absent; or, in other words, when some of the coefficients p_1, p_2, &c., are equal to zero.   The term p_n, which does not contain x, is called the *absolute term*.   An equation is *numerical,* or *algebraical,* according as its coefficients are numbers, or algebraical symbols.

2. **Numerical and Algebraical Equations.**—In many researches in both mathematical and physical science the final mathematical problem presents itself in the form of an equation on whose solution that of the problem depends.   It was natural,

therefore, that the attention of mathematicians should have been at an early stage in the history of the science directed towards inquiries of this nature. The science of the Theory of Equations, as it now stands, has grown out of the successive attempts of mathematicians to discover general methods for the solution of equations of any degree. When the coefficients of an equation are given numbers, the problem is to determine a numerical value, or perhaps several different numerical values, which will satisfy the equation. In this branch of the science very great progress has been made; and the best methods hitherto advanced for the discovery, either exactly or approximately, of the numerical values of the roots will be explained in their proper places in this work.

Equal progress has not been made in the general solution of equations whose coefficients are algebraical symbols. The student is aware that the root of an equation of the second degree, whose coefficients are such symbols, may be expressed in terms of these coefficients in a general formula; and that the numerical roots of any particular numerical equation may be obtained by substituting in this formula the particular numbers for the symbols. It was natural to inquire whether it was possible to discover any such formula in the case of equations of higher degrees. Such results have been attained in the case of equations of the third and fourth degrees. It will be shown that in certain cases these formulas fail to give us the solution of a numerical equation by substitution of the numerical coefficients for the general symbols, and are, therefore, in this respect, inferior to the corresponding algebraical solution of the quadratic.

Many attempts have been made to arrive at similar general formulas for equations of the fifth and higher degrees; but it may now be regarded as established by the researches of modern analysts that it is not possible by means of radical signs, and other signs of operation employed in common algebra, to express the root of an equation of the fifth or any higher degree in terms of the coefficients.

3. **Polynomials.**—One important object of the science of the Theory of Equations is thus the discovery of those values of the quantity x which give to the polynomial $f(x)$ the particular value zero. In attempting to discover such values of the variable we shall be led into many inquiries concerning the values assumed by the polynomial for other different values of x. We shall, in fact, see in the next Chapter that, corresponding to a continuous series of values of x varying from an infinitely great negative quantity $(-\infty)$ to an infinitely great positive quantity $(+\infty)$, $f(x)$ will assume also values continuously varying. The study of such variations is a very important part of the subject on which we are engaged. The general solution of numerical equations is, in fact, a tentative process; and by examining the values assumed by the polynomial for certain arbitrarily assumed values of the variable, we shall be led, if not to the root itself, at least to an indication of the neighbourhood in which it exists, and within which our further approximation must be carried on.

A polynomial is sometimes called a *quantic*. It is convenient to have distinct names for quantics of the 2nd, 3rd, 4th, 5th, &c., degrees. That of the 2nd degree is called a *quadratic* or *quadric*; that of the 3rd is called a *cubic*; that of the 4th a *quartic* or *biquadratic*; that of the 5th a *quintic*; and so on. The equations obtained by equating these quantics to zero are called *quadratic*, *cubic*, *biquadratic*, &c., *equations*, respectively.

CHAPTER I.

4. IN tracing the changes of value of a polynomial corresponding to changes in the variable, we shall first inquire what terms in the polynomial are most important when values very great or very small are assigned to x. This inquiry will form the subject of the present and succeeding Articles.

Writing the polynomial in the form

$$a_0 x^n \left\{ 1 + \frac{a_1}{a_0} \frac{1}{x} + \frac{a_2}{a_0} \frac{1}{x^2} + \ldots + \frac{a_{n-1}}{a_0} \frac{1}{x^{n-1}} + \frac{a_n}{a_0} \frac{1}{x^n} \right\},$$

it is plain that its value tends to become equal to $a_0 x^n$, as x tends towards ∞. We proceed, then, to inquire what is the value of x nearest to zero which will have the effect of making the term $a_0 x^n$ exceed the sum of all the others.

Theorem.—*If in the polynomial*

$$a_0 x^n + a_1 x^{n-1} + a_2 x^{n-2} + \ldots + a_{n-1} x + a_n$$

the value $\dfrac{a_k}{a_0} + 1$, *or any greater value, be substituted for* x, *where* a_k *is that one of the coefficients* a_1, a_2, $\ldots$ a_n *whose numerical value is greatest, irrespective of sign, the term containing the highest power of* x *will exceed the sum of all the terms which follow.*

The inequality

$$a_0 x^n > a_1 x^{n-1} + a_2 x^{n-2} + \ldots + a_{n-1} x + a_n$$

is satisfied by the following:—

$$a_0 x^n > a_k \left(x^{n-1} + x^{n-2} + \ldots + x + 1 \right),$$

where a_k is the greatest among the coefficients

$$a_1, a_2, \ldots a_{n-1}, a_n,$$

without regard to sign.

This leads, summing the geometric series, to the condition

$$a_0 x^n > a_k \frac{x^n - 1}{x - 1}, \quad \text{or} \quad x^n > \frac{a_k}{a_0 (x - 1)} (x^n - 1),$$

which is satisfied if $a_0 (x - 1)$ be > or $= a_k,$

$$\text{or} \qquad x > \text{or} = \frac{a_k}{a_0} + 1 ;$$

which proves the theorem.

Remark.—This theorem is useful in furnishing us, when the coefficients of the polynomial are given numbers, with a number such that when x receives values nearer to $+ \infty$ the polynomial will preserve constantly a positive sign. If we change the sign of x, the first term will retain its sign if n be even, and will become negative if n be odd; so that we are furnished by the same theorem with a negative value of x, such that for any value nearer to $- \infty$, the polynomial will retain constantly a positive sign if n be even, and a negative sign if n be odd. The constitution of the polynomial is, in general, such that limits much nearer to zero than those here arrived at can be found beyond which the function preserves the same sign; for in the above proof we have taken the most unfavourable case, *i.e.* where all the coefficients except the first are negative, and each equal to a_k; whereas in general the coefficients have varying values, positive, negative, or zero. Several theorems, having for their object the discovery of such closer limits, will be given in a subsequent Chapter.

5. We now proceed to determine what is the most important term in a polynomial when the value of x is indefinitely diminished, and what is the greatest value of x for which that term exceeds all the others.

It is evident that the polynomial tends to become equal to a_n as x tends towards zero. We inquire what is the greatest value of x which gives a_n the preponderance.

Theorem.—*If in the polynomial*

$$a_0 x^n + a_1 x^{n-1} + \ldots + a_{n-1} x + a_n$$

the value $\dfrac{a_n}{a_n + a_k}$, *or any smaller value, be substituted for* x, *where* a_k
is the greatest coefficient exclusive of a_n, *the term* a_n *will be numerically greater than the sum of all the others.*

To prove this, let $x = \dfrac{1}{y}$; then by the theorem of Art. 4, a_k being now the greatest among the coefficients a_0, a_1, ... a_{n-1}, without regard to sign, the value $\dfrac{a_k}{a_n} + 1$, or any greater value of y, will make

$$a_n y^n > a_{n-1} y^{n-1} + a_{n-2} y^{n-2} + \ldots + a_1 y + a_0,$$

that is, $\qquad a_n > a_{n-1} \dfrac{1}{y} + a_{n-2} \dfrac{1}{y^2} + \ldots a_0 \dfrac{1}{y^n} ;$

hence the value $\dfrac{a_n}{a_n + a_k}$, or any less value of x, will make

$$a_n > a_{n-1} x + a_{n-2} x^2 + \ldots + a_0 x^n.$$

This proposition is often stated in a different manner, as follows:—*Values so small may be assigned to* x *as to make the polynomial*

$$a_{n-1} x + a_{n-2} x^2 + \ldots + a_0 x^n$$

less than any assigned quantity.

This is evident, as in the above proof a_n may be taken to be the assigned quantity.

There is also another useful statement of the theorem, as follows:—*When the variable* x *receives a very small value, the sign of the polynomial*

$$a_{n-1} x + a_{n-2} x^2 + \ldots + a_0 x^n$$

is the same as the sign of its first term $a_{n-1} x$.

This appears by writing the expression in the form

$$x \left\{ a_{n-1} + a_{n-2} x + \ldots a_0 x^{n-1} \right\} ;$$

for when a value sufficiently small is given to x, the numerical value of the term a_{n-1} exceeds the sum of the other terms of the expression within the brackets, and the sign of that expression will consequently depend on the sign of a_{n-1}.

6. Change of form of a Polynomial corresponding to an increase or diminution of the Variable. Derived Functions.—We shall now examine the form assumed by the polynomial when $x + h$ is substituted for x.

If we suppose h essentially positive, the resulting form will correspond to an increase of the variable; and by changing the sign of h in the result we obtain the form corresponding to a diminution of x.

When x is changed to $x + h$, $f(x)$ becomes $f(x + h)$, or

$$a_0(x + h)^n + a_1(x+h)^{n-1} + a_2(x+h)^{n-2} + \ldots + a_{n-2}(x + h)^2 + a_{n-1}(x+h) + a_n.$$

Let each term of this expression be expanded by the binomial theorem, and the result arranged according to ascending powers of h. We then have

$$a_0 x^n + a_1 x^{n-1} + a_2 x^{n-2} + \ldots + a_{n-2} x^2 + a_{n-1} x + a_n$$

$$+ h \left\{ n a_0 x^{n-1} + (n - 1) a_1 x^{n-2} + (n - 2) a_2 x^{n-3} + \ldots + 2 a_{n-2} x + a_{n-1} \right\}$$

$$+ \frac{h^2}{1 \cdot 2} \left\{ n (n - 1) a_0 x^{n-2} + (n - 1)(n - 2) a_1 x^{n-3} + \ldots + 2 a_{n-2} \right\}$$

$$+ \quad \cdots \cdots$$

$$+ \frac{h^n}{1 \cdot 2 \cdot 3 \ldots n} \left\{ n \cdot n - 1 \ldots 2 \cdot 1 \right\} a_0.$$

It will be observed that the part of this expression independent of h is $f(x)$: a result which is obvious *à priori;* and that the subsequent coefficients of the different powers of h are functions of x of degrees diminishing by unity. It will further be observed that the coefficient of h may be derived from $f(x)$ in the following manner:—Let each term in $f(x)$ be multiplied by the exponent of x in that term, and let the exponent of x in the term be diminished by unity, the sign being retained; the sum of all the terms of $f(x)$ treated in this way will constitute a polynomial of dimensions one degree lower than those of $f(x)$. This polynomial is called the *first derived function* of $f(x)$. It is usual to represent this function by the notation $f'(x)$.

The coefficient of $\dfrac{h^2}{1.2}$ may be derived from $f'(x)$ by a process the same as that employed in deriving $f'(x)$ from $f(x)$, or by the operation twice performed on $f(x)$. It is represented by $f''(x)$. Thus $f''(x)$ is called the *second derived function* of $f(x)$; and in like manner the succeeding coefficients may all be derived by successive operations of this character; so that, employing the notation here indicated, we may write the result thus :—

$$f(x+h) = f(x) + f'(x)h + \frac{f''(x)}{1.2} h^2 + \frac{f'''(x)}{1.2.3} h^3 + \ldots + a_0 h^n.$$

We may observe that, since the interchange of x and h does not alter $f(x+h)$, the expansion may also be written in the form

$$f(x+h) = f(h) + f'(h) x + \frac{f''(h)}{1.2} x^2 + \frac{f'''(h)}{1.2.3} x^3 + \ldots + a_0 x^n.$$

We shall in general employ this notation; but on certain occasions it will be found more convenient when dealing with the successive derived functions to use suffixes instead of the accents here employed, the latter notation becoming cumbrous when we go beyond the first two or three derived functions. The above expansion will then be written as follows :—

$$f(x+h) = f(x) + f_1(x) h + f_2(x) \frac{h^2}{1.2} + \ldots + f_r(x) \frac{h^r}{1.2.3\ldots r} + \ldots$$

EXAMPLE.

Find what the polynomial $4x^3 + 6x^2 - 7x + 4$ becomes when x is changed into $x + h$.

Here

$$\begin{aligned}
f(x) &= 4x^3 + 6x^2 - 7x + 4,\\
f'(x) &= 12x^2 + 12x - 7,\\
f''(x) &= 24x + 12,\\
f'''(x) &= 24;
\end{aligned}$$

and the result is

$$4x^3 + 6x^2 - 7x + 4 + (12x^2 + 12x - 7) h + (24x + 12) \frac{h^2}{1.2} + 24 \frac{h^3}{1.2.3}.$$

This example shows how the absolute term of each function disappears when its derived is formed, the degree of the function diminishing, till finally $f_n(x)$ is reached, which is equal in general to $\{n . n-1 . n-2 \ldots 2 . 1\} a_0$; in this case $f_3(x) = \{3 . 2 . 1\} 4$.

7. Continuity of a Rational Integral Function of x.
—If the value of x be made to vary, by indefinitely small incre-
ments, from one quantity a to a greater quantity b, it becomes
an inquiry how the polynomial $f(x)$ varies at the same time.
The object of the present Article is to show that $f(x)$ passes at
the same time through all values between $f(a)$ and $f(b)$; in
other words, that it varies continuously along with x. Let x be
increased from a to $a + h$. The corresponding increment of
$f(x)$ is

$$f(a + h) - f(a),$$

which is equal, by Art. 6, to

$$f'(a)h + f''(a)\frac{h^2}{1 \cdot 2} + \ldots + a_0 h^n,$$

in which all the coefficients $f'(a)$, $f''(a)$, &c., are finite quantities.
Now, by the theorem of Art. 5, this latter expression may, by
taking h small enough, be made to assume a value less than any
assigned quantity; so that the difference between $f(a + h)$ and
$f(a)$ may be made as small as we please, and will ultimately
vanish with h. The same is true during all stages of the
variation of x from a to b; thus the continuity of the function
$f(x)$ is established.

It is to be observed that it is not here proved that $f(x)$
increases continuously from $f(a)$ to $f(b)$. It may either increase
or diminish, or at one time increase, and at another diminish;
but our proof shows that it cannot pass *per saltum* from one
value to another. The sign of $f'(a)$ will determine whether $f(x)$
is increasing or diminishing; for we know by Art. 5 that when
h is small enough the sign of the total increment will depend on
that of $f'(a)h$. We thus observe that *when $f'(a)$ is positive $f(x)$
is increasing with x; and when $f'(a)$ is negative $f(x)$ is diminishing
as x increases.*

**8. Form of the Quotient and Remainder when a
Polynomial is divided by a Binomial.**—Let the quotient,
when

$$a_0 x^n + a_1 x^{n-1} + a_2 x^{n-2} + \ldots + a_{n-1} x + a_n$$

is divided by $x - h$, be

$$b_0 x^{n-1} + b_1 x^{n-2} + \ldots + b_{n-2} x + b_{n-1}.$$

This we shall represent by Q, and the remainder by R. We have then the following equation:

$$f(x) \equiv (x - h)\, Q + R.$$

The meaning of this equation is, that when Q is multiplied by $x - h$, and R added, the result must be *identical*, term for term, with $f(x)$. In order to distinguish identical equations of this kind from others, it will often be found convenient to use the symbol here employed in place of the usual symbol of equality. The right-hand side of the identity is

$$b_0 x^n + \left.\begin{matrix} b_1 \\ -hb_0 \end{matrix}\right\} x^{n-1} + \left.\begin{matrix} b_2 \\ -hb_1 \end{matrix}\right\} x^{n-2} + \ldots + \left.\begin{matrix} b_{n-1} \\ -hb_{n-2} \end{matrix}\right\} x + \begin{matrix} R \\ -hb_{n-1} \end{matrix}.$$

Equating the coefficients of x on both sides, we get the following series of equations to determine $b_0,\ b_1,\ b_2, \ldots b_{n-1},\ R$:—

$$
\begin{aligned}
b_0 &= a_0, \\
b_1 &= b_0 h + a_1, \\
b_2 &= b_1 h + a_2, \\
b_3 &= b_2 h + a_3, \\
&\ \cdot \quad \cdot \quad \cdot \quad \cdot \\
b_{n-1} &= b_{n-2} h + a_{n-1}, \\
R &= b_{n-1} h + a_n.
\end{aligned}
$$

These equations supply a ready method of calculating in succession the coefficients $b_0,\ b_1,$ &c., of the quotient, and the remainder. For this purpose we write the series of operations in the following manner :—

$$
\begin{array}{ccccccc}
a_0, & a_1, & a_2, & a_3, & \ldots\ldots & a_{n-1}, & a_n, \\
 & b_0 h, & b_1 h, & b_2 h, & \ldots\ldots & b_{n-2} h, & b_{n-1} \cdot h \\
\hline
 & b_1, & b_2, & b_3, & \ldots\ldots & b_{n-1}, & R.
\end{array}
$$

In the first line are written down the successive coefficients of $f(x)$; the first term in the second line is obtained by multiplying a_0 (or b_0, which is equal to it) by h. The product $b_0 h$ is placed under a_1, and then added to it in order to obtain the term b_1 in the third line. This term, when obtained, is multiplied in its turn by h, and placed under a_2. The product is added to a_2 to obtain the second figure b_2 in the third line. The repetition of this process furnishes in succession all the coefficients of the quotient, the last figure thus obtained being the remainder. A few examples will make this plain.

EXAMPLES.

1. Find the quotient and remainder when $3x^4 - 5x^3 + 10x^2 + 11x - 61$ is divided by $x - 3$.

The calculation is arranged as follows :—

$$
\begin{array}{rrrrr}
3 & -5 & 10 & 11 & -61. \\
  & 9 & 12 & 66 & 231. \\
\hline
  & 4 & 22 & 77 & 170.
\end{array}
$$

Thus the quotient is $3x^3 + 4x^2 + 22x + 77$, and the remainder 170.

2. Find the quotient and remainder when $x^3 + 5x^2 + 3x + 2$ is divided by $x - 1$.
$$Ans.\quad Q = x^2 + 6x + 9,\quad R = 11.$$

3. Find Q and R when $x^5 - 4x^4 + 7x^3 - 11x - 13$ is divided by $x - 5$.

[N. B.—When any term in a polynomial is absent, care must be taken to supply the place of its coefficient by zero in writing down the coefficients of $f(x)$. In this example, therefore, the series in the first line will be

$$
\begin{array}{rrrrrr}
1 & -4 & 7 & 0 & -11 & -13.]
\end{array}
$$
$$Ans.\quad Q = x^4 + x^3 + 12x^2 + 60x + 289,\quad R = 1432.$$

4. Find the quotient and remainder when $x^9 + 3x^7 - 15x^2 + 2$ is divided by $x - 2$.
$$Ans.\quad Q = x^8 + 2x^7 + 7x^6 + 14x^5 + 28x^4 + 56x^3 + 112x^2 + 209x + 418,\quad R = 838.$$

5. Find the quotient and remainder when $x^5 + x^2 - 10x + 113$ is divided by $x + 4$.
$$Ans.\quad Q = x^4 - 4x^3 + 16x^2 - 63x + 242;\quad R = -855.$$

9. Tabulation of Functions.—The arithmetical operation explained in the preceding Article supplies a convenient practical method of calculating the value of a polynomial whose coefficients are given numbers when any number is substituted for x. For, the equation

$$f(x) = (x - h)\, Q + R,$$

since its two members are identically equal, must be satisfied when any quantity is substituted for x. Put $x = h$; then $f(h) = R$; $x - h$ being $= 0$, and Q remaining finite. Hence the result of substituting h for x in $f(x)$ is the remainder when $f(x)$ is divided by $x - h$, and can be calculated rapidly by the process of the last Article.

For example, the result of substituting 3 for x in the polynomial of Ex. 1, Art. 8, viz.,

$$3x^4 - 5x^3 + 10x^2 + 11x - 61,$$

is 170, this being the remainder after division by $x - 3$. The student can verify this by actual substitution.

Again, the result of substituting $- 4$ for x in

$$x^5 + x^2 - 10x + 113$$

is $- 855$, as appears from Ex. 5, Art. 8. We saw in Art. 7 that as x receives a continuous series of values increasing from $- \infty$ to $+ \infty$, $f(x)$ will pass through a corresponding continuous series. If we substitute in succession for x, in a polynomial whose coefficients are given numbers, a series of numbers such as

$$-\infty, \dots -3, -2, -1, \quad 0, \quad 1, \quad 2, \quad 3, \dots +\infty,$$

and calculate, and note down, the corresponding values of $f(x)$, the process may be called the *tabulation of the function*.

EXAMPLES.

1. Tabulate the trinomial $2x^2 + x - 6$, for the values of x

$$-4, -3, -2, -1, \quad 0, \quad 1, \quad 2, \quad 3, \quad 4.$$

Values of x,	-4	-3	-2	-1	0	1	2	3	4
Values of $f(x)$,	22	9	0	-5	-6	-3	4	15	30

2. Tabulate the polynomial $10x^3 - 17x^2 + x + 6$ for the values of x

$$-4, -3, -2, -1, \quad 0, \quad 1, \quad 2, \quad 3, \quad 4.$$

Values of x,	-4	-3	-2	-1	0	1	2	3	4
Values of $f(x)$	-910	-420	-144	-22	6	0	20	126	378

10. Graphic Representation of a Polynomial.—
Whenever we have to deal with a great number of values of any
varying quantity, it is important to be able to represent them in
some simple and expressive manner. This in the present instance can be effected, and the general character of the function
made apparent to the eye, by means of a graphic representation.

We proceed to explain such a representation of the function
$f(x)$.

Let two right lines OX, OY
(fig. 1), cut one another at right
angles, and be produced indefinitely
in both directions. These lines are
called the *axis of x* and *axis of y*,
respectively. Lines, such as OA,
measured on the axis of x at the
right-hand side of O, are regarded
as positive, and those, such as
OA', measured at the left-hand
side, as negative. Lines parallel

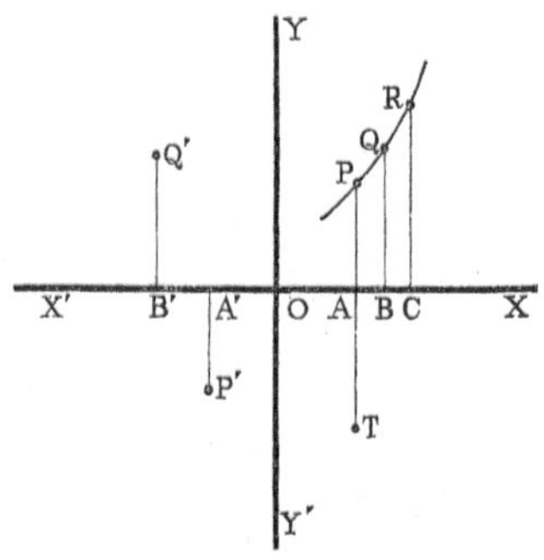

Fig. 1.

to OY which are above XX', such as AP or $B'Q'$, are positive;
and those below it, such as AT or $A'P'$, are negative. These
conventions are already familiar to the student acquainted with
Trigonometry.

We can now take any length we please on OX as unity,
and any number positive or negative will be represented by a
line measured on XX'; the series of numbers increasing from 0
to $+\infty$ in the direction OX, and diminishing from 0 to $-\infty$ in the
direction OX'. Let any number m be represented by OA; calculate $f(m)$; from A draw AP parallel to OY to represent $f(m)$
in magnitude on the same scale as that on which OA represents
m, and to represent by its position above or below the line OX
the sign of $f(m)$.

Corresponding to the different values of m represented by
OA, OB, OC, &c., we shall have a series of points P, Q, R, &c.,
which, when we suppose the number of values of m indefinitely
increased so as to include all numbers between $-\infty$ and $+\infty$, will

trace out a continuous curved line; and this curved line will, by the distances of its several points from the line OX, exhibit to the eye the several values of the function $f(x)$.

The student acquainted with analytic geometry will observe that the process we have here explained is that of tracing the plane curve whose equation is $y = f(x)$.

It is of course impossible to substitute for x all the numerical values between $-\infty$ and $+\infty$; and the labour of the calculation precludes our substitution of more than a limited number of values in any particular case. But the advantage of this graphic representation is, that in general we shall be able, from the calculation of a limited number of values of $f(x)$ corresponding to small integral values of x, to draw approximately the form of the curve representing the function, and thus obtain a general idea of its nature.

The process here described is also called *tracing the function.* We add examples :—

1. Let it be required to trace the trinomial $f(x) = 2x^2 + x - 6$.

From Ex. 1, Art. 9, we have, for the values of x

$$\ldots -4, -3, -2, -1, \quad 0, \quad 1, \quad 2, \quad 3, \quad 4, \ldots \infty$$

the corresponding values of $f(x)$

$$\ldots 22, \quad 9, \quad 0, -5, -6, -3, \quad 4, \quad 15, \quad 30, \ldots \infty$$

The unit of length taken is one-sixth of the line OD in fig. 2.

By means of these values we obtain the positions of nine points on the curve; seven of which, A, B, C, D, E, F, G, are here represented, the other two corresponding to values of $f(x)$ which lie out of the limits of our figure.

It may occur to the student that we have here exercised considerable imagination in drawing that part of the curve which lies between the points determined by calculation; and that much closer numerical values must be substituted for x in order to obtain the shape of the curve with any accuracy. He will learn, however, as he proceeds, that we are assisted in our approximation to the form of the curve by

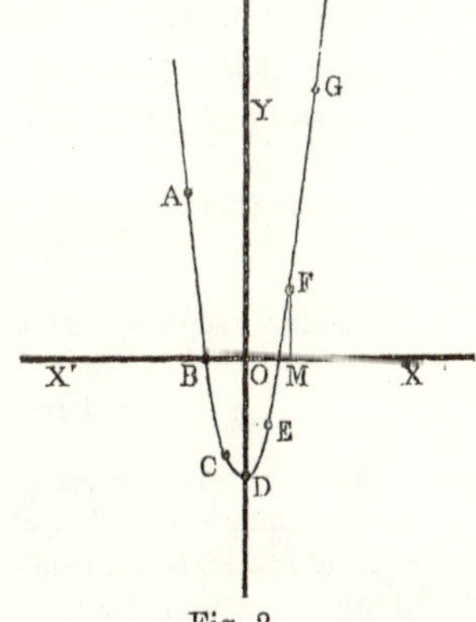

Fig. 2.

many other considerations besides the ascertained values of $f(x)$. Cases undoubtedly occur in which the portion of the curve between two values of x must be more closely examined, and then the substitution of nearer values of x will become necessary. The next example will furnish an illustration of such cases.

Remark.—The curve here traced cuts the axis of x in two points (a number equal to the degree of the polynomial): in other words, there are two values of x, for which the value of the given polynomial is zero; these are the two roots of the equation $2x^2 + x - 6 = 0$, viz., -2, and $\frac{3}{2}$. The curve corresponding to a given polynomial may not cut the axis of x at all, or may cut it in a number of points less than the degree of the polynomial. Such cases correspond to the imaginary roots of equations, as will appear more fully in the next Chapter. For example, the curve which represents the polynomial $2x^2 + x + 2$ will, when traced, lie entirely above the axis of x; in fact, since this function differs from the former only by the addition of the constant quantity 8, each value of $f(x)$ is obtained by adding 8 to the previously calculated value, and the entire curve can be obtained by simply supposing the previously traced curve to be moved up parallel to the axis of y through a distance of 8 of the units. It is evident, by the solution of the equation $2x^2 + x + 2 = 0$, that the two values of x which render the polynomial zero are in this case imaginary. Whenever, as here, the number of points in which the curve cuts the axis of x falls short of the degree of the polynomial, it is customary to speak of the curve as *cutting the line in imaginary points.*

2. Trace the polynomial

$$10x^3 - 17x^2 + x + 6.$$

This is already tabulated in Art. 9 for the values of x

$$-4, -3, -2, -1, \quad 0, \quad 1, \quad 2, \quad 3, \quad 4.$$

We may here remark, as an exercise on Art. 4, that this function retains positive values for all positive values of x greater than $2\cdot7$, and retains negative values for all values of x nearer to $-\infty$ than $-2\cdot7$. The curve will, then, if it cuts the axis of x at all, cut it at a point (or points) corresponding to some value (or values) of x between $-2{\cdot}7$ and $+2{\cdot}7$; so that if our object is to determine, or approximate to, the positions of the roots of the equation $f(x)=0$, our tabulation may be confined to the interval between $-2{\cdot}7$ and $2{\cdot}7$; indeed it may be confined to a closer interval, as will appear when we come to a more precise discussion of the limits of the roots of equations (cf. Remark, Art. 4). This is a case in which the substitution of integral values only of x gives very little help towards the tracing of the curve, and where, consequently, smaller intervals have to be examined. We give the tabulation of the function for intervals of one-tenth between the integers -1, 0; 0, 1; 1, 2. From these values the positions of the corresponding points on the curve may be approximately ascertained, and the curve traced as in fig. 3.

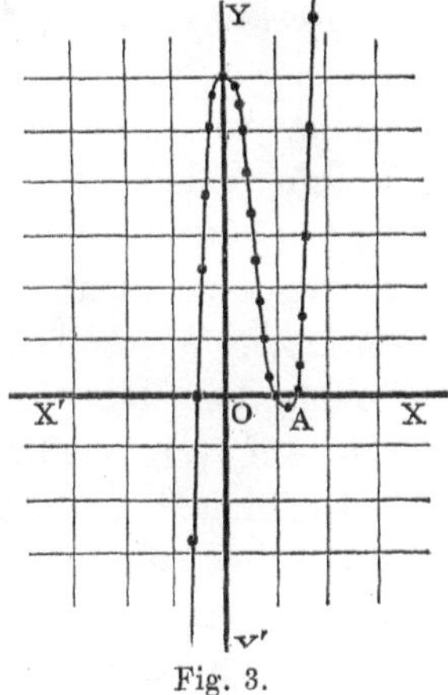

Fig. 3.

Values of x	-1	$-{\cdot}9$	$-{\cdot}8$	$-{\cdot}7$	$-{\cdot}6$	$-{\cdot}5$	$-{\cdot}4$	$-{\cdot}3$	$-{\cdot}2$	$-{\cdot}1$
Values of $f(x)$	-22	$-15{\cdot}96$	$-10{\cdot}8$	$-6{\cdot}46$	$-2{\cdot}88$	0	$2{\cdot}24$	$3{\cdot}9$	$5{\cdot}04$	$5{\cdot}72$

Values of x	0	${\cdot}1$	${\cdot}2$	${\cdot}3$	${\cdot}4$	${\cdot}5$	${\cdot}6$	${\cdot}7$	${\cdot}8$	${\cdot}9$
Values of $f(x)$	6	$5{\cdot}94$	$5{\cdot}6$	$5{\cdot}04$	$4{\cdot}32$	$3{\cdot}5$	$2{\cdot}64$	$1{\cdot}8$	$1{\cdot}04$	${\cdot}42$

Values of x	1	$1{\cdot}1$	$1{\cdot}2$	$1{\cdot}3$	$1{\cdot}4$	$1{\cdot}5$	$1{\cdot}6$	$1{\cdot}7$	$1{\cdot}8$	$1{\cdot}9$	2
Values of $f(x)$	0	$-{\cdot}16$	0	${\cdot}54$	$1{\cdot}52$	3	$5{\cdot}04$	$7{\cdot}7$	$11{\cdot}04$	$15{\cdot}12$	20

11. Maxima and Minima Values of Polynomials.— It is apparent from the considerations established in the preceding Articles, that as the variable x changes from $-\infty$ to $+\infty$, the function $f(x)$ may undergo many variations. It may go on for a certain period increasing, and then, ceasing to increase, may commence to diminish; it may then cease to diminish and commence again to increase; after which another period of diminution may arrive, or the function may (as in the last example of the preceding Art.) go on then continually increasing. At a stage where the function ceases to increase and commences to diminish, it is said to have attained a *maximum* value ; and when it ceases to diminish and commences to increase, it is said to have attained a *minimum* value. A polynomial may have several maxima or several minima values, or both, the number depending on its degree. Nothing exhibits so well as a graphic representation the occurrence of such a maximum or minimum value; as well as the various fluctuations of which the values of a polynomial are susceptible. We shall give in a subsequent Chapter the method of finding the values of x corresponding to the maxima or minima values of $f(x)$, together with criteria to decide between maxima and minima. These are among the considerations alluded to in Art. 10, as aiding us in the graphic construction of the polynomial. Another very material aid to such a construction would be a knowledge of the values of x corresponding to the points (if any) in which the line XX' is cut by the curve; that is to say, of the values of x which render $f(x) = 0$. Such a value of x is a root of the equation $f(x) = 0$.

We proceed in the next Chapter to a discussion of the roots, and general properties of equations.

CHAPTER II.

12. THE questions which we have now to discuss with respect to the equation

$$a_0 x^n + a_1 x^{n-1} + a_2 x^{n-2} + \ldots + a_{n-1} x + a_n = 0$$

are whether every such equation must have a root; whether, assuming the existence of roots, their number is definite or indefinite; what is their character; are they always real, or may they involve the imaginary expression $\sqrt{-1}$.

The following theorem enables us to establish the existence of a real root in many instances :—

Theorem.—*If two real quantities a and b be substituted for the unknown quantity x in any polynomial $f(x)$, and if they furnish results having different signs, one plus and the other minus; then the equation $f(x) = 0$ must have at least one real root intermediate in value between a and b.*

This theorem is an immediate consequence of the property of the continuity of the function $f(x)$ established in Art. 7; for since $f(x)$ changes continuously from $f(a)$ to $f(b)$, i. e. passes through all the intermediate values, while x changes from a to b; and since one of these quantities, $f(a)$ or $f(b)$, is positive, and the other negative, it follows that for some value of x intermediate between a and b, $f(x)$ must attain the value zero which is intermediate between $f(a)$ and $f(b)$.

The student will assist his conception of this theorem by reference to the graphic method of representation described in Art. 10. What is here proved, and what will appear obvious from the figure, is, that if there exist two points of the curved line representing the polynomial on opposite sides of the axis OX, then the curve joining these points must cut that axis at least once. It will also be evident from the figure that several values may exist between a and b for which $f(x) = 0$, *i.e.* for which the curve cuts the axis. For example, in fig. 3, Art. 10, $x = -2$ gives a negative value (-144), and $x = 2$ gives a positive value (20), and between these points of the curve there exist *three* points of section of the axis of x.

Corollary.—*If there exist no real quantity which, substituted for x, makes $f(x) = 0$, then $f(x)$ must be positive for every real value of x.*

For it is evident (Art. 4) that $x = \infty$ makes $f(x)$ positive; and no value of x, therefore, can make it negative; for if it did, the equation would by the theorem of this Article have a real root, which is contrary to our present hypothesis. In terms of the graphic representation this theorem may be expressed by saying that when the equation $f(x) = 0$ has no real root, the curve representing the polynomial $f(x)$ must lie entirely above the axis of x.

13. **Theorem.**—*Every equation of an odd degree has at least one real root of a sign opposite to that of its last term.*

This is an immediate consequence of the theorem in the last Article. Substitute in succession $-\infty$, 0, ∞ for x in the polynomial $f(x)$. The results are, n being odd (see Art. 4),

$$x = -\infty, \ f(x) \text{ is negative};$$

$$x = 0, \ \text{sign of } f(x) \text{ is the same as that of } a_n;$$

$$x = +\infty, \ f(x) \text{ is positive.}$$

If a_n is positive, the equation must have a real root between $-\infty$ and 0, *i.e.* a real negative root; and if a_n is negative, the equation must have a real root between 0 and ∞, *i.e.* a real positive root. The theorem is thus proved.

14. **Theorem.**—*Every equation of an even degree, whose last term is negative, has at least two real roots, one positive and the other negative.*

The results of substituting $-\infty$, 0, ∞ are in this case

$$-\infty, \qquad +,$$
$$0, \qquad -,$$
$$+\infty, \qquad +\,;$$

hence there is a real root between $-\infty$ and 0, and another between 0 and $+\infty$; *i. e.*, there exist at least one real negative, and one real positive root.

We have contented ourselves in both these Articles with proving the *existence* of roots, and for this purpose it is sufficient to substitute very large positive or negative values, as we have done, for x. We are of course able to narrow the limits within which the roots lie by the aid of the theorem of Art. (4), and still more by the aid of the theorems to be given subsequently, to which we have before made reference.

15. **Existence of a Root in the General Equation. Imaginary Roots.**—We have now proved the existence of a real root in the case of every equation except one of an even degree whose last term is positive. Such an equation may have no real root at all. It becomes then an inquiry whether, in the absence of real values, there may not be values involving the imaginary expression $\sqrt{-1}$ which, when substituted for x, reduce the polynomial to zero ; or whether there may not be in certain cases both real and imaginary values

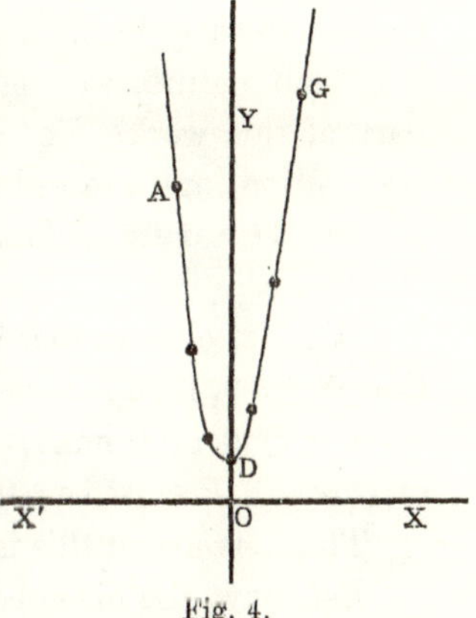

Fig. 4.

of the variable which satisfy the equation. We take a simple case to illustrate our meaning. As already remarked (see Ex. 1, Art. 10), the curve corresponding to the polynomial

$$f(x) = 2x^2 + x + 2$$

lies entirely above the axis of x, as in fig. 4. The equation

$f(x) = 0$ has no real roots; but it has the two imaginary roots

$$-\frac{1}{4} + \frac{\sqrt{15}}{4}\sqrt{-1}, \quad -\frac{1}{4} - \frac{\sqrt{15}}{4}\sqrt{-1},$$

as is evident by the solution of the quadratic.

In this simple instance we observe that, in the absence of any real values, there are two imaginary expressions which reduce the polynomial to zero. The general proposition of which this is a very particular illustration is, that *every rational integra equation*

$$a_0 x^n + a_1 x^{n-1} + a_2 x^{n-1} + \ldots + a_{n-1} x + a_n = 0$$

must have a root of the form

$$a + \beta \sqrt{-1},$$

a and β being real finite quantities. This proposition includes both real and imaginary roots, the former corresponding to the value $\beta = 0$.

As the proof of this proposition involves considerations somewhat advanced, and as we do not wish to interrupt the student so early in his study of the subject by investigations of a difficult nature, we shall defer its discussion to a subsequent part of this work. For the present, therefore, we assume the proposition, and proceed to derive certain consequences from it.

16. **Theorem.**—*Every equation of n dimensions has n roots, and no more.*

We first observe that if any quantity h is a root of the equation $f(x) = 0$, then $f(x)$ is divisible by $x - h$ without a remainder. This is evident from Art. 9; for if $f(h) = 0$, *i.e.* if h is a root of $f(x) = 0$, R must be $= 0$.

The converse of this is also obviously true.

Let, now, the given equation be

$$f(x) \equiv x^n + p_1 x^{n-1} + p_2 x^{n-2} + \ldots + p_{n-1} x + p_n = 0.$$

This equation must have a root, real or imaginary (see Art. 15), which we shall denote by the symbol a_1. Let the quotient, when $f(x)$ is divided by $x - a_1$, be $\phi_1(x)$; we have then the identical equation

$$f(x) \equiv (x - a_1)\,\phi_1(x).$$

Again, the equation $\phi_1(x) = 0$, which is of $n-1$ dimensions, must have a root, which we represent by a_2. Let the quotient obtained by dividing $\phi_1(x)$ by $x - a_2$ be $\phi_2(x)$. Hence

$$\phi_1(x) = (x - a_2)\,\phi_2(x),$$
$$\text{and} \quad \therefore\; f(x) = (x - a_1)\,(x - a_2)\,\phi_2(x),$$

where $\phi_2(x)$ is of $n-2$ dimensions.

Proceeding in this manner, we prove that $f(x)$ consists of the product of n factors, each containing x in the first degree, and a numerical factor $\phi_n(x)$. Comparing the coefficients of x^n, it is plain that $\phi_n(x) = 1$. Thus we prove the identical equation

$$f(x) = (x - a_1)(x - a_2)(x - a_3)\,\ldots\ldots\,(x - a_{n-1})(x - a_n).$$

It is evident that the substitution of any one of the quantities $a_1, a_2, \ldots a_n$ for x in the right-hand member of this equation will reduce that member to zero, and will therefore reduce $f(x)$ to zero; that is to say, the equation $f(x) = 0$ has for roots the n quantities $a_1, a_2, a_3 \ldots a_{n-1}, a_n$. And it can have no other roots; for if any quantity other than one of the quantities $a_1, a_2, \ldots a_n$ be substituted in the right-hand member of the above equation, the factors will be all different from zero, and therefore the product cannot vanish.

Corollary.—*Two polynomials of the nth degree cannot be equal to one another for more than n values of the variable without being completely identical.*

For if, in fact, we equate their difference to zero, we obtain an equation of the nth degree, which can be satisfied by n values only of the variable, unless each coefficient be separately equal to zero.

Remark.—The theorem of this Article, although of no assistance in the solution of the equation $f(x) = 0$, enables us to solve completely the converse problem, *i. e.* to find the equation whose roots are any n given quantities. The required equation is obtained by multiplying together the n simple factors formed by subtracting from x each of the given roots. It also enables us to obtain, when any one or more of the roots of a given equation are known, the equation containing the remaining roots. To effect this we have only to divide the given equation by the product of the given binomial factors. The quotient will be the required polynomial composed of the remaining factors.

Examples.

1. Find the equation whose roots are
$$-3, \quad -1, \quad 4, \quad 5.$$

Ans. $x^4 - 5x^3 - 13x^2 + 53x + 60 = 0.$

2. The equation
$$x^4 - 6x^3 + 8x^2 - 17x + 10 = 0$$
has a root 5; find the equation containing the remaining roots.

[N. B.—Use the method of division of Art. 8.]

Ans. $x^3 - x^2 + 3x - 2 = 0.$

3. Solve the equation
$$x^4 - 16x^3 + 86x^2 - 176x + 105 = 0,$$
two roots being 1 and 7.

Ans. The other two roots are 3, 5.

4. Form the equation whose roots are
$$-\frac{3}{2}, \quad 3, \quad \frac{1}{7}.$$

Ans. $14x^3 - 23x^2 - 60x + 9 = 0.$

5. Solve the cubic equation
$$x^3 - 1 = 0.$$

Here it is evident that $x = 1$ satisfies the equation. Divide by $x - 1$, and solve the resulting quadratic. The two roots are found to be
$$-\frac{1}{2} + \frac{1}{2}\sqrt{-3}, \quad -\frac{1}{2} - \frac{1}{2}\sqrt{-3}.$$

It can be easily shown that if either of these imaginary roots is squared, the other results. It is usual to represent these roots by ω and ω^2. They are called the two *imaginary cube roots of unity.* We have the identical equation
$$x^3 - 1 \equiv (x - 1)(x - \omega)(x - \omega^2).$$

6. Form an equation with rational coefficients which shall have for a root the irrational expression
$$\sqrt{p} + \sqrt{q}.$$

This expression has four different values according to the different combinations of the radical signs. These values are
$$\sqrt{p} + \sqrt{q}, \quad -\sqrt{p} - \sqrt{q}, \quad \sqrt{p} - \sqrt{q}, \quad -\sqrt{p} + \sqrt{q}.$$

The required equation is, therefore,

$$(x - \sqrt{p} - \sqrt{q})(x + \sqrt{p} + \sqrt{q})(x - \sqrt{p} + \sqrt{q})(x + \sqrt{p} - \sqrt{q}) = 0,$$

or

$$(x^2 - p - q - 2\sqrt{pq})(x^2 - p - q + 2\sqrt{pq}) = 0,$$

or, finally,

$$x^4 - 2(p + q)x^2 + (p - q)^2 = 0.$$

17. Equal Roots.—It must be observed that the n factors of which a polynomial $f(x)$ consists need not be all different from one another. The factor $x - a$, for example, may occur in the second or any other higher power not superior to n. In this case we speak of the equation $f(x) = 0$ still as having n roots, two or more being now equal to one another; and the root a is called a multiple root of the equation; double, triple, &c., according to the number of times the factor is repeated.

The propriety of this mode of speaking will be apparent, and the nature of multiple roots of equations made clearer, by a reference to the graphic construction in Art. 10, fig. 3. We see by an inspection of the figure that the two positive roots of the equation $10x^3 - 17x^2 + x + 6 = 0$ are nearly equal, and we can conceive that a slight addition to the absolute term of this polynomial, which is as we have seen in the remark, Ex. 1, Art. 10, equivalent to a small parallel movement upwards of the whole curve, would have the effect of rendering equal the roots of the equation thus altered. In that case the line OX would no longer cut the curve in two distinct points, but would *touch* it. Now, when a line touches a curve it is properly said to meet the curve, not once, but in *two coincident points*. The occurrence of a triple or higher multiple root can be illustrated in a similar manner. Our principal object here is to indicate by a figure the manner in which double roots enter into equations, and to point out how these double roots form as it were the connecting link between real and imaginary roots. Let us suppose that the above polynomial is further altered by another small addition to the absolute term. We shall then have a graphic representation in which the axis OX cuts the curve in only one real point, viz., that corresponding to the negative root, the two points of section corresponding to the two positive roots having now disappeared.

Consider, for example, the polynomial $10x^3 - 17x^2 + x + 28$, which is obtained from that of Ex. 2, Art. 10, by the addition of 22. The student can easily construct the figure: the point corresponding to A in fig. 3 will now lie much above the axis of x. Divide by $x + 1$, and obtain the trinomial $10x^2 - 27x + 28$ which contains the remaining two roots. They are easily found to be

$$\frac{27}{20} + \frac{\sqrt{391}}{20}\sqrt{-1}, \quad \frac{27}{20} - \frac{\sqrt{391}}{20}\sqrt{-1}.$$

In the examples we have studied, in both this Art. and Art. 15, we observe that a change in the form of a polynomial may convert it from one having real roots into another in which two of the real roots become equal, and a further change may convert it into a form where the two roots become imaginary. We also observe in these examples that when a change of form of the polynomial causes one real root to disappear, a second also disappears, and the two are replaced by a pair of imaginary roots. This is true in general, as will be established in the next Article.

18. **Imaginary Roots enter Equations in Pairs.**— The proposition we have to prove may be stated as follows :— *If an equation $f(x) = 0$, whose coefficients are all real quantities, have for a root the imaginary expression $\alpha + \beta\sqrt{-1}$, it must also have for a root the conjugate imaginary expression $\alpha - \beta\sqrt{-1}$.* The product

$$(x - \alpha - \beta\sqrt{-1})(x - \alpha + \beta\sqrt{-1}) \equiv (x - \alpha)^2 + \beta^2.$$

Let the polynomial $f(x)$ be divided by the second member of this identity, and if possible let there be a remainder $Rx + R'$. We have then the identical equation

$$f(x) \equiv \{(x - \alpha)^2 + \beta^2\}\,Q + Rx + R',$$

where Q is the quotient, of $n - 2$ dimensions in x. Substitute in this identity $a + \beta \sqrt{-1}$ for x. This, by hypothesis, causes $f(x)$ to vanish. It also causes $(x - a)^2 + \beta^2$ to vanish. Hence

$$R \left(a + \beta \sqrt{-1} \right) + R' = 0.$$

From this we obtain the two equations

$$R a + R' = 0, \qquad R \beta = 0,$$

since the real and imaginary parts cannot destroy one another ; hence

$$R = 0, \qquad R' = 0.$$

Thus the remainder $R x + R'$ vanishes ; and, therefore, $f(x)$ is divisible without remainder by the product of the two factors

$$x - a - \beta \sqrt{-1}, \qquad x - a + \beta \sqrt{-1}.$$

The equation has, consequently, the root $a - \beta \sqrt{-1}$ as well as the root $a + \beta \sqrt{-1}$.

Thus the total number of imaginary roots in an equation with real coefficients will always be even ; and every polynomial may be regarded as composed of real factors, each pair of imaginary roots producing a real quadratic factor, and each real root producing a real simple factor. The actual resolution of the polynomial into these factors constitutes the complete solution of the equation.

We observed in Art. 17 that equal roots may be considered as the connecting link between real and imaginary roots. We may now regard this statement from another point of view. Suppose a polynomial has the quadratic factor $(x - a)^2 + k$, and let its form be altered by means of slight alterations in the value of k. When k is negative, the quadratic factor gives a pair of *real* roots ; when $k = 0$, this factor has two *equal* roots, a ; when k is positive, the factor has two *imaginary* roots.

Remark.—A proof exactly similar to that above given shows that *surd roots*, of the form $a \pm \sqrt{\gamma}$, enter equations whose coefficients are rational in pairs.

EXAMPLES.

1. Form the cubic equation which shall contain the two roots

$$1, \quad 3 + 2\sqrt{-1}.$$

$$Ans. \quad x^3 - 7x^2 + 19x - 13 = 0.$$

2. Form a rational equation which shall have for two of its roots

$$1 + 5\sqrt{-1}, \quad 5 - \sqrt{-1}.$$

$$Ans. \quad x^4 - 12x^3 + 72x^2 - 312x + 676 = 0.$$

3. Solve the equation

$$x^4 + 2x^3 - 5x^2 + 6x + 2 = 0,$$

which has a root

$$-2 + \sqrt{3}.$$

$$Ans. \text{ The roots are } -2 \pm \sqrt{3}, \quad 1 \pm \sqrt{-1}.$$

4. Solve the equation

$$3x^3 - 4x^2 + x + 88 = 0,$$

one root being

$$2 + \sqrt{-7}.$$

$$Ans. \text{ The roots are } 2 \pm \sqrt{-7}, \quad -\frac{8}{3}.$$

19. Descartes' Rule of Signs—Positive Roots.—This rule, which enables us, by the mere inspection of a given equation, to assign a superior limit to the number of its positive roots, may be enunciated as follows :—*No equation can have more positive roots than it has changes of sign from + to −, and from − to +, in the terms of its first member.*

We shall content ourselves for the present with the proof which is usually given, and which is more a verification than a general demonstration of this celebrated theorem of Descartes. It will be subsequently shown that this rule of Descartes, and other similar rules which were discovered by early investigators relative to the number of the positive, negative, and imaginary roots of equations, are immediate deductions from the more general theorems of Budan and Fourier.

Let the signs of a polynomial taken at random succeed each other in the following order :—

$$+ \; + \; - \; + \; - \; - \; - \; + \; + \; - \; + \; -.$$

In this there are in all seven changes of sign, including

those from + to −, and from − to +.   It is proposed to show that if this polynomial be multiplied by a binomial whose signs, corresponding to a positive root, are + −, the resulting polynomial will have at least one more change of sign than the original.

We write down only the signs which occur in the operation as follows :—

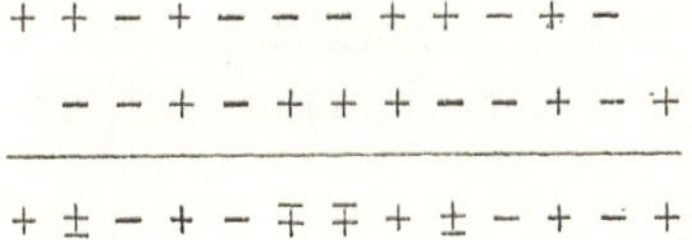

We have placed here in the result the ambiguous sign ± wherever there are two terms with different signs to be added. We observe in this case, and it will readily appear also for every other arrangement, that the effect of the process is to introduce the ambiguous sign wherever the sign + follows +, or − follows −, in the original polynomial.   The number of variations of sign is never diminished.   There is, moreover, always one variation added at the end. This is obvious in the above instance, where the original polynomial terminates with a variation; if it terminate with a continuation of sign, it will equally appear that the corresponding ambiguity in the resulting polynomial must furnish one additional variation either with the preceding or with the superadded sign.   Thus, in even the most unfavourable case, that, namely, in which the continuations of sign in the original remain continuations in the resulting polynomial, there is one variation added ; and we may conclude in general that the effect of the multiplication of a polynomial by a binomial factor $x - a$ is to introduce at least one additional change of sign.

Suppose now a polynomial formed of the product of the factors corresponding to the negative and imaginary roots of an equation; the effect of multiplying this by each of the factors $x - a$, $x - \beta$, $x - \gamma$, &c., corresponding to the positive roots a, β, γ, &c., . . . is to introduce at least one change of sign for each ; so that when the complete product is formed containing all the roots, we are sure that the resulting polynomial has at least as many changes of sign as it has positive roots. This is Descartes' proposition.

20. **Descartes' Rule of Signs—Negative Roots.**—In order to give the most advantageous statement to Descartes' rule in the case of negative roots, we first prove that if $-x$ be substituted for x in the equation $f(x) = 0$, the resulting equation will have the same roots as the original except that their signs will be changed. This follows from the identical equation of Art. 16

$$f(x) \equiv (x - a_1)\ (x - a_2)\ (x - a_3) \ldots (x - a_n),$$

from which we derive

$$f(-x) \equiv (-1)^n\ (x + a_1)\ (x + a_2)\ (x + a_3) \ldots (x + a_n).$$

From this it is evident that the roots of $f(-x) = 0$ are

$$-a_1,\ \ -a_2,\ \ -a_3, \ldots \ldots -a_n.$$

Hence the negative roots of $f(x)$ are positive roots of $f(-x)$, and we may enunciate Descartes' rule for negative roots as follows :—*No equation can have a greater number of negative roots than there are changes of sign in the terms of the polynomial $f(-x)$.*

21. **Use of Descartes' Rule in proving the existence of Imaginary Roots.**—We are often able to detect the existence of imaginary roots in equations by the application of Descartes' rule ; for if it should happen that the sum of the greatest possible number of positive roots, added to the greatest possible number of negative roots, is less than the degree of the equation, we are sure of the existence of imaginary roots. Take, for example, the equation

$$x^8 + 10\,x^3 + x - 4 = 0.$$

This, having only one variation, cannot have more than one positive root. And, changing x into $-x$, we get

$$x^8 - 10\,x^3 - x - 4 = 0 ;$$

which, having only one variation, the original equation cannot have more than one negative root. Hence, in the proposed equation there cannot exist more than two real roots. It has, therefore, at least six imaginary roots. This application of Descartes' rule is available only in the case of incomplete equations; for it is easily seen that the sum of the number of variations in $f(x)$ and $f(-x)$ is exactly equal to the degree of the equation when it is complete.

22. **Theorem.**—We shall close this chapter with the following theorem, which defines fully the conclusions which can be drawn as to the roots of an equation from the signs furnished by its first member when two given numbers are substituted for x:—*If two numbers a and b, substituted for x in the polynomial $f(x)$, give results with contrary signs, an odd number of real roots of the equation $f(x) = 0$ lies between them; and if they give results with the same sign, either no real root or an even number of real roots lies between them.*

We proceed to prove the first part of this proposition: the second is proved in an exactly similar manner.

Let the following m roots $a_1, a_2, \ldots a_m$, and no others, of the equation $f(x) = 0$ lie between the quantities a and b, of which, as usual, we take a to be the lesser.

Let $\phi(x)$ be the quotient when $f(x)$ is divided by the product of the m factors $(x - a_1)(x - a_2) \ldots (x - a_m)$. We have, then, the identical equation

$$f(x) = (x - a_1)(x - a_2) \ldots (x - a_m)\, \phi(x).$$

Putting in this successively $x = a$, $x = b$, we obtain

$$f(a) = (a - a_1)(a - a_2) \ldots (a - a_m)\, \phi(a),$$
$$f(b) = (b - a_1)(b - a_2) \ldots (b - a_m)\, \phi(b).$$

Now $\phi(a)$ and $\phi(b)$ have the same sign; for if they had different signs there would be, by Art. 12, one root at least of the equation $\phi(x) = 0$ between them. By hypothesis, $f(a)$ and $f(b)$ have different signs; hence the signs of the products

$$(a - a_1)(a - a_2) \ldots (a - a_m),$$
$$(b - a_1)(b - a_2) \ldots (b - a_m),$$

are different; but the sign of the second is positive, since all its factors are positive; hence the sign of the first is negative; but all the factors of the first are negative; therefore their number must be odd; which proves the proposition.

In this proposition it is to be understood that multiple roots are counted a number of times equal to the degree of their multiplicity.

It is instructive to regard this proposition by the light of the graphic method of construction, from which point of view it appears almost intuitively true; for if a point be taken on a curve at one side of the axis, we must cross the axis an odd number of times to reach a point at the other side; and we must cross it an even number of times, or not at all, to reach any other point at the same side of the axis.

EXAMPLES.

1. If the signs of the terms of an equation be all positive, it cannot have a positive root.

2. If the signs of the terms of any complete equation be alternately positive and negative, it cannot have a negative root.

3. If an equation consist of a number of terms connected by + signs followed by a number of terms connected by − signs, it has one positive root and no more.

[Apply Art. 12, substituting 0 and ∞; and Art. 19.]

4. If an equation involve only even powers of x, and if all the coefficients have positive signs, it cannot have a real root.

[Apply Arts. 19 and 20.]

5. If an equation involve only odd powers of x, and if the coefficients have all positive signs, it has the root zero and no other real root.

6. If an equation be complete, the number of continuations of sign in $f(x)$ is the same as the number of variations of sign in $f(-x)$.

7. When an equation is complete; if all its roots are real, the number of positive roots is equal to the number of variations, and the number of negative roots is equal to the number of continuations of sign.

8. An equation having an even number of variations of sign must have its last sign positive, and one having an odd number of variations must have its last sign negative.

[N. B.—The sign + is always given to the highest power of x.]

9. Hence prove that if an equation has an even number of variations it must have an equal or less even number of positive roots; and if it has an odd number of variations it must have an equal or less odd number of positive roots; in other

words, the number of positive roots when less than the number of variations must differ from it by an even number.

[Substitute 0 and ∞, and apply Art. 22.]

10. Find an inferior limit to the number of imaginary roots of the equation

$$x^6 - 3x^2 - x + 1 = 0.$$

Ans. At least two imaginary roots.

11. Find the nature of the roots of the equation

$$x^4 + 15x^2 + 7x - 11 = 0.$$

[Apply Arts. 14, 19, 20.]

Ans. One positive, one negative, two imaginary.

12. Show that the equation

$$x^3 + qx + r = 0,$$

where q and r are essentially positive, has one negative and two imaginary roots.

13. Show that the equation

$$x^3 - qx + r = 0,$$

where q and r are essentially positive, has one negative root; and that the other two roots are either imaginary or both positive.

14. Show that the equation

$$\frac{A^2}{x-a} + \frac{B^2}{x-b} + \frac{C^2}{x-c} + \ldots \frac{L^2}{x-l} = x - m,$$

where a, b, c, $\ldots$ l are numbers all different from one another, cannot have an imaginary root.

Substitute $\alpha + \beta \sqrt{-1}$ and $\alpha - \beta \sqrt{-1}$ in succession for x, and subtract. We get an expression which can vanish only on the supposition $\beta = 0$.

15. Show that the equation

$$x^n - 1 = 0$$

has, when n is even, two real roots 1 and -1, and no other real root; and, when n is odd, the real root 1, and no other real root.

[This and the next example follow readily from Arts. 19 and 20.]

16. Show that the equation

$$x^n + 1 = 0$$

has, when n is even, no real root; and, when n is odd, the real root -1, and no other real root.

17. Solve the equation

$$x^4 + 2qx^3 + 3q^2x^2 + 2q^3x - r^4 = 0.$$

This is equivalent to

$$(x^2 + qx + q^2)^2 - q^4 - r^4 = 0.$$

$$\textit{Ans. } -\frac{1}{2}q + \sqrt{-\frac{3}{4}q^2 + \sqrt{q^4 + r^4}}.$$

The different signs of the radicals give four combinations, and the expression here written involves the four roots.

18. Form the equation which has for roots the different values of the expression

$$2 + \theta \sqrt{7} + \sqrt{11 + \theta \sqrt{7}},$$

where $\theta^2 = 1$.

If no restriction had been made by the introduction of θ, this expression would have 8 values. The $\sqrt{7}$ must now be taken with the same sign where it occurs under the second radical and free from it. There are, therefore, only four values in all. *Ans.* $x^4 - 8x^3 - 12x^2 + 84x - 63 = 0$.

19. Form the equation which has for roots the four values of

$$-9 + \theta \sqrt{137} + 3 \sqrt{34 - 2\theta \sqrt{137}},$$

where $\theta^2 = 1$. *Ans.* $x^4 + 36x^3 - 400x^2 - 3168x + 7744 = 0$.

20. Form an equation with rational coefficients which shall have for roots all the values of the expression

$$\theta_1 \sqrt{p} + \theta_2 \sqrt{q} + \theta_3 \sqrt{r},$$

where $$\theta_1{}^2 = 1, \quad \theta_2{}^2 = 1, \quad \theta_3{}^2 = 1.$$

There are eight different values of this expression, viz.,

$$\sqrt{p} + \sqrt{q} + \sqrt{r}, \qquad -\sqrt{p} - \sqrt{q} - \sqrt{r},$$

$$\sqrt{p} - \sqrt{q} - \sqrt{r}, \qquad -\sqrt{p} + \sqrt{q} + \sqrt{r},$$

$$-\sqrt{p} + \sqrt{q} - \sqrt{r}, \qquad \sqrt{p} - \sqrt{q} + \sqrt{r},$$

$$-\sqrt{p} - \sqrt{q} + \sqrt{r}, \qquad \sqrt{p} + \sqrt{q} - \sqrt{r}.$$

Assume $$x = \theta_1 \sqrt{p} + \theta_2 \sqrt{q} + \theta_3 \sqrt{r}.$$

Squaring this, we have

$$x^2 = p + q + r + 2(\theta_2 \theta_3 \sqrt{qr} + \theta_3 \theta_1 \sqrt{rp} + \theta_1 \theta_2 \sqrt{pq}).$$

Transposing, and squaring again,

$$(x^2 - p - q - r)^2 = 4(qr + rp + pq) + 8\theta_1 \theta_2 \theta_3 \sqrt{pqr} \,(\theta_1 \sqrt{p} + \theta_2 \sqrt{q} + \theta_3 \sqrt{r}). \quad (1)$$

Transposing, substituting x for $\theta_1 \sqrt{p} + \theta_2 \sqrt{q} + \theta_3 \sqrt{r}$, and squaring, we obtain the final equation free from radicals

$$\{x^4 - 2x^2(p + q + r) + p^2 + q^2 + r^2 - 2qr - 2rp - 2pq\}^2 = 64\,pqr\,x^2.$$

This is an equation of the eighth degree, whose roots are the values above written. Since θ_1, θ_2, θ_3 have disappeared, it is indifferent which of the eight roots $\pm\sqrt{p} \pm \sqrt{q} \pm \sqrt{r}$ is assumed equal to x in the first instance. The final equation is that which would have been obtained if each of the 8 roots had been subtracted from x, and the continued product formed, as in Ex. 6, Art. 16.

CHAPTER III.

RELATIONS BETWEEN THE ROOTS AND COEFFICIENTS OF EQUA-
TIONS, WITH APPLICATIONS TO SYMMETRIC FUNCTIONS OF THE
ROOTS.

23. Relations between the Roots and Coefficients.—
Taking for simplicity the coefficient of the highest power of x
as unity, and representing, as in Art. 16, the n roots of an equation by $a_1, a_2, a_3, \ldots a_n$, we have the following identity:—

$$x^n + p_1 x^{n-1} + p_2 x^{n-2} + \ldots + p_{n-1} x + p_n$$

$$\equiv (x - a_1)(x - a_2)(x - a_3) \ldots (x - a_n). \tag{1}$$

When the factors of the second member of this identity are
multiplied together, the product will consist, as is proved in
elementary works on Algebra, of a highest term x^n; plus a
term x^{n-1} multiplied by the factor

$$-(a_1 + a_2 + a_3 + \ldots + a_n),$$

i. e. the sum of the roots with their signs changed; plus a term
x^{n-2} multiplied by the factor

$$a_1 a_2 + a_1 a_3 + a_2 a_3 + \ldots + a_{n-1} a_n,$$

i. e. the sum of the products of the roots taken in pairs; plus a
term x^{n-3} multiplied by the factor

$$-(a_1 a_2 a_3 + a_1 a_2 a_4 + \ldots + a_{n-2} a_{n-1} a_n),$$

i. e. the sum of the products of the roots with their signs
changed taken three by three; and so on. It is plain that the
sign of each coefficient will be negative or positive according as
the number of roots in each product is odd or even. The last
term is

$$\pm a_1 a_2 a_3 \ldots a_{n-1} a_n,$$

the sign being $-$ if n is odd, and $+$ if n is even. Equating the coefficients of x on each side of the identity (1), we have the following series of equations :—

$$
\left.
\begin{aligned}
p_1 &= -\,(a_1 + a_2 + a_3 + \ldots + a_n),\\
p_2 &= \quad (a_1 a_2 + a_1 a_3 + a_2 a_3 + \ldots + a_{n-1} a_n),\\
p_3 &= -\,(a_1 a_2 a_3 + a_1 a_3 a_4 + \ldots + a_{n-2} a_{n-1} a_n),\\
&\quad\cdot\quad\cdot\quad\cdot\quad\cdot\quad\cdot\quad\cdot\quad\cdot\\
p_n &= (-1)^n\, a_1 a_2 a_3 \ldots a_{n-1} a_n,
\end{aligned}
\right\} \qquad (2)
$$

which furnish us with the following

Theorem.—*In every algebraic equation, the coefficient of whose highest term is unity, the coefficient p_1 of the second term with its sign changed is equal to the sum of the roots.*

The coefficient p_2 of the third term is equal to the sum of the products of the roots taken two by two.

The coefficient p_3 of the fourth term with its sign changed is equal to the sum of the products of the roots taken three by three ; and so on, the signs of the coefficients being taken alternately negative and positive, and the number of roots multiplied together in each term of the corresponding function of the roots increasing by unity, till finally that function is reached which consists of the product of the n roots.

When the coefficient a_0 of x^n is not unity (see Art. 1), we must divide each term of the equation by it. The sum of the roots is then equal to $-\dfrac{a_1}{a_0}$; the sum of their products in pairs is equal to $\dfrac{a_2}{a_0}$; and so on.

Cor. 1.—Every root of an equation is a divisor of the last term.

Cor. 2.—If the roots of an equation be all positive, the coefficients will be alternately positive and negative; and if the roots be all negative, the coefficients will be all positive. This is obvious from the equations (2) [cf. Arts. 19 and 20].

24. **Applications of the Theorem.**—Since the equations (2) of the preceding Article furnish us with n distinct relations

between the n roots and the coefficients, it might perhaps be supposed that by their means some advantage is gained in the general solution of the equation. Such, however, is not the case ; for suppose it were attempted to determine by means of these equations a root, a_1, of the original equation, this could be effected only by the elimination of the other roots by means of the given equations, and the consequent determination of a final equation of which a_1 is one of the roots. Now, in whatever way this final equation is obtained, it must have for solution not only a_1, but each of the other roots a_2, a_3 ... a_n; for, since all the roots enter in the same manner in the equations (2), if it had been proposed to determine a_2 (or any other root) by the elimination of the rest, our final equation could differ from that obtained for a_1 only by the substitution of a_2 (or that other root) for a_1. The final equation arrived at, therefore, by the process of elimination must have the n quantities a_1, a_2, ... a_n for roots ; and cannot, consequently, be easier of solution than the given equation. This final equation is, in fact, the original equation itself, with the root we are seeking substituted for x. This we shall show for the particular case of a cubic. The process is perfectly general, and may be applied to an equation of any degree. Let a, β, γ be the roots of the equation

$$x^3 + p_1 x^2 + p_2 x + p_3 = 0.$$

We have, by Art. 23,

$$p_1 = - (a + \beta + \gamma),$$
$$p_2 = \quad a\beta + a\gamma + \beta\gamma,$$
$$p_3 = - a\beta\gamma.$$

Add these three equations, after multiplying the first by a^2, and the second by a. We find

$$p_1 a^2 + p_2 a + p_3 = - a^3,$$

or $\qquad a^3 + p_1 a^2 + p_2 a + p_3 = 0,$

which is the given cubic with a in the place of x.

The student can take as an exercise to prove the same result in the case of an equation of the fourth degree. In general the successive equations of Art. 23 are to be multiplied by a^{n-1}, a^{n-2}, a^{n-3}, &c., and added.

Although the equations (2) furnish, as we have seen, no aid in the general solution of the given equation, they are of use, when some (one or more) of the roots are known, in facilitating the finding of the others; they may also be of use in finding the roots when certain particular relations among them are known to exist.

EXAMPLES.

1. Solve the equation

$$x^3 - 5x^2 - 16x + 80 = 0,$$

the sum of two of its roots being equal to nothing.

Let the roots be α, β, γ. We have, then,

$$\alpha + \beta + \gamma = 5,$$

$$\alpha\beta + \alpha\gamma + \beta\gamma = -16,$$

$$\alpha\beta\gamma = -80.$$

Taking $\beta + \gamma = 0$, we have, from the first of these, $\alpha = 5$, and from either the second or third we obtain $\beta\gamma = -16$. We find for β and γ the values 4 and -4. Thus the three roots are 5, 4, -4.

2. Solve the equation

$$x^3 - 3x^2 + 4 = 0,$$

two of its roots being equal.

Let the roots be α, α, β. We have

$$2\alpha + \beta = 3,$$

$$\alpha^2 + 2\alpha\beta = 0,$$

from which we find $\alpha = 2$, and $\beta = -1$. The roots are 2, 2, -1.

3. The equation

$$x^4 + 4x^3 - 2x^2 - 12x + 9 = 0$$

has two pairs of equal roots; find them.

Let the roots be α, α, β, β; we have

$$2\alpha + 2\beta = -4,$$

$$\alpha^2 + \beta^2 + 4\alpha\beta = -2,$$

from which we obtain for α and β the values 1 and -3.

4. Solve the equation

$$x^3 - 9x^2 + 14x + 24 = 0,$$

two of whose roots are in the ratio of 3 to 2.

Let the roots be α, β, γ, with the relation $2\alpha = 3\beta$. By elimination of α we easily obtain

$$5\beta + 2\gamma = 18,$$

$$3\beta^2 + 5\beta\gamma = 28,$$

from which we have the following quadratic for β :—

$$19\beta^2 - 90\beta + 56 = 0.$$

The roots of this are 4, and $\dfrac{14}{19}$; the former gives for α and γ the values 6 and -1. The three roots are 6, 4, -1. The student will here ask what is the significance of the value $\dfrac{14}{19}$ of β; and the same difficulty may have presented itself in the previous examples. It will be observed that in all examples of this nature we never require all the relations between the roots and coefficients in order to determine the required unknown quantities. The reason of this is, that the given condition establishes one or more relations among the roots. Whenever the equations employed appear to furnish more than one system of values for the roots, the actual roots are easily determined by the condition that they must satisfy the equation (or equations) between the roots and coefficients which we have not employed in determining them. Thus, in the present example, the value $\beta = 4$ gives a system satisfying the omitted equation

$$\alpha\beta\gamma = -24 ;$$

while the value $\beta = \dfrac{14}{19}$ gives a system not satisfying this equation, and is therefore to be rejected.

5. The roots of the equation

$$x^3 - 9x^2 + 23x - 15 = 0$$

are in arithmetical progression; find them.

Let the roots be $\alpha - \delta$, α, $\alpha + \delta$; we have at once

$$3\alpha = 9,$$

$$3\alpha^2 - \delta^2 = 23,$$

from which we obtain the three roots 1, 3, 5.

6. The roots of

$$x^4 + 2x^3 - 21x^2 - 22x + 40 = 0$$

are in arithmetical progression; find them.

Assume for the roots $\alpha - 3\delta$, $\alpha - \delta$, $\alpha + \delta$, $\alpha + 3\delta$.

Ans. -5, -2, 1, 4.

7. The roots of

$$27x^3 + 42x^2 - 28x - 8 = 0$$

are in geometric progression; find them.

Assume for the roots $\alpha\rho$, α, $\dfrac{\alpha}{\rho}$. From the third of the equations (2), Art. 23, we have $\alpha^3 = \dfrac{8}{27}$, or $\alpha = \dfrac{2}{3}$. We get a quadratic for ρ, and either value gives

$$\textit{Ans.}\ -2,\quad \frac{2}{3},\quad \frac{-2}{9}.$$

8. Solve the equation

$$3x^4 - 40x^3 + 130x^2 - 120x + 27 = 0,$$

whose roots are in geometric progression.

Assume for the roots $\dfrac{\alpha}{\rho^3}$, $\dfrac{\alpha}{\rho}$, $\alpha\rho$, $\alpha\rho^3$. Using the second and fourth of the equa-

tions (2), Art. 23, we easily find $\textit{Ans.}\ \dfrac{1}{3},\quad 1,\quad 3,\quad 9.$

9. Solve the equation

$$x^4 + 15x^3 + 70x^2 + 120x + 64 = 0,$$

whose roots are in geometric progression. $\textit{Ans.}\ -1,\ -2,\ -4,\ -8.$

10. Solve the equation

$$6x^3 - 11x^2 + 6x - 1 = 0,$$

whose roots are in harmonic progression.

Take the roots α, β, γ. We have the relation

$$\frac{1}{\alpha} + \frac{1}{\gamma} = \frac{2}{\beta},$$

hence

$$\beta\gamma + \gamma\alpha + \alpha\beta = 3\gamma\alpha\ ;$$

$$\textit{Ans.}\ 1,\quad \frac{1}{2},\quad \frac{1}{3}.$$

11. Solve the equation

$$81x^3 - 18x^2 - 36x + 8 = 0,$$

whose roots are in harmonic progression.

$$\textit{Ans.}\ \frac{2}{9},\quad \frac{2}{3},\quad -\frac{2}{3}.$$

12. If the roots of the equation

$$x^3 - px^2 + qx - r = 0$$

be in harmonic progression, show that the mean root is $\dfrac{3r}{q}$.

13. The equation

$$x^4 - 2x^3 + 4x^2 + 6x - 21 = 0$$

has two roots equal in magnitude and opposite in sign; determine all the roots.

Take $\alpha + \beta = 0$, and employ the first and third of equations (2), Art. 23.

$$\textit{Ans.}\ \sqrt{3},\ -\sqrt{3},\ 1 \pm \sqrt{-6}.$$

14. The equation
$$3x^4 - 25x^3 + 50x^2 - 50x + 12 = 0$$
has two roots whose product is 2 ; find all the roots.

$$\text{\emph{Ans.} } 6, \quad \frac{1}{3}, \quad 1 \pm \sqrt{-1}.$$

15. One of the roots of the cubic
$$x^3 - px^2 + qx - r = 0$$
is double another ; show that it may be found from a quadratic equation.

16. Show that all the roots of the equation
$$x^n + p_1 x^{n-1} + p_2 x^{n-2} + \ldots + p_{n-1} x + p_n = 0$$
can be obtained when they are in arithmetical progression.

Let the roots be $\alpha, \; \alpha + \delta, \; \alpha + 2\delta, \; \ldots \alpha + (n-1)\delta$. The first of equations (2) gives

$$-p_1 = n\alpha + \{1 + 2 + 3 + \ldots + (n-1)\}\delta$$
$$= n\alpha + \frac{n(n-1)}{2}\delta. \tag{1}$$

Again, since the sum of the squares of any number of quantities is equal to the square of their sum minus twice the sum of their products in pairs, we have the equation

$$p_1{}^2 - 2p_2 = \alpha^2 + (\alpha + \delta)^2 + (\alpha + 2\delta)^2 + \ldots$$
$$= n\alpha^2 + n(n-1)\alpha\delta + \frac{n(n-1)(2n-1)}{6}\delta^2. \tag{2}$$

Subtracting the square of (1) from n times the equation (2), we find δ^2 in terms of p_1 and p_2. We can then find α from equation (1). Thus all the roots can be expressed in terms of the coefficients p_1 and p_2.

17. Find the condition which must be satisfied by the coefficients of the equation
$$x^3 - px^2 + qx - r = 0$$
when two of its roots α, β are connected by a relation $\alpha + \beta = 0$. *Ans.* $pq - r = 0$.

18. Find the condition that the cubic
$$x^3 - px^2 + qx - r = 0$$
should have its roots in geometric progression. *Ans.* $p^3 r - q^3 = 0$.

19. Find the condition that the same cubic should have its roots in harmonic progression (see Ex. 12). *Ans.* $27r^2 - 9pqr + 2q^3 = 0$.

20. Find the condition that the equation
$$x^4 + px^3 + qx^2 + rx + s = 0$$
should have two roots connected by the relation $\alpha + \beta = 0$; and find in that case the two quadratics from which α, β and γ, δ are obtained.

$$\text{\emph{Ans.} } pqr - p^2 s - r^2 = 0, \quad px^2 + r = 0, \quad x^2 + px + \frac{ps}{r} = 0.$$

21. Find the condition that the biquadratic of Ex. 20 should have its roots connected by the relation $\beta + \gamma = \alpha + \delta$. *Ans.* $p^3 - 4pq + 8r = 0$.

22. Find the condition that the roots α, β, γ, δ of

$$x^4 + px^3 + qx^2 + rx + s = 0$$

should be connected by the relation $\alpha\beta = \gamma\delta$.   *Ans.* $p^2 s - r^2 = 0$.

23. Show that the condition obtained in Ex. 22 is satisfied when the roots of the biquadratic are in geometric progression.

25. Depression of an Equation when a relation exists between two of its Roots.—The examples given under the preceding Article illustrate the use of the equations connecting the roots and coefficients in determining the roots in particular cases when known relations exist among them. The object of the present Article is to show that, in general, *if a relation of the form $\beta = \phi(\alpha)$ exist between two of the roots of an equation $f(x) = 0$, the equation may be depressed two dimensions.*

Let $\phi(x)$ be substituted for x in the identity

$$f(x) \equiv a_0 x^n + a_1 x^{n-1} + \ldots + a_n,$$

then $f(\phi(x)) \equiv a_0(\phi(x))^n + a_1(\phi(x))^{n-1} + \ldots + a_{n-1}\phi(x) + a_n.$

We represent, for convenience, the second member of this identity by $F(x)$. Substitute α for x, then

$$F(\alpha) = f(\phi(\alpha)) = f(\beta) = 0 ;$$

hence α satisfies the equation $F(x) = 0$, and it also satisfies the equation $f(x) = 0$; hence the polynomials $f(x)$ and $F(x)$ have a common measure $x - \alpha$; thus α can be determined, and from it $\phi(\alpha)$ or β, and the given equation can be depressed two dimensions.

EXAMPLES.

1. The equation

$$x^3 - 5x^2 - 4x + 20 = 0$$

has two roots whose difference $= 3$: find them.

Here $\beta - \alpha = 3$, $\beta = 3 + \alpha$; substitute $x + 3$ for x in the given polynomial $f(x)$; it becomes $x^3 + 4x^2 - 7x - 10$; the common measure of this and $f(x)$ is $x - 2$; from which $\alpha = 2$, $\beta = 5$; the third root is -2.

2. The equation

$$x^4 - 5x^3 + 11x^2 - 13x + 6 = 0$$

has two roots connected by the relation $2\beta + 3\alpha = 7$: find all the roots.

Ans. 1,   2, $1 \pm \sqrt{-2}$.

In connexion with this subject it may be observed, that when two polynomials $f(x)$ and $F(x)$ have one or more common factors, this factor or factors may be found by the ordinary process for obtaining their common measure. Thus, if we know that two given equations have common roots, we can obtain these roots by equating to zero the greatest common measure of the given polynomials.

EXAMPLES.

1. The equations

$$2x^3 + 5x^2 - 6x - 9 = 0,$$

$$3x^3 + 7x^2 - 11x - 15 = 0,$$

have two common roots, find them.   *Ans.* $-1, \ -3.$

2. The equations

$$x^3 + px^2 + qx + r = 0,$$

$$x^3 + p'x^2 + q'x + r' = 0,$$

have two common roots; find the quadratic which furnishes them, and also the 3rd root of each.

$$Ans. \ \ x^2 + \frac{q-q'}{p-p'}\, x + \frac{r-r'}{p-p'} = 0, \quad \frac{-r\,(p-p')}{r-r'}, \quad \frac{-r'\,(p-p')}{r-r'}.$$

26. **The Cube Roots of Unity.**—Equations of the forms

$$x^n - 1 = 0, \quad x^n + 1 = 0$$

are called *binomial*. The roots of the former are called the n n^{th} *roots of unity*. A general discussion of these forms will be given in a subsequent Chapter. We confine ourselves at present to the simple case of the binomial cubic, for which certain useful properties of the roots can be easily established. It has been already shown (see Ex. 5, Art. 16), that the roots of the cubic

$$x^3 - 1 = 0$$

are $1, \ \ -\frac{1}{2} + \frac{1}{2}\sqrt{-3}, \ \ -\frac{1}{2} - \frac{1}{2}\sqrt{-3}.$

If either of the imaginary roots be represented by ω, the other is easily seen to be ω^2, by actually squaring ; or we may see the same thing as follows :—

If ω be a root of the cubic, ω^2 must also be a root; for, since $\omega^3 = 1$, we get, by squaring, $\omega^6 = 1$, which is $(\omega^2)^3 = 1$, thus showing that ω^2 satisfies the cubic $x^3 - 1 = 0$. We have then the identity

$$x^3 - 1 \equiv (x - 1)(x - \omega)(x - \omega^2).$$

Changing x into $-x$, we get the following identity also :—

$$x^3 + 1 \equiv (x + 1)(x + \omega)(x + \omega^2),$$

which furnishes the roots of

$$x^3 + 1 = 0.$$

Whenever in any product of quantities involving the imaginary cube roots of unity any power higher than the second presents itself, it can be replaced by ω, or ω^2, or by unity; for example,

$$\omega^4 = \omega^3 . \omega = \omega, \quad \omega^5 = \omega^3 . \omega^2 = \omega^2, \quad \omega^6 = \omega^3 . \omega^3 = 1, \ \&c.$$

The first or second of equations (2), Art. 23, gives the following property of the imaginary cube roots :—

$$1 + \omega + \omega^2 = 0.$$

By the aid of this equation any expression involving real quantities and the imaginary cube roots can be written in either of the forms $P + \omega Q$, $P + \omega^2 Q$.

EXAMPLES.

1. Show that the product

$$(\omega m + \omega^2 n)(\omega^2 m + \omega n)$$

is rational.   *Ans.* $m^2 - mn + n^2$.

2. Prove the following identities :—

$$m^3 + n^3 \equiv (m + n)(\omega m + \omega^2 n)(\omega^2 m + \omega n),$$

$$m^3 - n^3 \equiv (m - n)(\omega m - \omega^2 m)(\omega^2 m - \omega n).$$

3. Show that the product

$$(\alpha + \omega \beta + \omega^2 \gamma)(\alpha + \omega^2 \beta + \omega \gamma)$$

is rational.   *Ans.* $\alpha^2 + \beta^2 + \gamma^2 - \beta\gamma - \gamma\alpha - \alpha\beta$.

4. Prove the identity

$$(\alpha + \beta + \gamma)(\alpha + \omega\beta + \omega^2\gamma)(\alpha + \omega^2\beta + \omega\gamma) \equiv \alpha^3 + \beta^3 + \gamma^3 - 3\alpha\beta\gamma.$$

5. Prove the identity

$$(\alpha + \omega\beta + \omega^2\gamma)^3 + (\alpha + \omega^2\beta + \omega\gamma)^3 \equiv (2\alpha - \beta - \gamma)(2\beta - \gamma - \alpha)(2\gamma - \alpha - \beta).$$

[Apply Ex. 2.]

6. Prove the identity

$$(\alpha + \omega\beta + \omega^2\gamma)^3 - (\alpha + \omega^2\beta + \omega\gamma)^3 \equiv -3\sqrt{-3}\,(\beta - \gamma)(\gamma - \alpha)(\alpha - \beta).$$

[Apply Ex. 2, and substitute for $\omega - \omega^2$ its value $\sqrt{-3}$.]

7. Prove the identity

$$\alpha'^3 + \beta'^3 + \gamma'^3 - 3\alpha'\beta'\gamma' \equiv (\alpha^3 + \beta^3 + \gamma^3 - 3\alpha\beta\gamma)^2,$$

where

$$\alpha' \equiv \alpha^2 + 2\beta\gamma, \quad \beta' \equiv \beta^2 + 2\gamma\alpha, \quad \gamma' \equiv \gamma^2 + 2\alpha\beta.$$

8. Find the equation whose roots are

$$m + n, \quad \omega m + \omega^2 n, \quad \omega^2 m + \omega n.$$

$$Ans. \ x^3 - 3\,mnx - (m^3 + n^3) = 0.$$

9. Find the equation whose roots are

$$l + m + n, \quad l + \omega m + \omega^2 n, \quad l + \omega^2 m + \omega n.$$

$$Ans. \ x^3 - 3\,lx^2 + 3\,(l^2 - mn)\,x - (l^3 + m^3 + n^3 - 3\,lmn) = 0.$$

Remark.—Corresponding to the n n^{th} roots of unity there will be n n^{th} roots of any quantity. The n roots of the equation

$$x^n - a = 0$$

are the n n^{th} roots of a.

The three cube roots, for example, of a are

$$\sqrt[3]{a}, \quad \omega\sqrt[3]{a}, \quad \omega^2\sqrt[3]{a},$$

where $\sqrt[3]{a}$ represents the real cube root according to the ordinary arithmetical interpretation. Each of these satisfies the cubic

$$x^3 - a = 0.$$

It is to be observed that the three cube roots may be obtained by multiplying *any one* of the three above written by 1, ω, ω^2.

In addition, therefore, to the real cube root there are two imaginary cube roots obtained by multiplying the real cube root by the imaginary cube roots of unity. Thus, besides the ordinary cube root 3, the number 27 has the two imaginary cube roots

$$-\frac{3}{2} + \frac{3}{2}\sqrt{-3}, \quad -\frac{3}{2} - \frac{3}{2}\sqrt{-3},$$

as the student can easily verify.

10. Form a rational equation which shall have

$$\omega \sqrt[3]{Q + \sqrt{Q^2 + P^3}} + \omega^2 \sqrt[3]{Q - \sqrt{Q^2 + P^3}}$$

for a root; where $\omega^3 = 1$.

Compare Ex. 8.   *Ans.* $x^3 + 3Px - 2Q = 0$.

11 Form an equation with rational coefficients which shall have

$$\theta_1 \sqrt[3]{P} + \theta_2 \sqrt[3]{Q}$$

for a root, where $\theta_1{}^3 = 1$, and $\theta_2{}^3 = 1$.

Cubing the equation

$$x = \theta_1 \sqrt[3]{P} + \theta_2 \sqrt[3]{Q},$$

and substituting x for its value on the right-hand side, we get

$$x^3 - P - Q = 3\theta_1\theta_2 \sqrt[3]{PQ} \cdot x.$$

Cubing again, we have

$$(x^3 - P - Q)^3 = 27 PQ x^3.$$

Since θ_1 and θ_2 may each have any one of the values 1, ω, ω^2, the nine roots of this equation are

$$\sqrt[3]{P} + \sqrt[3]{Q}, \qquad \omega \sqrt[3]{P} + \omega \sqrt[3]{Q}, \qquad \omega^2 \sqrt[3]{P} + \omega^2 \sqrt[3]{Q},$$

$$\omega \sqrt[3]{P} + \omega^2 \sqrt[3]{Q}, \qquad \omega^2 \sqrt[3]{P} + \sqrt[3]{Q}, \qquad \omega \sqrt[3]{P} + \sqrt[3]{Q},$$

$$\omega^2 \sqrt[3]{P} + \omega \sqrt[3]{Q}, \qquad \sqrt[3]{P} + \omega^2 \sqrt[3]{Q}, \qquad \sqrt[3]{P} + \omega \sqrt[3]{Q}.$$

We see also that, since θ_1 and θ_2 have disappeared from the final equation, it is indifferent which of these nine roots is assumed equal to x in the first instance. The resulting equation is that which would have been obtained by multiplying together the nine factors of the form $x - \sqrt[3]{P} - \sqrt[3]{Q}$ obtained from the nine roots above written.

12. Form separately the three cubics whose roots are the groups in three (written in vertical columns in Ex. 11) of the roots of the equation of the preceding example. We can write them down from Ex. 8, taking first m and n equal to $\sqrt[3]{P}, \sqrt[3]{Q}$; then equal to $\omega\sqrt[3]{P}, \omega\sqrt[3]{Q}$; and finally equal to $\omega^2\sqrt[3]{P}, \omega^2\sqrt[3]{Q}$.

$$\textit{Ans.} \quad x^3 - 3\sqrt[3]{PQ}\,x - P - Q = 0,$$

$$x^3 - 3\omega^2\sqrt[3]{PQ}\,x - P - Q = 0,$$

$$x^3 - 3\omega\sqrt[3]{PQ}\,x - P - Q = 0.$$

27. Symmetric Functions of the Roots.—Symmetric functions of the roots of an equation are those in which all the roots are alike involved, so that the function is unaltered in value when any two of the roots are interchanged. For example, those functions of the roots (the sum, the sum of the products in pairs, &c.) with which we were concerned in Art. 23 are functions of this nature; for, as the student will readily perceive, if in any of these expressions the root a_1, let us say, be written in every place where a_2 occurs, and a_2 in every place where a_1 occurs, the value of the expression will be unchanged.

The functions discussed in Art. 23 are the simplest symmetric functions of the roots, each root entering in the first degree only in any term of any one of them.

By means of the equations (2) of Art. 23 the values of an infinite variety of symmetric functions of the roots can, without knowing the roots themselves, be obtained in terms of the coefficients. It will in fact be shown when we come to the more general discussion of this subject, that any rational symmetric function of the roots can be so expressed. We content ourselves in the present Chapter with some simple examples, most of which will have reference to equations of the third and fourth degrees. It is usual to represent a symmetric function by the Greek letter Σ attached to one term of it from which the entire expression may be written down. Thus, if a, β, γ be the roots of a cubic, $\Sigma a^2 \beta^2$ represents the symmetric function

$$a^2 \beta^2 + a^2 \gamma^2 + \beta^2 \gamma^2,$$

where the products in pairs are taken, and each term squared. Again, $\Sigma a^2 \beta$ represents

$$a^2 \beta + a^2 \gamma + \beta^2 \gamma + \beta^2 a + \gamma^2 a + \gamma^2 \beta,$$

where all possible permutations of the roots two by two are taken, and the first root in each term then squared.

For a biquadratic whose roots are a, β, γ, δ, $\Sigma a^2 \beta^2$ represents

$$a^2 \beta^2 + a^2 \gamma^2 + a^2 \delta^2 + \beta^2 \gamma^2 + \beta^2 \delta^2 + \gamma^2 \delta^2,$$

and so on.

EXAMPLES.

1. Find the value of $\Sigma \alpha^2 \beta$ of the roots of the cubic

$$x^3 + px^2 + qx + r = 0.$$

Multiplying together the equations

$$\alpha + \beta + \gamma = -p,$$

$$\beta\gamma + \gamma\alpha + \alpha\beta = q,$$

we obtain

$$\Sigma \alpha^2 \beta + 3\alpha\beta\gamma = -pq;$$

hence

$$\Sigma \alpha^2 \beta = 3r - pq.$$

2. Find for the same cubic the value of

$$\alpha^2 + \beta^2 + \gamma^2. \hspace{3cm} Ans. \ \Sigma \alpha^2 = p^2 - 2q.$$

3. Find for the same cubic the value of

$$\alpha^3 + \beta^3 + \gamma^3.$$

Multiplying the values of $\Sigma \alpha$ and $\Sigma \alpha^2$, we obtain

$$\alpha^3 + \beta^3 + \gamma^3 + \Sigma \alpha^2 \beta = -p^3 + 2pq;$$

hence, by Ex. 1,

$$\Sigma \alpha^3 = -p^3 + 3pq - 3r.$$

4. Find for the same cubic the value of

$$\beta^2 \gamma^2 + \gamma^2 \alpha^2 + \alpha^2 \beta^2.$$

We easily obtain

$$\beta^2 \gamma^2 + \gamma^2 \alpha^2 + \alpha^2 \beta^2 + 2\alpha\beta\gamma (\alpha + \beta + \gamma) = q^2,$$

from which

$$\Sigma \alpha^2 \beta^2 = q^2 - 2pr.$$

5. Find for the same cubic the value of

$$(\beta + \gamma)(\gamma + \alpha)(\alpha + \beta).$$

This is equal to

$$2\alpha\beta\gamma + \Sigma \alpha^2 \beta. \hspace{3cm} Ans. \ r - pq.$$

6. Find the value of the symmetric function

$$\alpha^2 \beta\gamma + \alpha^2 \beta\delta + \alpha^2 \gamma\delta + \beta^2 \alpha\gamma + \beta^2 \alpha\delta + \beta^2 \gamma\delta$$

$$+ \gamma^2 \alpha\beta + \gamma^2 \alpha\delta + \gamma^2 \beta\delta + \delta^2 \alpha\beta + \delta^2 \alpha\gamma + \delta^2 \beta\gamma$$

of the roots of the biquadratic

$$x^4 + px^3 + qx^2 + rx + s = 0.$$

Multiplying together

$$\alpha + \beta + \gamma + \delta = -p,$$

$$\alpha\beta\gamma + \alpha\beta\delta + \alpha\gamma\delta + \beta\gamma\delta = -r,$$

we obtain

$$\Sigma \alpha^2 \beta\gamma + 4\alpha\beta\gamma\delta = pr;$$

hence

$$\Sigma \alpha^2 \beta\gamma = pr - 4s.$$

7. Find for the same biquadratic the symmetric function

$$\alpha^2 + \beta^2 + \gamma^2 + \delta^2.$$

Squaring $\Sigma\alpha$, we easily obtain

$$\Sigma\alpha^2 = p^2 - 2q.$$

8. Find for the same biquadratic the symmetric function

$$\alpha^2\beta^2 + \alpha^2\gamma^2 + \alpha^2\delta^2 + \beta^2\gamma^2 + \beta^2\delta^2 + \gamma^2\delta^2.$$

Squaring the equation

$$\Sigma\alpha\beta = q,$$

we obtain

$$\Sigma\alpha^2\beta^2 + 2\Sigma\alpha^2\beta\gamma + 6\alpha\beta\gamma\delta = q^2;$$

hence

$$\Sigma\alpha^2\beta^2 = q^2 - 2pr + 2s.$$

9. Find for the same biquadratic $\Sigma\alpha^3\beta$.

To form this symmetric function, we take the two permutations $\alpha\beta$ and $\beta\alpha$ of the letters α, β; these give two terms $\alpha^3\beta$ and $\beta^3\alpha$ of Σ. We have similarly two terms from every other pair of the letters α, β, γ, δ; so that the symmetric function consists of 12 terms in all.

Multiply together the two equations

$$\Sigma\alpha\beta = q,$$

$$\Sigma\alpha^2 = p^2 - 2q,$$

$$\Sigma\alpha^2\, \Sigma\alpha\beta = \Sigma\alpha^3\beta + \Sigma\alpha^2\beta\gamma,$$

as can be easily seen.

It is convenient to remark here, that results of the kind expressed by this last equation can be verified by the consideration that the number of terms in both members of the equation must be the same. Thus, in the present instance, since $\Sigma\alpha^2$ contains 4 terms, and $\Sigma\alpha\beta$ 6 terms, their product must contain 24; and these are in fact the 12 terms which form $\Sigma\alpha^3\beta$, together with the 12 which form $\Sigma\alpha^2\beta\gamma$ (see Ex. 6).

Using results of previous examples, we have

$$\Sigma\alpha^3\beta = p^2q - 2q^2 - pr + 4s.$$

10. Find for the same biquadratic the value of

$$\alpha^4 + \beta^4 + \gamma^4 + \delta^4.$$

Squaring $\Sigma\alpha^2$, and employing results already obtained,

$$\Sigma\alpha^4 = p^4 - 4p^2q + 2q^2 + 4pr - 4s.$$

11. Find in terms of the coefficients the sum of the squares of the roots of the equation

$$x^n + p_1 x^{n-1} + p_2 x^{n-2} + \ldots + p_n = 0.$$

Squaring $\Sigma\alpha_1$, we easily find

$$p_1^2 = \Sigma\alpha_1^2 + 2\Sigma\alpha_1\alpha_2:$$

hence

$$\Sigma\alpha_1^2 = p_1^2 - 2p_2.$$

E

12. Find the sum of the reciprocals of the roots of the equation in the preceding example.

From the second last, and last of the equations of Art. 23, we have

$$\alpha_2\alpha_3\ldots\alpha_n + \alpha_1\alpha_3\ldots\alpha_n + \ldots + \alpha_1\alpha_2\ldots\alpha_{n-1} = (-1)^{n-1}p_{n-1},$$
$$\alpha_1\alpha_2\alpha_3\ldots\alpha_n = (-1)^n p_n;$$

dividing the former by the latter, we have

$$\frac{1}{\alpha_1} + \frac{1}{\alpha_2} + \frac{1}{\alpha_3} + \ldots + \frac{1}{\alpha_n} = \frac{-p_{n-1}}{p_n},$$

or

$$\Sigma\frac{1}{\alpha_1} = \frac{-p_{n-1}}{p_n}.$$

In a similar manner the sum of the products in pairs, in threes, &c. of the reciprocals of the roots can be found by dividing the 3rd last, or 4th last, &c. coefficient by the last.

13. Find for the cubic equation

$$a_0 x^3 + 3a_1 x^2 + 3a_2 x + a_3 = 0$$

the values, in terms of the coefficients, of the following three functions of the roots α, β, γ :—

$$(\beta-\gamma)^2 + (\gamma-\alpha)^2 + (\alpha-\beta)^2,$$
$$\alpha(\beta-\gamma)^2 + \beta(\gamma-\alpha)^2 + \gamma(\alpha-\beta)^2,$$
$$\alpha^2(\beta-\gamma)^2 + \beta^2(\gamma-\alpha)^2 + \gamma^2(\alpha-\beta)^2.$$

It will be often found convenient to write, as in the present example, an equation with *binomial coefficients*, that is, numerical coefficients corresponding to those in the expansion by the binomial theorem, in addition to the literal coefficients a_0, a_1, &c.

We easily obtain

$$a_0^2\{(\beta-\gamma)^2 + (\gamma-\alpha)^2 + (\alpha-\beta)^2\} = 18(a_1^2 - a_0 a_2),$$
$$a_0^2\{\alpha(\beta-\gamma)^2 + \beta(\gamma-\alpha)^2 + \gamma(\alpha-\beta)^2\} = 9(a_0 a_3 - a_1 a_2),$$
$$a_0^2\{\alpha^2(\beta-\gamma)^2 + \beta^2(\gamma-\alpha)^2 + \gamma^2(\alpha-\beta)^2\} = 18(a_2^2 - a_1 a_3).$$

14. Find in terms of the coefficients of the cubic in the preceding example the quadratic

$$(x-\alpha)^2(\beta-\gamma)^2 + (x-\beta)^2(\gamma-\alpha)^2 + (x-\gamma)^2(\alpha-\beta)^2 = 0,$$

where α, β, γ are the roots of the cubic.

$$\textit{Ans.} \quad (a_0 a_2 - a_1^2)x^2 + (a_0 a_3 - a_1 a_2)x + (a_1 a_3 - a_2^2) = 0.$$

15. Find for the cubic of Example 13 the value of

$$(2\alpha-\beta-\gamma)(2\beta-\gamma-\alpha)(2\gamma-\alpha-\beta).$$

Since

$$2\alpha-\beta-\gamma = 3\alpha-(\alpha+\beta+\gamma) = 3\alpha + \frac{3a_1}{a_0},$$

the required value is easily obtained by substituting $-\dfrac{a_1}{a_0}$ for x in the identity

$$a_0 x^3 + 3a_1 x^2 + 3a_2 x + a_3 \equiv a_0(x-\alpha)(x-\beta)(x-\gamma).$$

$$\textit{Ans.} \quad a_0^3(2\alpha-\beta-\gamma)(2\beta-\gamma-\alpha)(2\gamma-\alpha-\beta) = -27(a_0^2 a_3 - 3a_0 a_1 a_2 + 2a_1^3).$$

16. Find, for the biquadratic equation

$$a_0 x^4 + 4a_1 x^3 + 6a_2 x^2 + 4a_3 x + a_4 = 0,$$

the value of the symmetric function of the roots

$$(\beta - \gamma)^2 (\alpha - \delta)^2 + (\gamma - \alpha)^2 (\beta - \delta)^2 + (\alpha - \beta)^2 (\gamma - \delta)^2.$$

Here the equation is written with binomial coefficients corresponding to the expansion of the binomial to the 4th power. The symmetric function in question is easily seen to be identical with

$$2 \Sigma \alpha^2 \beta^2 - 2 \Sigma \alpha^2 \beta\gamma + 12 \alpha\beta\gamma\delta.$$

Employing the results of examples 6 and 8, we find

$$a_0^2 \{ (\beta - \gamma)^2 (\alpha - \delta)^2 + (\gamma - \alpha)^2 (\beta - \delta)^2 + (\alpha - \beta)^2 (\gamma - \delta)^2 \} = 24 \, (a_0 a_4 - 4 a_1 a_3 + 3 a_2^2).$$

17. Taking the six products in pairs of the four roots of the equation of Ex. 16, and adding each product, *e.g.* $\alpha\beta$, to that which contains the remaining two roots, *e.g.* $\gamma\delta$, we have the three sums in pairs

$$\beta\gamma + \alpha\delta, \qquad \gamma\alpha + \beta\delta, \qquad \alpha\beta + \gamma\delta \, ;$$

it is required to form the two following symmetric functions of the roots of the biquadratic :—

$$(\gamma\alpha + \beta\delta)(\alpha\beta + \gamma\delta) + (\alpha\beta + \gamma\delta)(\beta\gamma + \alpha\delta) + (\beta\gamma + \alpha\delta)(\gamma\alpha + \beta\delta),$$

$$(\beta\gamma + \alpha\delta)(\gamma\alpha + \beta\delta)(\alpha\beta + \gamma\delta).$$

The former of these is the sum of the products in pairs, and the latter the continued product, of the three expressions above given. As these three functions of the roots are important in the theory of the biquadratic, we shall represent them uniformly by the letters λ, μ, ν. We have to find expressions in terms of the coefficients for $\mu\nu + \nu\lambda + \lambda\mu$ and $\lambda\mu\nu$.

The former is $\Sigma \alpha^2 \beta\gamma$, and is easily found to be (see Ex. 6)

$$a_0^2 \Sigma \mu\nu = 4 \, (4 a_1 a_3 - a_0 a_4).$$

The latter is, when multiplied out, equal to

$$\alpha\beta\gamma\delta \left(\alpha^2 + \beta^2 + \gamma^2 + \delta^2 \right) + \alpha^2 \beta^2 \gamma^2 \delta^2 \left(\frac{1}{\alpha^2} + \frac{1}{\beta^2} + \frac{1}{\gamma^2} + \frac{1}{\delta^2} \right),$$

and gives after easy calculations the following :—

$$a_0^3 \lambda\mu\nu = 8 \, (2 a_0 a_3^2 - 3 a_0 a_2 a_4 + 2 a_1^2 a_4).$$

18. Find, for the biquadratic of Ex. 16, the following symmetric function :—

$$\{ (\gamma - \alpha)(\beta - \delta) - (\alpha - \beta)(\gamma - \delta) \} \{ (\alpha - \beta)(\gamma - \delta) - (\beta - \gamma)(\alpha - \delta) \}$$
$$\{ (\beta - \gamma)(\alpha - \delta) - (\gamma - \alpha)(\beta - \delta) \}.$$

This is an important symmetric function in the theory of the biquadratic. To prevent any ambiguity in writing this, or corresponding functions in which the differences of the roots of the biquadratic enter, we explain the notation we uniformly employ.

Taking in circular order the three roots α, β, γ, we have the three differences $\beta - \gamma$, $\gamma - \alpha$, $\alpha - \beta$; and subtracting δ from each root in turn, we have the three other differences $\alpha - \delta$, $\beta - \delta$, $\gamma - \delta$. We combine these in pairs as follows:—

$$(\beta - \gamma)(\alpha - \delta),$$

$$(\gamma - \alpha)(\beta - \delta),$$

$$(\alpha - \beta)(\gamma - \delta).$$

The symmetric function in question is the product of the differences of these three.

Employing the values of λ, μ, ν in the preceding example, we have

$$-\mu + \nu \equiv (\beta - \gamma)(\alpha - \delta), \quad -\nu + \lambda \equiv (\gamma - \alpha)(\beta - \delta), \quad -\lambda + \mu \equiv (\alpha - \beta)(\gamma - \delta).$$

We have, then, to find the value of

$$(2\lambda - \mu - \nu)(2\mu - \nu - \lambda)(2\nu - \lambda - \mu),$$

or

$$(3\lambda - \Sigma\,\alpha\beta)(3\mu - \Sigma\,\alpha\beta)(3\nu - \Sigma\,\alpha\beta),$$

in terms of the coefficients of the biquadratic.

Multiplying this out, substituting the value of $\Sigma\,\alpha\beta$, and attending to the results of Ex. 17, we obtain

$$a_0{}^3(2\lambda - \mu - \nu)(2\mu - \nu - \lambda)(2\nu - \lambda - \mu) = -432\{a_0 a_2 a_4 + 2a_1 a_2 a_3 - a_0 a_3{}^2 - a_1{}^2 a_4 - a_2{}^3\}.$$

The functions of the coefficients arrived at in Examples 15, 16, and the present example, will be found to be of great importance in the theory of the cubic and biquadratic equations.

19. Find, for the biquadratic of Ex. 16, the symmetric function

$$(\alpha - \beta)^2 + (\alpha - \gamma)^2 + (\alpha - \delta)^2 + (\beta - \gamma)^2 + (\beta - \delta)^2 + (\gamma - \delta)^2.$$

This may be written $\Sigma(\alpha - \beta)^2$.

$$Ans. \quad a_0{}^2\,\Sigma(\alpha - \beta)^2 = 48(a_1{}^2 - a_0 a_2).$$

20. Find, for the biquadratic of Example 6, the symmetric function $\Sigma(\alpha - \beta)^4$.

$$Ans. \quad \Sigma(\alpha - \beta)^4 = 3p^4 - 16p^2 q + 20q^2 + 4pr - 16s.$$

21. Show that, for the biquadratic of Example 16, the value of the same symmetric function can be written in the form

$$a_0{}^4\,\Sigma(\alpha - \beta)^4 = 16\{48(a_0 a_2 - a_1{}^2)^2 - a_0{}^2(a_0 a_4 - 4a_1 a_3 + 3a_2{}^2)\}.$$

22. Prove the following relation between the roots and coefficients of the biquadratic of Ex. 16:—

$$a_0{}^3(\beta + \gamma - \alpha - \delta)(\gamma + \alpha - \beta - \delta)(\alpha + \beta - \gamma - \delta) = 32(a_0{}^2 a_3 - 3a_0 a_1 a_2 + 2a_1{}^3).$$

28. *Remark.*—We close this chapter with certain observations which will be found useful in verifying the results of the calculation of symmetric functions. The first is, that *the degree of any symmetric function in the roots is always equal to the sum of the suffixes in each term of its value in terms of the coefficients.* The student will observe that this is true in the case of the results of Examples 13, 15, 16, 17, 18, 19, 21, 22; and that it must be so in general appears from the equations (2) of Art. 23, for the suffix of each coefficient in those equations is equal to the degree in the roots of the corresponding function of the roots; hence in any product of any powers of the coefficients the sum of the suffixes must be equal to the degree of the corresponding function of the roots.

The second observation is, that *for any symmetric function of the differences of the roots of an equation written with binomial coefficients, the value in terms of the coefficients must be such that the algebraic sum of the numerical factors in it is equal to cipher.* This is true in general; for if in the equation with binomial coefficients

$$a_0 x^n + n a_1 x^{n-1} + \frac{n \cdot n - 1}{1 \cdot 2} a_2 x^{n-2} + \ldots + a_n = 0$$

we make

$$a_0 = a_1 = a_2 = \ldots = a_n = 1,$$

it becomes

$$(x + 1)^n = 0,$$

i.e. all the roots become equal; hence any function of the differences of the roots must in that case vanish, and therefore also the function of the coefficients which is equal to it; but this consists of the algebraic sum of the numerical factors after putting

$$a_0 = a_1 = a_2 = \ldots = 1.$$

In examples 15, 16, 18, 19, 21, 22, of Art. 27, we have instances of this.

Miscellaneous Examples.

1. Find the value of the symmetric function

$$\frac{\beta^2 + \gamma^2}{\beta\gamma} + \frac{\gamma^2 + \alpha^2}{\gamma\alpha} + \frac{\alpha^2 + \beta^2}{\alpha\beta}$$

of the roots of the cubic equation

$$x^3 + px^2 + qx + r = 0.$$

Ans. $\dfrac{pq}{r} - 3.$

2. Find for the same cubic the value of

$$(\beta + \gamma - \alpha)^3 + (\gamma + \alpha - \beta)^3 + (\alpha + \beta - \gamma)^3.$$

Ans. $24\,r - p^3.$

3. Find the value of $\Sigma\,\alpha^3\beta^3$ of the roots of the same equation.

Here $\Sigma\,\alpha\beta\,\Sigma\,\alpha^2\beta^2 = \Sigma\,\alpha^3\beta^3 + \alpha\beta\gamma\,\Sigma\,\alpha^2\beta$; hence &c.

Ans. $q^3 - 3pqr + 3r^2.$

4. Find for the same cubic the symmetric function

$$(\beta^3 - \gamma^3)^2 + (\gamma^3 - \alpha^3)^2 + (\alpha^3 - \beta^3)^2.$$

$\Sigma\,\alpha^6$ is easily obtained by squaring $\Sigma\,\alpha^3$ (see Ex. 3, Art. 27).

Ans. $2p^6 - 12p^4q + 12p^3r + 18p^2q^2 - 18pqr - 6q^3.$

5. Find for the same cubic the value of

$$\frac{\beta^2 + \gamma^2}{\beta + \gamma} + \frac{\gamma^2 + \alpha^2}{\gamma + \alpha} + \frac{\alpha^2 + \beta^2}{\alpha + \beta}.$$

Ans. $\dfrac{2p^2q - 4pr - 2q^2}{r - pq}.$

6. Find for the same cubic the value of

$$\frac{\alpha^2 + \beta\gamma}{\beta + \gamma} + \frac{\beta^2 + \gamma\alpha}{\gamma + \alpha} + \frac{\gamma^2 + \alpha\beta}{\alpha + \beta}.$$

Ans. $\dfrac{p^4 - 3p^2q + 5pr + q^2}{r - pq}.$

7. Find for the same cubic the value of

$$\frac{2\beta\gamma - \alpha^2}{\beta + \gamma - \alpha} + \frac{2\gamma\alpha - \beta^2}{\gamma + \alpha - \beta} + \frac{2\alpha\beta - \gamma^2}{\alpha + \beta - \gamma}.$$

Ans. $\dfrac{p^4 - 2p^2q + 14pr - 8q^2}{4pq - p^3 - 8r}.$

8. Find the symmetric function $\Sigma\left(\dfrac{\alpha - \beta}{\alpha + \beta}\right)^2$ for the same cubic.

Ans. $\dfrac{-p^2q^2 - 4p^3r + 8q^3 - 2pqr - 9r^2}{(r - pq)^2}.$

9. Find in terms of the coefficients the value of $\Sigma \dfrac{\alpha\beta}{\gamma^2}$ of the biquadratic equation

$$x^4 + px^3 + qx^2 + rx + s = 0.$$

Here $\Sigma \alpha\beta \, \Sigma \dfrac{1}{\alpha^2} = \Sigma \dfrac{\alpha}{\beta} + \Sigma \dfrac{\alpha\beta}{\gamma^2}$; and $\Sigma \alpha \, \Sigma \dfrac{1}{\alpha} = 4 + \Sigma \dfrac{\alpha}{\beta}$.

$$Ans. \quad \frac{qr^2 - 2q^2 s - prs + 4s^2}{s^2}.$$

10. Find the value of $\Sigma \dfrac{\alpha}{\beta^2}$ for the equation

$$x^n + p_1 x^{n-1} + p_2 x^{n-2} + \ldots + p_{n-1} x + p_n = 0.$$

$$Ans. \quad \frac{p_{n-1} p_n - p_1 p_{n-1}^2 + 2 p_1 p_{n-2} p_n}{p_n^2}.$$

11. Find for the biquadratic of Question 9 the value of

$$(\beta\gamma - \alpha\delta)(\gamma\alpha - \beta\delta)(\alpha\beta - \gamma\delta).$$

Compare Ex. 22, Art. 24. $\qquad\qquad\qquad Ans. \quad r^2 - p^2 s.$

12. Find the value of $\Sigma (a_0 \alpha + a_1)^2 (\beta - \gamma)^2$ of the roots of the cubic

$$a_0 x^3 + 3 a_1 x^2 + 3 a_2 x + a_3 = 0.$$

$$Ans. \quad \frac{18}{a_0^2} (a_0 a_2 - a_1^2)^2.$$

13. Find the symmetric function $\Sigma \dfrac{a_1^2 + a_2^2}{a_1 a_2}$ of the roots of the equation

$$x^n + p_1 x^{n-1} + p_2 x^{n-2} + \ldots + p_{n-1} x + p_n = 0.$$

The given function may be written in the form

$$a_1 \left\{ \frac{1}{a_1} + \frac{1}{a_2} + \ldots + \frac{1}{a_n} \right\} - 1$$

$$+ a_2 \left\{ \frac{1}{a_1} + \frac{1}{a_2} + \ldots + \frac{1}{a_n} \right\} - 1$$

$$+ \ldots\ldots\ldots$$

$$+ a_n \left\{ \frac{1}{a_1} + \frac{1}{a_2} + \ldots + \frac{1}{a_n} \right\} - 1,$$

or $\Sigma a_1 \, \Sigma \dfrac{1}{a_1} - n$; hence $\qquad\qquad Ans. \quad \dfrac{p_1 p_{n-1}}{p_n} - n.$

14. Clear of radicals the equation

$$\sqrt{t - \alpha^2} + \sqrt{t - \beta^2} + \sqrt{t - \gamma^2} = 0;$$

and express the coefficients of the resulting equation in t in terms of the coefficients of the cubic of Ex. 1.

$$Ans. \quad 3t^2 - 2(p^2 - 2q)t - p^4 + 4p^2 q - 8pr = 0.$$

15. If α, β, γ, δ be the roots of the biquadratic of Question 9, prove

$$(\alpha^2 + 1)\,(\beta^2 + 1)\,(\gamma^2 + 1)\,(\delta^2 + 1) = (1 - q + s)^2 + (p - r)^2.$$

Substitute in turn each of the roots of the equation $x^2 + 1 = 0$ in the identity of Art. 16, and multiply.

16. For the general equation prove

$$(\alpha_1^2 + 1)\,(\alpha_2^2 + 1)\ldots\ldots(\alpha_n^2 + 1) = (1 - p_2 + p_4 - \ldots)^2 + (p_1 - p_3 + \ldots)^2.$$

17. If α, β, γ, δ be the roots of the equation

$$a_0 x^4 + 4\,a_1 x^3 + 6\,a_2 x^2 + 4\,a_3 x + a_4 = 0\,;$$

prove

$$a_0^3\,(\beta + \gamma)\,(\gamma + \alpha)\,(\alpha + \beta)\,(\alpha + \delta)\,(\beta + \delta)\,(\gamma + \delta) = 16\,\{\,6\,a_1 a_2 a_3 - a_0 a_3^2 - a_1^2 a_4\,\}.$$

The symmetric function in question is $(\mu + \nu)\,(\nu + \lambda)\,(\lambda + \mu)$, or $\Sigma\lambda\,\Sigma\mu\nu - \lambda\mu\nu$, where λ, μ, ν have the values of Ex. 17, Art. 27.

18. The distances on a right line of two pairs of points from a fixed origin are the roots (α, β) and (α', β') of the two quadratic equations

$$ax^2 + 2\,bx + c = 0,\quad a' x^2 + 2\,b' x + c' = 0\,;$$

prove that when one pair of the points are the harmonic conjugates of the other pair, the following relation exists :—

$$ac' + a' c - 2\,bb' = 0.$$

19. The distances of three points A, B, C on a right line from a fixed origin O on the line are the roots of the equation

$$ax^3 + 3\,bx^2 + 3\,cx + d = 0\,;$$

find the condition that one of the points A, B, C should bisect the distance between the other two.

Compare Ex. 15, Art. 27.   *Ans.* $a^2 d - 3\,abc + 2\,b^3 = 0.$

20. Retaining the notation of the preceding question, find the condition that the four points O, A, B, C should form a harmonic division.

$$\textit{Ans.}\quad ad^2 - 3\,bcd + 2\,c^3 = 0.$$

This can be derived from the result of Ex. 19 by changing the roots into their reciprocals, or it can be easily calculated independently.

21. If the roots $(\alpha, \beta, \gamma, \delta)$ of the equation

$$ax^4 + 4\,bx^3 + 6\,cx^2 + 4\,dx + e = 0$$

are so related that $\alpha - \delta$, $\beta - \delta$, $\gamma - \delta$ are in harmonic progression ; prove the relation among the coefficients

$$ace + 2\,bcd - ad^2 - b^2 e - c^3 = 0.$$

Compare Ex. 18, Art. 27.

22. Express

$$(2\,\beta\gamma - \gamma\alpha - \alpha\beta)\,(2\,\gamma\alpha - \alpha\beta - \beta\gamma)\,(2\,\alpha\beta - \beta\gamma - \gamma\alpha)$$

as the sum of two cubes.

$$\textit{Ans.}\quad (\beta\gamma + \omega\gamma\alpha + \omega^2 \alpha\beta)^3 + (\beta\gamma + \omega^2 \gamma\alpha + \omega\alpha\beta)^3.$$

Compare Ex. 5, Art. 26.

23. Form the equation whose roots are

$$-\frac{\beta\gamma + \omega\gamma\alpha + \omega^2\alpha\beta}{\alpha + \omega\beta + \omega^2\gamma}, \qquad -\frac{\beta\gamma + \omega^2\gamma\alpha + \omega\alpha\beta}{\alpha + \omega^2\beta + \omega\gamma},$$

where $\omega^3 = 1$, and α, β, γ are the roots of the cubic

$$ax^3 + 3bx^2 + 3cx + d = 0.$$

Ans. $(ac - b^2)\, x^2 + (ad - bc)\, x + (bd - c^2) = 0.$

Compare Exs. 13 and 14, Art. 27.

24. Clear of radicals the following equation, where $\omega^3 = 1$:—

$$x = \sqrt{a(1-\omega) + b(1-\omega^2)} + \sqrt{a(1-\omega^2) + b(1-\omega)}.$$

Ans. $x^4 - 6(a+b)\, x^2 - 3(a-b)^2 = 0.$

25. Express

$$(x + y + z)^3 + (x + \omega y + \omega^2 z)^3 + (x + \omega^2 y + \omega z)^3$$

in terms of $x^3 + y^3 + z^3$ and xyz, where $\omega^3 = 1$.

Ans. $3(x^3 + y^3 + z^3) + 18\,xyz.$

26. If

$$(x^3 + y^3 + z^3 - 3xyz)(x'^3 + y'^3 + z'^3 - 3x'y'z') \equiv X^3 + Y^3 + Z^3 - 3XYZ,$$

find X, Y, Z in terms of x, y, z ; x', y', z'.

Apply Example 4, Art. 26.

Ans. $X = xx' + yy' + zz', \quad Y = xy' + yz' + zx', \quad Z = xz' + yx' + zy'.$

27. Resolve

$$(\alpha + \beta + \gamma)^3 \alpha\beta\gamma - (\beta\gamma + \gamma\alpha + \alpha\beta)^3$$

into three factors, each of the second degree in α, β, γ.

Ans. $(\alpha^2 - \beta\gamma)(\beta^2 - \gamma\alpha)(\gamma^2 - \alpha\beta).$

Compare Ex. 18, Art. 24.

28. Resolve

$$l^2 + m^2 + n^2 - 2mn - 2nl - 2lm$$

into four factors.

Ans. $(\sqrt{l} + \sqrt{m} + \sqrt{n})(\sqrt{l} - \sqrt{m} - \sqrt{n})(-\sqrt{l} + \sqrt{m} - \sqrt{n})(-\sqrt{l} - \sqrt{m} + \sqrt{n}).$

29. Resolve

$$(m - n)^2 + (n - l)^2 + (l - m)^2$$

into factors.

Ans. $2(l + \omega m + \omega^2 n)(l + \omega^2 m + \omega n).$

30. If

$$a + m + n = 0, \quad b + \omega m + \omega^2 n = 0, \quad c + \omega^2 m + \omega n = 0,$$

where $\omega^3 = 1$, prove

$$(b - c)^2 (c - a)^2 (a - b)^2 = -27(m^3 - n^3)^2.$$

31. Find the value of

$$(\alpha^2 + 2)(\beta^2 + 2)(\gamma^2 + 2)(\delta^2 + 2),$$

where α, β, γ, δ are the roots of the equation

$$x^4 - 7x^3 + 8x^2 - 5x + 10 = 0.$$

Substitute in succession for x the two roots of the equation $x^2 + 2 = 0$, and multiply.   (Compare Ex. 15.)

Ans.   166.

32. Resolve into simple factors each of the following expressions :—

(1). $(\beta - \gamma)^2 (\beta + \gamma - 2\alpha) + (\gamma - \alpha)^2 (\gamma + \alpha - 2\beta) + (\alpha - \beta)^2 (\alpha + \beta - 2\gamma)$.

(2). $(\beta - \gamma)(\beta + \gamma - 2\alpha)^2 + (\gamma - \alpha)(\gamma + \alpha - 2\beta)^2 + (\alpha - \beta)(\alpha + \beta - 2\gamma)^2$.

Ans.   (1). $(2\alpha - \beta - \gamma)(2\beta - \gamma - \alpha)(2\gamma - \alpha - \beta)$.
(2). $-9(\beta - \gamma)(\gamma - \alpha)(\alpha - \beta)$.

33. Find the condition that the cubic equation

$$x^3 - px^2 + qx - r = 0$$

should have a pair of roots of the form $a \pm a\sqrt{-1}$; and show how to determine the roots in that case.

If the real root is b, we easily find, by forming the sum of the squares of the roots, $p^2 - 2q = b^2$.   The required condition is

$$(p^2 - 2q)(q^2 - 2pr) - r^2 = 0.$$

34. Solve the equation

$$x^3 - 7x^2 + 20x - 24 = 0$$

whose roots are of the form indicated in **Ex.** 33.

Ans.   Roots 3, and $2 \pm 2\sqrt{-1}$.

35. Find the conditions that the biquadratic equation

$$x^4 - px^3 + qx^2 - rx + s = 0$$

should have roots of the form $a \pm a\sqrt{-1}$, $b \pm b\sqrt{-1}$.   Here there must be two conditions among the coefficients, as there are only two independent quantities involved in the roots.

Ans. $p^2 - 2q = 0$; $r^2 - 2qs = 0$.

36. Solve the biquadratic

$$x^4 - 4x^3 + 8x^2 - 120x + 900 = 0$$

whose roots are of the form in **Ex.** 35.

Ans. $3 \pm 3\sqrt{-1}$, $-5 \mp 5\sqrt{-1}$.

37. If $\alpha + \beta\sqrt{-1}$ be a root of the equation

$$x^3 + qx + r = 0,$$

prove that 2α will be a root of the equation

$$x^3 + qx - r = 0.$$

38. Find the condition that the cubic equation

$$x^3 + px^2 + qx + r = 0$$

should have two roots α, β connected by the relation $\alpha\beta + 1 = 0$.

Ans. $1 + q + pr + r^2 = 0$.

39. Find the condition that the biquadratic

$$x^4 + px^3 + qx^2 + rx + s = 0$$

should have two roots connected by the relation $\alpha\beta + 1 = 0$.

The condition arranged according to powers of s is

$$1 + q + pr + r^2 + (p^2 + pr - 2q - 1)s + (q - 1)s^2 + s^3 = 0.$$

40. Solve the equation

$$x^5 - 11x^4 + 32x^3 + 8x^2 - 144x + 144 = 0,$$

three roots of which are in Harmonic Progression, and two of the form $\pm a$.

Ans. Roots $\pm 2, 2, 3, 6$.

41. Find the value of $\Sigma(a_1 - a_2)^2 a_3 a_4 \ldots a_n$ of the roots of the equation

$$x^n + p_1 x^{n-1} + p_2 x^{n-2} + \ldots + p_n = 0.$$

This is readily reducible to Ex. 13.

Ans. $(-1)^n \{ p_1 p_{n-1} - n^2 p_n \}$.

42. If the roots of the equation

$$a_0 x^n + n a_1 x^{n-1} + \frac{n(n-1)}{1 \cdot 2} a_2 x^{n-2} + \ldots + a_n = 0$$

be in Arithmetical Progression, show that they can be obtained from the expression

$$-\frac{a_1}{a_0} \pm \frac{r}{a_0} \sqrt{\frac{3(a_1^2 - a_0 a_2)}{n+1}}$$

by giving to r all the values $1, 3, 5, \ldots n-1$, when n is even; and all the values $0, 2, 4, 6 \ldots n-1$, when n is odd.

43. Representing the differences of three quantities α, β, γ by $\alpha_1, \beta_1, \gamma_1$, as follows :—

$$\alpha_1 = \beta - \gamma, \quad \beta_1 = \gamma - \alpha, \quad \gamma_1 = \alpha - \beta;$$

prove the relations

$$\alpha_1^3 + \beta_1^3 + \gamma_1^3 = 3\alpha_1 \beta_1 \gamma_1,$$
$$\alpha_1^4 + \beta_1^4 + \gamma_1^4 = \tfrac{1}{2} \{ \alpha_1^2 + \beta_1^2 + \gamma_1^2 \}^2,$$
$$\alpha_1^5 + \beta_1^5 + \gamma_1^5 = \tfrac{5}{2} \{ \alpha_1^2 + \beta_1^2 + \gamma_1^2 \} \alpha_1 \beta_1 \gamma_1.$$

These can be derived by taking $\alpha_1, \beta_1, \gamma_1$ to be roots of the equation

$$x^3 + qx - r = 0$$

(where the second term is absent since the sum of the roots $= 0$), and calculating the symmetric functions $\Sigma \alpha_1^3, \Sigma \alpha_1^4, \Sigma \alpha_1^5$ in terms of q and r. The process can be extended to form $\Sigma \alpha_1^6, \Sigma \alpha_1^7$, &c. They are all capable of being expressed in terms of the product $\alpha_1 \beta_1 \gamma_1$ and the sum of squares $\alpha_1^2 + \beta_1^2 + \gamma_1^2$; the former being equal to r, and the latter to $-2(\beta_1 \gamma_1 + \gamma_1 \alpha_1 + \alpha_1 \beta_1)$, or $-2q$.

44. If α, β, γ be the roots of the cubic

$$a_0 x^3 + 3a_1 x^2 + 3a_2 x + a_3 = 0,$$

find the value of $\Sigma(\beta - \gamma)^4$ in terms of the coefficients. This is easily obtained by means of the second relation in Ex. 43.

Ans. $a_0^4 \Sigma(\beta - \gamma)^4 = 162(a_0 a_2 - a_1^2)^2$.

CHAPTER IV.

29. Transformation of Equations.—We can, without knowing the roots of an equation, transform it into another whose roots shall have certain assigned relations to those of the proposed. The utility of this process consists in the fact that the discussion of the transformed equation will often be more simple than that of the original. We proceed to explain the most important transformations of equations.

30. **Roots with Signs changed.**—To transform an equation into another whose roots are those of the given equation with contrary signs, let a_1, a_2, a_3, ... a_n be the roots of

$$x^n + p_1 x^{n-1} + p_2 x^{n-2} + \ldots . p_{n-1}x + p_n = 0 ;$$

then

$$x^n + p_1 x^{n-1} + p_2 x^{n-2} + \ldots + p_{n-1}x + p_n \equiv (x - a_1)(x - a_2) \ldots (x - a_n) ;$$

change x into $-y$; we have, then, whether n be even or odd,

$$y^n - p_1 y^{n-1} + p_2 y^{n-2} - \ldots \pm p_{n-1}y \mp p_n \equiv (y + a_1)(y + a_2) \ldots (y + a_n).$$

The polynomial in y equated to zero is an equation whose roots are $-a_1$, $-a_2$, ... $-a_n$; and to effect the required transformation we have only to *change the signs of every alternate term of the given equation beginning with the second.*

Examples.

1. Find the equation whose roots are those of

$$x^5 + 7x^4 + 7x^3 - 8x^2 + x + 1 = 0$$

with their signs changed. *Ans.* $x^5 - 7x^4 + 7x^3 + 8x^2 + x - 1 = 0.$

2. Change the signs of the roots of the equation

$$x^7 + 3x^5 + x^3 - x^2 + 7x + 2 = 0.$$

[Supply the missing terms with zero coefficients.]

Ans. $x^7 + 3x^5 + x^3 + x^2 + 7x - 2 = 0.$

31. Roots Multiplied by a Given Quantity.—To transform an equation whose roots are $a_1, a_2, \ldots a_n$ into another whose roots are $ma_1, ma_2, \ldots ma_n$, we change x into $\dfrac{y}{m}$ in the identity of the preceding article. We have then, after multiplication by m^n,

$$y^n + mp_1 y^{n-1} + m^2 p_2 y^{n-2} + \ldots + m^{n-1} p_{n-1} y + m^n p_n$$
$$\equiv (y - ma_1)(y - ma_2) \ldots \ldots (y - ma_n).$$

Hence, to multiply the roots of an equation by a given quantity m, *we have only to multiply the successive coefficients, beginning with the second, by m, m^2, m^3, $\ldots m^n$.*

The present transformation is useful for getting rid of the coefficient of the first term of an equation when it is not unity; and generally for removing fractional coefficients from an equation. If there is a coefficient a_0 of the first term, we form the equation whose roots are $a_0 a_1$, $a_0 a_2$, $\ldots a_0 a_n$; the transformed equation will be divisible by a_0, and after such division the coefficient of x^n will be unity.

When there are fractional coefficients, we can get rid of them by multiplying the roots by a quantity m, which is the least common multiple of all the denominators of the fractions. In many cases, multiplication by a quantity less than the least common multiple will be sufficient for this purpose, as will appear in the following examples :—

EXAMPLES.

1. Change the equation
$$3x^4 - 4x^3 + 4x^2 - 2x + 1 = 0$$
into another the coefficient of whose highest term will be unity. We multiply the roots by 3.

Ans. $x^4 - 4x^3 + 12x^2 - 18x + 27 = 0.$

2. Remove the fractional coefficients from the equation

$$x^3 - \frac{1}{2}x^2 + \frac{2}{3}x - 1 = 0.$$

Multiply the roots by 6.

Ans. $x^3 - 3x^2 + 24x - 216 = 0.$

3. Remove the fractional coefficients from the equation

$$x^3 - \frac{5}{2}x^2 - \frac{7}{18}x + \frac{1}{108} = 0.$$

By noting the factors which occur in the denominators of these fractions, we easily observe that a number much smaller than the least common multiple will suffice to remove the fractions. If the required multiplier is m, we write the transformed equation thus:—

$$x^3 - m\frac{5}{2}x^2 - m^2\frac{7}{3^2 \cdot 2}x + \frac{m^3}{3^3 \cdot 2^2} = 0 \; ;$$

it is evident that the value 6 for m will make each coefficient integral; hence we have only to multiply the roots by 6.

$$\textit{Ans. } x^3 - 15x^2 - 14x + 2 = 0.$$

4. Remove the fractional coefficients from the equation

$$x^4 + \frac{3}{10}x^2 + \frac{13}{25}x + \frac{77}{1000} = 0.$$

The student must be careful in examples of this kind to supply the missing terms with zero coefficients. The required multiplier is 10.

$$\textit{Ans. } x^4 + 30x^2 + 520x + 770 = 0.$$

5. Remove the fractional coefficients from the equation

$$x^4 - \frac{5}{6}x^3 + \frac{5}{12}x^2 - \frac{13}{900} = 0.$$

$$\textit{Ans. } x^4 - 25x^3 + 375x^2 - 11700 = 0.$$

32. Reciprocal Roots.—To transform an equation into one whose roots are the reciprocals of the roots of the proposed equation, we change x into $\dfrac{1}{y}$ in the identity of Art. 30. This gives, after certain easy reductions,

$$\frac{1}{y^n} + \frac{p_1}{y^{n-1}} + \frac{p_2}{y^{n-2}} + \ldots + \frac{p_{n-1}}{y} + p_n = \frac{p_n}{y^n}\left(y - \frac{1}{a_1}\right)\left(y - \frac{1}{a_2}\right)\ldots\left(y - \frac{1}{a_n}\right),$$

or

$$y^n + \frac{p_{n-1}}{p_n}y^{n-1} + \frac{p_{n-2}}{p_n}y^{n-2} + \ldots + \frac{p_1}{p_n}y + \frac{1}{p_n} = \left(y - \frac{1}{a_1}\right)\left(y - \frac{1}{a_2}\right)\ldots\left(y - \frac{1}{a_n}\right);$$

hence, if in the given equation we replace x by $\dfrac{1}{y}$, and multiply by y^n, the resulting polynomial in y equated to zero will have for roots $\dfrac{1}{a_1}, \dfrac{1}{a_2}, \ldots \dfrac{1}{a_n}$.

Example.

Find the equation whose roots are the reciprocals of those of

$$x^4 - 3x^3 + 7x^2 + 5x - 2 = 0.$$

$$\textit{Ans. } 2y^4 - 5y^3 - 7y^2 + 3y - 1 = 0.$$

33. Reciprocal Equations.—There is a certain class of equations which remain unaltered when x is changed into $\dfrac{1}{x}$. These are called *reciprocal equations*. The conditions which must obtain among the coefficients of an equation in order that it should be one of this class are, from the preceding Article, plainly the following :—

$$\frac{p_{n-1}}{p_n} = p_1, \quad \frac{p_{n-2}}{p_n} = p_2, \quad \&c. \ldots \frac{p_1}{p_n} = p_{n-1}, \quad \frac{1}{p_n} = p_n.$$

The last of these conditions gives $p_n^2 = 1$, or $p_n = \pm 1$. Reciprocal equations are divided into two classes, according as p_n is equal to $+1$, or to -1.

(1). In the first case

$$p_{n-1} = p_1, \quad p_{n-2} = p_2, \quad \ldots \quad p_1 = p_{n-1} ;$$

and we have the *first class of reciprocal equations, in which the coefficients of the corresponding terms taken from the beginning and end are equal in magnitude and have the same signs.*

(2). In the second case, when $p_n = -1$,

$$p_{n-1} = -p_1, \quad p_{n-2} = -p_2, \quad \&c. \ldots p_1 = -p_{n-1} ;$$

and we have the *second class of reciprocal equations, in which corresponding terms counting from the beginning and end are equal in magnitude but different in sign.* It is to be observed that in this case when the degree of the equation is even, say $n = 2m$, one of the conditions becomes $p_m = -p_m$, or $p_m = 0$; so that in reciprocal equations of the second class, whose degree is even, the middle term is absent.

If a be a root of a reciprocal equation, $\dfrac{1}{a}$ must also be a root, for it is a root of the transformed equation, and the transformed equation is identical with the proposed ; hence the roots of a reciprocal equation occur in pairs, $a, \dfrac{1}{a}; \beta, \dfrac{1}{\beta}; \&c.$ When the degree is odd there must be a root which is its own reciprocal ; and it is in fact obvious from the form of the equation that -1, or $+1$ is then a root, according as the equation is of the first or second of the above classes. In either case we can divide off by the known factor $(x + 1$ or $x - 1)$, and we are left a reciprocal equation of even degree and of the first class. In equations of the second class of even degree $x^2 - 1$ is a factor, since the equation may be written in the form

$$x^n - 1 + p_1 x (x^{n-2} - 1) + \ldots = 0.$$

By dividing by $x^2 - 1$, this also is reducible to a reciprocal equation of the first class of even degree. Hence all reciprocal equations may be reduced to *those of the first class whose degree is even*, and this consequently may be regarded as *the standard form of reciprocal equations*.

Reduce to a reciprocal equation of even degree and of first class

$$x^6 + \frac{5}{6} x^5 - \frac{22}{3} x^4 + \frac{22}{3} x^2 - \frac{5}{6} x - 1 = 0.$$

$$Ans. \quad x^4 + \frac{5}{6} x^3 - \frac{19}{3} x^2 + \frac{5}{6} x + 1 = 0.$$

34. To Increase or Diminish the Roots by a Given Quantity.—We change the variable in the polynomial $f(x)$ by the substitution $x = y + h$; the resulting equation in y will have roots each less or greater by h than the given equation in x, according as h is positive or negative. The resulting equation is (see Art. 6)

$$f(h) + f'(h) y + \frac{f''(h)}{1 \cdot 2} y^2 + \frac{f'''(h)}{1 \cdot 2 \cdot 3} y^3 + \ldots = 0.$$

There is a mode of formation of this equation which for practical purposes is much more convenient than the direct calculation of the derived functions, and the substitution in them of the given quantity h. This we proceed to explain. Let the proposed equation be

$$a_0 x^n + a_1 x^{n-1} + a_2 x^{n-2} + \ldots + a_{n-1} x + a_n = 0 ;$$

and suppose the transformed polynomial in y to be

$$A_0 y^n + A_1 y^{n-1} + A_2 y^{n-2} + \ldots + A_{n-1} y + A_n:$$

since $y = x - h$, this is equivalent to

$$A_0 (x - h)^n + A_1 (x - h)^{n-1} + \ldots + A_{n-1} (x - h) + A_n,$$

which must be identical with the given polynomial: we conclude that if the given polynomial be divided by $x - h$, the remainder is A_n, and the quotient

$$A_0 (x - h)^{n-1} + A_1 (x - h)^{n-2} + \ldots + A_{n-2} (x - h) + A_{n-1}:$$

if this again be divided by $x - h$, the remainder is A_{n-1}, and the quotient

$$A_0 (x - h)^{n-2} + A_1 (x - h)^{n-3} + \ldots + A_{n-2}.$$

Proceeding in this way, we are able by a repetition of arithmetical processes, of the kind explained in Art. 8, to calculate in succession the several coefficients A_n, A_{n-1}, &c., of the transformed equation; the last, A_0, being equal to a_0.

EXAMPLES.

1. Find the equation whose roots are those of

$$x^4 - 5x^3 + 7x^2 - 17x + 11 = 0,$$

each diminished by 4.

The operation is best exhibited as follows :—

1	− 5	7	− 17	11
	4	− 4	12	− 20
	− 1	3	− 5	− 9
	4	12	60	
	3	15	**55**	
	4	28		
	7	**43**		
	4			
	11			

Here the first division of the given polynomial by $x - 4$ gives the remainder $- 9$ ($= A_4$), and the quotient $x^3 - x^2 + 3x - 5$ (cf. Art. 8). Dividing this again by

F

$x-4$, we get the remainder 55 ($=A_3$), and the quotient $x^2+3x+15$. Dividing this again, we get the remainder 43 ($=A_2$), and quotient $x+7$; and dividing this we get $A_1=11$, and $A_0=1$; hence the required transformed equation is

$$y^4+11y^3+43y^2+55y-9=0.$$

2. Find the equation whose roots are those of

$$x^5+4x^3-x^2+11=0,$$

each diminished by 3.

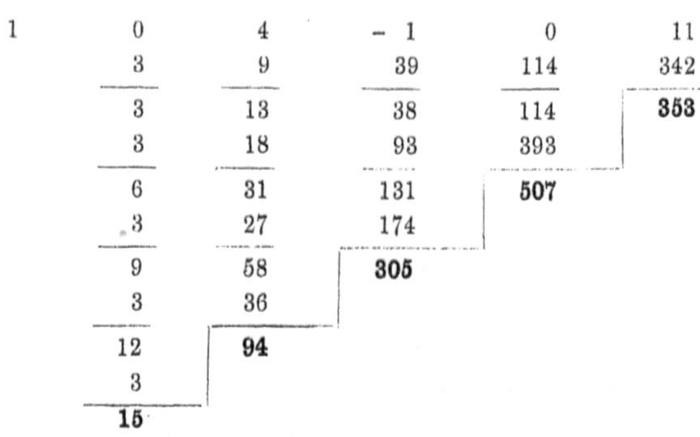

The transformed equation is, therefore,

$$y^5+15y^4+94y^3+305y^2+507y+353=0.$$

The student is recommended to attend carefully to these examples; as he will find, when we come to a discussion of the general solution of numerical equations, that the most convenient method of effecting such solution is only a repetition of the process here described.

3. Find the equation whose roots are those of

$$4x^5-2x^3+7x-3=0,$$

each increased by 2.

[The multiplier in this operation is, of course, -2.]

$$Ans. \quad 4y^5-40y^4+158y^3-308y^2+303y-129=0.$$

4. Increase by 7 the roots of the equation

$$3x^4+7x^3-15x^2+x-2=0.$$

$$Ans. \quad 3y^4-77y^3+720y^2-2876y+4058=0.$$

5. Diminish by 23 the roots of the equation

$$5x^3-13x^2-12x+7=0.$$

The operation may be conveniently performed by first diminishing the roots by 20, and then diminishing the roots of the transformed equation again by 3. The process may be exhibited in two stages, as follows, the broken lines marking the conclusion of each stage :—

$$
\begin{array}{cccc}
5 & -\ 13 & -\ 12 & 7 \\
 & 100 & 1740 & 34560 \\
\hline
 & 87 & 1728 & \mathbf{34567} \\
 & 100 & 3740 & 19122 \\
\hline
 & 187 & \mathbf{5468} & \mathbf{53689} \\
 & 100 & 906 & \\
\hline
 & \mathbf{287} & 6374 & \\
 & 15 & 951 & \\
\hline
 & 302 & \mathbf{7325} & \\
 & 15 & & \\
\hline
 & 317 & & \\
 & 15 & & \\
\hline
 & \mathbf{332} & &
\end{array}
$$

$$Ans.\ \ 5y^3 + 332y^2 + 7325y + 53689 = 0.$$

35. Removal of Terms.—One of the chief uses of the transformation of the preceding Article is to remove a certain specified term from an equation. Such a step often facilitates its solution. Writing the transformed equation in descending powers of y, we have

$$a_0 y^n + (n a_0 h + a_1)\, y^{n-1} + \left\{ \frac{n(n-1)}{1\,.\,2} a_0 h^2 + (n-1)\, a_1 h + a_2 \right\} y^{n-2} + \ldots = 0.$$

If h be such as to satisfy the equation $n a_0 h + a_1 = 0$, the transformed equation will want the second term. If h be either of the values which satisfy the equation

$$\frac{n(n-1)}{1\,.\,2} a_0 h^2 + (n-1)\, a_1 h + a_2 = 0,$$

the transformed equation will want the third term; the removal of the fourth term will require the solution of a cubic for h; and so on. To remove the last term we must solve the equation $f(h) = 0$, which is the original equation itself.

EXAMPLES.

1. Transform the equation

$$x^3 - 6x^2 + 4x - 7 = 0$$

into one which shall want the second term.

$$na_0h + a_1 = 0 \quad \text{gives } h = 2.$$

Diminish the roots by 2.   *Ans.* $y^3 - 8y - 15 = 0.$

2. Transform the equation

$$x^4 + 8x^3 + x - 5 = 0$$

into one which shall want the second term.

Increase the roots by 2.   *Ans.* $y^4 - 24y^2 + 65y - 55 = 0.$

3. Transform the equation

$$x^4 - 4x^3 - 18x^2 - 3x + 2 = 0$$

into one which shall want the third term.

The quadratic for h is

$$6h^2 - 12h - 18 = 0, \quad \text{giving } h = 3, \ h = -1.$$

Thus there are two ways of effecting the transformation.

Diminishing the roots by 3, we obtain

$$(1) \quad y^4 + 8y^3 - 111y - 196 = 0.$$

Increasing the roots by 1, we obtain

$$(2) \quad y^4 - 8y^3 - 17y - 8 = 0.$$

36. Binomial Coefficients.—In many algebraical processes it is found convenient to write the polynomial $f(x)$ in the following form :—

$$a_0 x^n + na_1 x^{n-1} + \frac{n(n-1)}{1 \cdot 2} a_2 x^{n-2} + \ldots + \frac{n(n-1)}{1 \cdot 2} a_{n-2} x^2 + na_{n-1} x + a_n,$$

in which each term is affected, in addition to the literal coefficient, with the numerical coefficient of the corresponding term in the expansion of $(x + 1)^n$ by the binomial theorem. The student will find examples of equations written in this way on referring to Article 27, Examples 13 and 16. The form is one to which any given polynomial can be at once reduced.

We adopt the following notation :—

$$U_n = a_0 x^n + na_1 x^{n-1} + \frac{n(n-1)}{1 \cdot 2} a_2 x^{n-2} + \ldots + na_{n-1} x + a_n,$$

thus using U with the suffix n to represent the polynomial of the nth degree written with binomial coefficients.

We have, changing n into $n-1$, &c.,

$$U_{n-1} \equiv a_0 x^{n-1} + (n-1) a_1 x^{n-2} + \ldots + (n-1) a_{n-2} x + a_{n-1},$$

$$\ldots \ldots \ldots \ldots \ldots \ldots \ldots$$

$$U_3 \equiv a_0 x^3 + 3a_1 x^2 + 3a_2 x + a_3,$$

$$U_2 \equiv a_0 x^2 + 2a_1 x + a_2,$$

$$U_1 \equiv a_0 x + a_1,$$

$$U_0 \equiv a_0.$$

One advantage of the binomial form is, that the derived functions can be immediately written down. The first derived function of U_n is, plainly,

$$n \left\{ a_0 x^{n-1} + (n-1) a_1 x^{n-2} + \frac{(n-1)(n-2)}{1.2} a_2 x^{n-3} + \ldots + a_{n-1} \right\};$$

or $n U_{n-1}$; so that the first derived function of a polynomial represented in this way can be formed by applying to the suffix of U the rule given in Art. 6, with respect to the exponent of the variable. Thus, for example, the first derived of U_4 is formed by multiplying the function by 4, and diminishing the suffix by unity; it is, then, $4 U_3$, as the student can easily verify.

We proceed now to prove that the substitution of $y + h$ for x transforms the polynomial U_n, or

$$a_0 x^n + n a_1 x^{n-1} + \frac{n(n-1)}{1.2} a_2 x^{n-1} + \ldots + n a_{n-1} x + a_n,$$

into

$$A_0 y^n + n A_1 y^{n-1} + \frac{n(n-1)}{1.2} A_2 y^{n-2} + \ldots + n A_{n-1} y + A_n,$$

where

$$A_0, \; A_1, \; \ldots \; A_{n-1}, \; A_n,$$

are the functions which result by substituting h for x in

$$U_0, \; U_1, \; U_2, \; \ldots \; U_{n-1}, \; U_n;$$

i.e. $A_0 = a_0, \quad A_1 = a_0 h + a_1, \quad A_2 = a_0 h^2 + 2a_1 h + a_2,$ &c. $\ldots$

Representing the derived functions of $f(h)$ by suffixes, as explained in Art. 6, we may write the result of the transformation, *i.e.* $f(y+h)$, in the following form:—

$$f(h) + f_1(h)\, y + \frac{f_2(h)}{1 \cdot 2}\, y^2 + \ldots + \frac{f_{n-1}(h)}{1 \cdot 2 \ldots n-1}\, y^{n-1} + \frac{f_n(h)}{1 \cdot 2 \ldots n}\, y^n :$$

$f(h)$ is the result of substituting h for x in U_n; it is, therefore, A_n: its first derived $f_1(h)$ is, by the above rule, nA_{n-1}; the first derived of this again is $n(n-1)A_{n-2}$; and so on. Making these substitutions, we have the result above stated, which enables us to write down without any calculation the transformed equation.

EXAMPLES.

1. Find the result of substituting $y+h$ for x in the polynomial

$$a_0 x^3 + 3a_1 x^2 + 3a_2 x + a_3.$$

Ans. $a_0 y^3 + 3(a_0 h + a_1) y^2 + 3(a_0 h^2 + 2a_1 h + a_2) y + a_0 h^3 + 3a_1 h^2 + 3a_2 h + a_3.$

The student will find it a useful exercise to verify this result by the process of operation explained in Art. 34, which may often be employed with advantage in the case of algebraical as well as numerical examples.

2. Remove the second term from the equation

$$a_0 x^3 + 3a_1 x^2 + 3a_2 x + a_3 = 0.$$

We must diminish the roots by a quantity h obtained from the equation

$$a_0 h + a_1 = 0, \quad i.e. \ \ h = \frac{-a_1}{a_0}.$$

Substituting this value of h in A_2, and A_3, the resulting equation in y is

$$y^3 + \frac{3(a_0 a_2 - a_1^2)}{a_0^2}\, y + \frac{a_0^2 a_3 - 3a_0 a_1 a_2 + 2a_1^3}{a_0^3} = 0.$$

3. Find the condition that the second and third terms of the equation $U_n = 0$ should be capable of being removed by the same substitution.

Here A_1 and A_2 must vanish for the same value of h; and eliminating h between them we find the required condition.

Ans. $a_0 a_2 - a_1^2 = 0$

4. Solve the equation

$$x^3 + 6x^2 + 12x - 19 = 0$$

by removing its second term.

The third term is removed by the same substitution, which gives

$$y^3 - 27 = 0.$$

The required roots are obtained by subtracting 2 from each root of the latter equation.

5. Find the condition that the second and fourth terms of the equation $U_n = 0$ should be capable of being removed by the same transformation. Here the coefficients A_1 and A_3 must vanish for the same value of h: eliminating h between the equations

$$a_0 h + a_1 = 0, \quad a_0 h^3 + 3a_1 h^2 + 3a_2 h + a_3 = 0,$$

we obtain the required condition

$$a_0{}^2 a_3 - 3a_0 a_1 a_2 + 2a_1{}^3 = 0.$$

. *Note.*—When this condition holds among the coefficients of a biquadratic equation its solution is reducible to that of a quadratic; for when the second term is removed the resulting equation is a quadratic for y^2 ; and from the values of y those of x can be obtained.

6. Solve the equation

$$x^4 + 16x^3 + 72x^2 + 64x - 129 = 0$$

by removing its second term.

The equation in y is

$$y^4 - 24y^2 - 1 = 0.$$

7. Solve in the same manner the equation

$$x^4 + 20x^3 + 143x^2 + 430x + 462 = 0.$$

Ans. The roots are $-7, \ -3, \ -5 \pm \sqrt{3}$.

8. Find the condition that the same transformation should remove the second and fifth terms of the equation $U_n = 0$.

Ans. $a_0{}^3 a_4 - 4a_0{}^2 a_1 a_3 + 6a_0 a_1{}^2 a_2 - 3a_1{}^4 = 0.$

37. **The Cubic.**—On account of their peculiar interest, we shall consider in this and the next following Articles the equations of the third and fourth degrees, in connexion with the transformation of the preceding Article. When $y + h$ is substituted for x in the equation

$$a_0 x^3 + 3a_1 x^2 + 3a_2 x + a_3 = 0, \tag{1}$$

we obtain

$$a_0 y^3 + 3A_1 y^2 + 3A_2 y + A_3 = 0,$$

where A_1, A_2, A_3 have the values of Art. 36.

If the transformed equation wants the second term,

$$A_1 = 0, \quad \text{or } h = -\frac{a_1}{a_0};$$

substituting this value for h in A_2 and A_3, we find, as in Ex. 2, Art. 36,

$$a_0 A_2 = a_0 a_2 - a_1^2, \quad a_0^2 A_3 = a_0^2 a_3 - 3a_0 a_1 a_2 + 2a_1^3;$$

hence the transformed cubic, wanting the second term, is

$$y^3 + \frac{3}{a_0^2}(a_0 a_2 - a_1^2) y + \frac{1}{a_0^3}(a_0^2 a_3 - 3a_0 a_1 a_2 + 2a_1^3) = 0.$$

The functions of the coefficients here involved are of such importance in the theory of algebraic equations, that it is customary to represent them by single letters. We accordingly adopt the notation

$$a_0 a_2 - a_1^2 = H, \quad a_0^2 a_3 - 3a_0 a_1 a_2 + 2a_1^3 = G;$$

and then write the transformed equation in the form

$$y^3 + \frac{3H}{a_0^2} y + \frac{G}{a_0^3} = 0. \tag{2}$$

We here observe that if the roots of this equation be multiplied by a_0 it becomes

$$z^3 + 3Hz + G = 0. \tag{3}$$

This is the form of the cubic we shall employ when we come to treat of its algebraical solution. The variable

$$z = a_0 y = a_0 x + a_1.$$

The original cubic is in fact identical with

$$(a_0 x + a_1)^3 + 3H(a_0 x + a_1) + G = 0,$$

after the factor a_0^2 is removed from this, as the student can easily verify.

If the roots of the original equation be α, β, γ, those of the transformed equation (2) will be

$$\alpha + \frac{a_1}{a_0}, \quad \beta + \frac{a_1}{a_0}, \quad \gamma + \frac{a_1}{a_0};$$

or, since

$$\alpha + \beta + \gamma = -\frac{3a_1}{a_0},$$

they will be

$$\tfrac{1}{3}(2\alpha - \beta - \gamma), \quad \tfrac{1}{3}(2\beta - \gamma - \alpha), \quad \tfrac{1}{3}(2\gamma - \alpha - \beta).$$

We can write down immediately the values of the symmetric functions

$$\Sigma\,(2a - \beta - \gamma)(2\beta - \gamma - a)\ ;\ (2a - \beta - \gamma)(2\beta - \gamma - a)(2\gamma - a - \beta)$$

of the roots of the original cubic. The latter will be found to agree with the value already found in Ex. 15, Art. 27.

Remark.—We now make with regard to the general equation an important observation : that any symmetric function of the *differences* of the roots a, β, γ, δ, &c., can be expressed by the functions of the coefficients which occur in the transformed equation wanting the second term. This is obvious, since the difference of any two roots a', β' of the transformed equation is equal to the difference of the two corresponding roots a, β of the original equation ; and the symmetric function of the differences $a' - \beta'$, &c., can be expressed in terms of the coefficients of the transformed equation. For example, in the case of the cubic, all symmetric functions of the differences of the roots can be expressed as functions of a_0, H, and G. Illustrations of this principle will be found among the examples of Art. 27.

38. **The Biquadratic.**—The transformed equation, wanting the second term, is

$$a_0 y^4 + 6A_2 y^2 + 4A_3 y + A_4 = 0,$$

where A_2 and A_3 have the same values as in the preceding Article ; and for A_4 we have

$$a_0^3 A_4 = a_0^3 a_4 - 4a_0^2 a_1 a_3 + 6a_0 a_1^2 a_2 - 3a_1^4.$$

The transformed equation is then

$$y^4 + \frac{6}{a_0^2}Hy^2 + \frac{4}{a_0^3}Gy + \frac{1}{a_0^4}(a_0^3 a_4 - 4a_0^2 a_1 a_3 + 6a_0 a_1^2 a_2 - 3a_1^4) = 0.$$

We might represent the absolute term of this equation by a symbol like H and G, and have thus three functions of the coefficients, in terms of which all symmetric functions of the differences of the roots of the biquadratic could be expressed. It is more convenient, however, to regard this absolute term as composed of H and another function of the coefficients determined in the following manner :—

$$a_0^3 a_4 - 4a_0^2 a_1 a_3 + 6a_0 a_1^2 a_2 - 3a_1^4 \equiv a_0^2 (a_0 a_4 - 4a_1 a_3 + 3a_2^2) - 3(a_0 a_2 - a_1^2)^2.$$

This identity is obvious ; and enables us to introduce a function of the coefficients

$$a_0 a_4 - 4a_1 a_3 + 3a_2^2,$$

which is of great importance in the theory of the biquadratic. This function is represented by the letter I, and we have

$$a_0^3 a_4 - 4a_0^2 a_1 a_3 + 6a_0 a_1^2 a - 3a_1^4 \equiv a_0^2 I - 3H^2.$$

The transformed equation may now be written

$$y^4 + \frac{6H}{a_0^2} y^2 + \frac{4G}{a_0^3} y + \frac{a_0^2 I - 3H^2}{a_0^4} = 0. \tag{1}$$

We can multiply the roots of this equation, as in the case of the cubic of Art. 37, by a_0 ; and obtain

$$z^4 + 6Hz^2 + 4Gz + a_0^2 I - 3H^2 = 0. \tag{2}$$

This form will be found convenient in the treatment of the algebraical solution of the biquadratic. The variable is the same as in the case of the cubic, i.e. $a_0 x + a_1$. The original biquadratic is in fact identical with

$$(a_0 x + a_1)^4 + 6H(a_0 x + a_1)^2 + 4G(a_0 x + a_1) + a_0^2 I - 3H^2 = 0,$$

after the factor a_0^3 is removed from this latter equation.

Any symmetric function of the differences of the roots of the original biquadratic can therefore be expressed by a_0, H, G, and I.

If the roots of the original equation be α, β, γ, δ, those of the transformed (1) will be, as is easily seen,

$$\tfrac{1}{4}(3\alpha - \beta - \gamma - \delta), \quad \tfrac{1}{4}(3\beta - \gamma - \delta - \alpha), \quad \tfrac{1}{4}(3\gamma - \delta - \alpha - \beta), \quad \tfrac{1}{4}(3\delta - \alpha - \beta - \gamma).$$

The sum of these $= 0$; the sum of their products in pairs $= \dfrac{6H}{a_0^2}$; the sum of their products in threes $= \dfrac{-4G}{a_0^3}$; and for their continued product we have

$$a_0^4 (3a - \beta - \gamma - \delta)(3\beta - \gamma - \delta - a)(3\gamma - \delta - a - \beta)(3\delta - a - \beta - \gamma)$$
$$= 256 (a_0^2 I - 3H^2).$$

There is another function of the coefficients to which we wish now to call attention, as it will be found to be of great importance in the subsequent discussion of the biquadratic. It is the function which presents itself in Ex. 18, Art. 27, *i.e.*

$$a_0 a_2 a_4 + 2a_1 a_2 a_3 - a_0 a_3^2 - a_1^2 a_4 - a_2^3.$$

This is denoted by the letter J. The example in question shows that it is a function of the differences of the roots. We ought, therefore, to be able to express it in terms of a_0, H, G, and I. We have, in fact, the identity

$$a_0^3 J \equiv a_0^2 HI - G^2 - 4H^3,$$

which the student can easily verify.

Or this relation can be derived as follows:—Whenever a function of the coefficients a_0, a_1, a_2, &c. is the expression of a function of the differences of the roots, it must be unaltered by the transformation which removes the second term of the equation; hence its value is unaltered when we change a_1 into zero, a_2 into A_2, a_3 into A_3, &c. Thus

$$a_0 a_2 a_4 + 2a_1 a_2 a_3 - a_0 a_3^2 - a_1^2 a_4 - a_2^3 \equiv a_0 A_2 A_4 - a_0 A_3^2 - A_2^3 ;$$

substituting for A_2, A_3, A_4 their values in terms of H, G, I, we easily obtain the above identity, which will usually be written in the form

$$G^2 + 4H^3 = a_0^2 (HI - a_0 J).$$

39. Homographic Transformation.—The transformation considered in Art. 34 is a particular case of the following, in which x is connected with the new variable y by the equation

$$y = \frac{\lambda x + \mu}{\lambda' x + \mu'}.$$

If $\lambda = 1$, $\mu = -h$, $\lambda' = 0$, $\mu' = 1$, we have $y = x - h$, as in Art. 34. Solving for x in terms of y, we have

$$x = \frac{\mu - \mu' y}{\lambda' y - \lambda}.$$

This value can be substituted for x in the given equation, and the resulting equation of the nth degree in y obtained.

Let a, β, γ, δ, &c., be the roots of the original equation, and a', β', γ', δ', &c., the corresponding roots of the transformed equation.   We easily obtain from the equations

$$a' = \frac{\lambda a + \mu}{\lambda' a + \mu'}, \quad \beta' = \frac{\lambda \beta + \mu}{\lambda' \beta + \mu'}, \text{ &c.,}$$

the relation

$$a' - \beta' = \frac{(\lambda \mu' - \mu \lambda')(a - \beta)}{(\lambda' a + \mu')(\lambda' \beta + \mu')} ;$$

with corresponding relations for the differences of any other pair of roots.   If we take four roots, and the four corresponding roots, we get

$$\frac{(a' - \beta')(\gamma' - \delta')}{(a' - \gamma')(\beta' - \delta')} = \frac{(a - \beta)(\gamma - \delta)}{(a - \gamma)(\beta - \delta)}.$$

Thus, if the roots of the proposed equation represent the distances of a number of points on a right line from a fixed origin on the line, the roots of the transformed will represent the distances of a corresponding system of points, so related to the former that the anharmonic ratio of any four of one system is the same as that of their four conjugates in the other system.   It is in consequence of this property that the transformation is called *homographic.*

It is important to observe that the transformation here considered, in which the variables x and y are connected by a relation of the form

$$Axy + Bx + Cy + D = 0,$$

is the most general transformation in which to one value of either variable corresponds one, and only one, value of the other.

40. **Transformation in General.**—In the general problem of transformation we have to form a new equation in y, whose roots are connected by a given relation $\phi(x, y) = 0$ with the roots of the proposed equation $f(x) = 0$.   The transformed equation will then be obtained by substituting in the given equation the value of x in terms of y obtained from the given

relation $\phi(x, y) = 0$; or, in other words, by eliminating x between the two equations $f(x) = 0$, and $\phi(x, y) = 0$. For example, suppose it were required to form the equation whose roots are the sums of every two of the roots (α, β, γ) of the cubic

$$x^3 - px^2 + qx - r = 0;$$

here

$$y = \beta + \gamma = \alpha + \beta + \gamma - \alpha = p - \alpha.$$

The equation $\phi(x, y) = 0$ is here $y = p - x$; for when x takes the value α, y takes one of the proposed values; and when x takes the values β and γ, y takes the other proposed values. The transformed equation is then obtained by substituting $p - y$ for x in the given equation.

EXAMPLES.

1. If α, β, γ be the roots of the cubic

$$x^3 - px^2 + qx - r = 0,$$

form the equation whose roots are

$$\beta\gamma + \frac{1}{\alpha}, \quad \gamma\alpha + \frac{1}{\beta}, \quad \alpha\beta + \frac{1}{\gamma};$$

here

$$y = \beta\gamma + \frac{1}{\alpha} = \frac{\alpha\beta\gamma + 1}{\alpha} = \frac{1 + r}{\alpha};$$

and the given relation is $xy = 1 + r$; the transformed equation is then obtained by substituting $\dfrac{1 + r}{y}$ for x in $f(x) = 0$.

$$Ans. \quad ry^3 - q(1 + r)y^2 + p(1 + r)^2 y - (1 + r)^3 = 0.$$

2. Form, for the same cubic, the equation whose roots are

$$\alpha\beta + \alpha\gamma, \quad \alpha\beta + \beta\gamma, \quad \beta\gamma + \alpha\gamma.$$

Substitute $\dfrac{r}{q - y}$ for x.
$\qquad Ans. \quad y^3 - 2qy^2 + (pr + q^2)y + r^2 - pqr = 0.$

3. Form, for the same cubic, the equation whose roots are

$$\frac{\alpha}{\beta + \gamma - \alpha}, \quad \frac{\beta}{\gamma + \alpha - \beta}, \quad \frac{\gamma}{\alpha + \beta - \gamma}.$$

Substitute $\dfrac{py}{1 + 2y}$ for x.
$\qquad Ans. \quad (p^3 - 4pq + 8r)y^3 + (p^3 - 4pq + 12r)y^2 + (6r - pq)y + r = 0.$

41. Equation of Differences of a Cubic.—We shall now apply the transformation explained in the preceding Article to an important problem, *i.e.* the formation of the equation whose roots are the squares of the differences of those of a given cubic. We shall do this in the first instance for the cubic

$$x^3 + qx + r = 0, \tag{1}$$

in which the second term is absent, and to which the general equation is easily reducible. Let the roots be a, β, γ. We have to form the equation in y whose roots are

$$(\beta - \gamma)^2, \ (\gamma - a)^2, \ (a - \beta)^2 \, ;$$

$$y = (\beta - \gamma)^2 = a^2 + \beta^2 + \gamma^2 - a^2 - \frac{2a\beta\gamma}{a} \, ;$$

$$a^2 + \beta^2 + \gamma^2 = -2q, \quad a\beta\gamma = -r.$$

The equation $\phi(x, y)$ of Art. 40 takes here the form

$$y = -2q - x^2 + \frac{2r}{x},$$

or

$$x^3 + (y + 2q)\,x - 2r = 0 \, ;$$

subtracting from this the proposed equation, we get

$$(y + q)\,x - 3r = 0, \quad \text{or } x = \frac{3r}{y + q} \, ;$$

hence the transformed equation in y is

$$y^3 + 6qy^2 + 9q^2y + 4q^3 + 27r^2 = 0. \tag{2}$$

If it be proposed to form the equation whose roots are the squares of the differences of those (a, β, γ) of the cubic

$$a_0 x^3 + 3a_1 x^2 + 3a_2 x + a_3 = 0, \tag{3}$$

we first remove the second term ; the resulting equation (see Art. 37) is

$$y^3 + \frac{3H}{a_0{}^2}\,y + \frac{G}{a_0{}^3} = 0 \, ;$$

and the required equation is the same as the equation of differences of this latter, since the difference of any two roots is unaltered by removing the second term. We can thus write down the required equation by putting

$$q = \frac{3H}{a_0{}^2}, \quad r = \frac{G}{a_0{}^3}$$

n the above.   The result is

$$x^3 + \frac{18H}{a_0{}^2} x^2 + \frac{81H^2}{a_0{}^4} x + \frac{27}{a_0{}^6} (G^2 + 4H^3) = 0, \qquad (4)$$

which has for roots

$$(\beta - \gamma)^2, \ (\gamma - a)^2, \ (a - \beta)^2.$$

The equation (4) can be written in a form free from fractions by multiplying the roots by $a_0{}^2$.   It becomes then

$$x^3 + 18Hx^2 + 81H^2x + 27\,(G^2 + 4H^3) = 0, \qquad (5)$$

whose roots are

$$a_0{}^2 (\beta - \gamma)^2 \quad a_0{}^2 (\gamma - a)^2, \quad a_0{}^2 (a - \beta)^2.$$

We can write down from this an important function of the roots of the cubic (3), *i.e.* the *product of the squares of the diffe-rences*, in terms of the coefficients :—

$$a_0{}^6 (\beta - \gamma)^2 (\gamma - a)^2 (a - \beta)^2 = -27\,(G^2 + 4H^3). \qquad (6)$$

It is evident from the identity of Art. 38 that $G^2 + 4H^3$ contains $a_0{}^2$ as a factor.   We have in fact the equation

$$G^2 + 4H^3 \equiv a_0{}^2 \{ a_0{}^2 a_3{}^2 - 6a_0 a_1 a_2 a_3 + 4a_0 a_2{}^3 + 4a_1{}^3 a_3 - 3a_1{}^2 a_2{}^2 \}.$$

The expression in brackets is called the *discriminant* of the cubic, and is represented by Δ ; giving the identities

$$G^2 + 4H^3 \equiv a_0{}^2 \Delta, \quad HI - a_0 J \equiv \Delta.$$

EXAMPLES.

1. Form the equation whose roots are the squares of the differences of those of

$$x^3 - 7x + 6 = 0.$$

Ans. $y^3 - 42y^2 + 441y - 400 = 0.$

2. Form the equation of differences of

$$x^3 + 6x^2 + 7x + 2 = 0.$$

First remove the second term.

Ans. $y^3 - 30y^2 + 225y - 68 = 0.$

3. Form the equation of differences of

$$x^3 + 6x^2 + 9x + 4 = 0.$$

Ans. $y^3 - 18y^2 + 81y = 0.$

42. Criterion of the Nature of the Roots of a Cubic.
—We can from the form of the equation of differences obtained in Art. 41 derive criteria in terms of the coefficients of the nature of the roots of the algebraical cubic. For, if the equation (5) of Art. 41 has a negative root, the cubic ((3) Art. 41) must have a pair of imaginary roots, in order that the square of their difference should be negative; and if (5) has no negative root, the cubic (3) has all its roots real, since a pair of imaginary roots of (3) would give rise to a negative root of (5).

The following four cases exist : —

(1). *When $G^2 + 4H^3$ is negative, the roots of the cubic are all real.*—For, to make this negative H must be negative (and $4H^3 > G^2$) ; the signs of the equation (5) are then alternately positive and negative, and, therefore (Art. 20), (5) has no negative root ; and consequently the given cubic has all its roots real.

(2). *When $G^2 + 4H^3$ is positive, the cubic has two imaginary roots.*—For the equation (5) must then (Art. 13) have a negative root.

(3). *When $G^2 + 4H^3 = 0$, the cubic has two equal roots.*—For the equation (5) has then one root equal to cipher. This is called the *condition for equal roots.*

(4). *When $G = 0$, and $H = 0$, the cubic has its three roots equal.*—For the roots of (5) are then all equal to cipher. These are the *conditions that the cubic should be a perfect cube.* They may also be expressed, as can be easily seen, in the form

$$\frac{a_0}{a_1} = \frac{a_1}{a_2} = \frac{a_2}{a_3}.$$

43. Transformation of Equations by Symmetric Functions.—Suppose it is required to transform an equation into another whose roots shall be given rational functions of the roots of the proposed. Let the given function be $\phi(\alpha, \beta, \gamma \ldots)$, where ϕ may involve all the roots, or any number of them. We form all possible combinations $\phi(\alpha\beta\gamma)$, $\phi(\alpha\beta\delta)$, &c., of the roots of this type, and write down the transformed equation as follows :—

$$\{y - \phi(\alpha\beta\gamma \ldots)\}\{y - \phi(\alpha\beta\delta \ldots)\}\ldots = 0.$$

When this is expanded, the successive coefficients of y will be symmetric functions of the roots a, β, γ, &c., of the given equation; and may therefore be expressed in terms of the coefficients of that equation.

EXAMPLES.

1. The roots of
$$x^3 + px^2 + qx + r = 0$$
are a, β, γ; find the equation whose roots are a^2, β^2, γ^2.

Suppose the transformed equation to be
$$y^3 + Py^2 + Qy + R = 0;$$
then
$$-P = a^2 + \beta^2 + \gamma^2, \quad Q = \Sigma a^2 \beta^2, \quad -R = a^2 \beta^2 \gamma^2;$$
and we have to form the symmetric functions Σa^2, $\Sigma a^2 \beta^2$, $a^2 \beta^2 \gamma^2$, of the given equation. We easily obtain
$$\Sigma a^2 = p^2 - 2q, \quad \Sigma a^2 \beta^2 = q^2 - 2pr, \quad a^2 \beta^2 \gamma^2 = r^2;$$
and the transformed equation is
$$y^3 - (p^2 - 2q) y^2 + (q^2 - 2pr) y - r^2 = 0.$$

2. Find for the same cubic the equation whose roots are a^3, β^3, γ^3.

$$\textit{Ans.} \quad y^3 + (p^3 - 3pq + 3r) y^2 + (q^3 - 3pqr + 3r^2) y + r^3 = 0.$$

3. If a, β, γ, δ be the roots of
$$x^4 + px^3 + qx^2 + rx + s = 0;$$
find the equation whose roots are a^2, β^2, γ^2, δ^2.

Let the transformed equation be
$$y^4 + Py^3 + Qy^2 + Ry + S = 0,$$
then
$$-P = \Sigma a^2, \quad Q = \Sigma a^2 \beta^2, \quad -R = \Sigma a^2 \beta^2 \gamma^2, \quad S = a^2 \beta^2 \gamma^2 \delta^2.$$
We have then the required equation (cf. Examples 8 and 17, Art. 27).

$$\textit{Ans.} \quad y^4 - (p^2 - 2q) y^3 + (q^2 - 2pr + 2s) y^2 - (r^2 - 2qs) y + s^2 = 0.$$

4. If a, β, γ, δ be the roots of
$$a_0 x^4 + 4a_1 x^3 + 6a_2 x^2 + 4a_3 x + a_4 = 0;$$
find the equation whose roots are λ, μ, ν; viz.,
$$\beta\gamma + a\delta, \quad \gamma a + \beta\delta, \quad a\beta + \gamma\delta.$$
See Ex. 17, Art. 27.

$$\textit{Ans.} \quad y^3 - \frac{6a_2}{a_0} y^2 + \frac{4}{a_0^2} (4a_1 a_3 - a_0 a_4) y - \frac{8}{a_0^3} (2a_0 a_3^2 - 3a_0 a_2 a_4 + 2a_1^2 a_4) = 0.$$

5. Show that the transformed equation, when the roots of the resulting cubic of Ex. 4 are multiplied by $\tfrac{1}{4} a_0$, and the second term of the equation then removed, is
$$z^3 - Iz + 2J = 0.$$

G

44. Formation of the Equation whose Roots are any Powers of those of the Proposed.—The method of effecting this transformation by symmetric functions, as explained in Art. 43, is often laborious. A much simpler process, involving only multiplication, can be employed. It depends on a knowledge of the solution of the binomial equation

$$x^n - 1 = 0.$$

This will be given in the following Chapter. The general process will be sufficiently obvious to the student from the application to the 2nd and 3rd degrees which will be found among the following examples :—

EXAMPLES.

1. To form the equation whose roots are the squares of the roots of

$$x^n + p_1 x^{n-1} + p_2 x^{n-2} + \ldots + p_{n-1} x + p_n = 0.$$

To effect this transformation, we have the identity

$$x^n + p_1 x^{n-1} + p_2 x^{n-2} + \ldots + p_{n-1} x + p_n \equiv (x - a_1)(x - a_2) \ldots (x - a_n) ;$$

change x into $-x$, and obtain, as in Art. 30,

$$x^n - p_1 x^{n-1} + p_2 x^{n-2} - \ldots \pm p_{n-1} x \mp p_n \equiv (x + a_1)(x + a_2) \ldots (x + a_n) ;$$

multiplying these equations, we have

$$(x^n + p_2 x^{n-2} + p_4 x^{n-4} + \ldots)^2 - (p_1 x^{n-1} + p_3 x^{n-3} + \ldots)^2 \equiv (x^2 - a_1{}^2)(x^2 - a_2{}^2) \ldots (x^2 - a_n{}^2) ;$$

it is evident that the first member of this identity contains, when expanded, only even powers of x ; we may then replace x^2 by y ; and get

$$y^n + (2p_2 - p_1{}^2) y^{n-1} + (p_2{}^2 - 2p_1 p_3 + 2p_4) y^{n-2} + \ldots \equiv (y - a_1{}^2)(y - a_2{}^2) \ldots (y - a_n{}^2).$$

The first member of this equated to zero is the required transformed equation.

N.B.—This transformation will often enable us to determine a limit to the number of real roots of the proposed equation. For, the square of a real root must be positive ; and therefore the original equation cannot have more real roots than the transformed has positive roots.

2. Find the equation whose roots are the squares of those of

$$x^3 - x^2 + 8x - 6 = 0.$$

Ans. $y^3 + 15y^2 + 52y - 36 = 0.$

The latter equation, by Descartes' rule of signs, cannot have more than one positive root ; hence the former must have a pair of imaginary roots.

3. Find the equation whose roots are the squares of the roots of the equation

$$x^5 + x^3 + x^2 + 2x + 3 = 0.$$

$$\text{Ans. } y^5 + 2y^4 + 5y^3 + 3y^2 - 2y - 9 = 0.$$

It follows from Descartes' rule of signs that the original equation must have four imaginary roots.

4. Verify by the method of Ex. 1 the Examples 1 and 3 of Art. 43.

5. To form the equation whose roots are the cubes of the roots of

$$x^n + p_1 x^{n-1} + p_2 x^{n-2} + \ldots + p_{n-1} x + p_n = 0.$$

It will be observed that in Ex. 1 the process consists in multiplying together $f(x)$, the given polynomial, and $f(-x)$; the variables involved in these being those obtained by multiplying x by the two roots of the equation $x^2 - 1 = 0$. In the present case we must multiply together $f(x)$, $f(\omega x)$, $f(\omega^2 x)$; the variables involved being obtained by multiplying x by the roots of the equation $x^3 - 1 = 0$. The transformation may be conveniently represented as follows:—

Write the polynomial $f(x)$ in the form

$$(p_n + p_{n-3} x^3 + \ldots) + x(p_{n-1} + p_{n-4} x^3 + \ldots) + x^2(p_{n-2} + p_{n-5} x^3 + \ldots),$$

which we represent, for brevity, by

$$P + xQ + x^2 R,$$

in which P, Q, and R are all functions of x^3.

We have then

$$P + xQ + x^2 R \equiv (x - \alpha_1)(x - \alpha_2) \ldots (x - \alpha_n). \tag{1}$$

Changing, in this identity, x into ωx and $\omega^2 x$ successively, we obtain

$$P + \omega x Q + \omega^2 x^2 R \equiv (\omega x - \alpha_1)(\omega x - \alpha_2) \ldots (\omega x - \alpha_n), \tag{2}$$

$$P + \omega^2 x Q + \omega x^2 R \equiv (\omega^2 x - \alpha_1)(\omega^2 x - \alpha_2) \ldots (\omega^2 x - \alpha_n), \tag{3}$$

since P, Q, and R, being functions of x^3, are unaltered.

Multiplying together the equations (1), (2), (3), and attending to the results of Art. 26, we obtain

$$P^3 + x^3 Q^3 + x^6 R^3 - 3x^3 PQR \equiv (x^3 - \alpha_1^3)(x^3 - \alpha_2^3) \ldots (x^3 - \alpha_n^3).$$

The first member of this identity contains x in powers which are multiples of 3 only. We can, therefore, substitute y for x^3 and obtain the required transformed equation.

6. Find the equation whose roots are the cubes of the roots of

$$x^4 - x^3 + 2x^2 + 3x + 1 = 0.$$

$$\text{Ans. } y^4 + 14y^3 + 50y^2 + 6y + 1 = 0.$$

7. Verify by the method of Ex. 5 the result of Ex. 2 of Art 43.

8. Form the equation whose roots are the cubes of the roots of

$$ax^3 + 3bx^2 + 3cx + d = 0.$$

$$\text{Ans. } a^3 y^3 + 3(a^2 d + 9b^3 - 9abc)y^2 + 3(ad^2 + 9c^3 - 9bcd)y + d^3 = 0.$$

MISCELLANEOUS EXAMPLES.

1. The roots of the equation

$$x^3 - 6x^2 + 11x - 6 = 0$$

are α, β, γ; form the equation whose roots are

$$\beta^2 + \gamma^2, \quad \gamma^2 + \alpha^2, \quad \alpha^2 + \beta^2.$$

Ans. $y^3 - 28y^2 + 245y - 650 = 0.$

2. The roots of the cubic

$$x^3 + 2x^2 + 3x + 1 = 0$$

are α, β, γ; form the equation whose roots are

$$\frac{1}{\beta^3} + \frac{1}{\gamma^3} - \frac{1}{\alpha^3}, \quad \frac{1}{\gamma^3} + \frac{1}{\alpha^3} - \frac{1}{\beta^3}, \quad \frac{1}{\alpha^3} + \frac{1}{\beta^3} - \frac{1}{\gamma^3}.$$

Ans. $y^3 + 12y^2 - 172y - 2072 = 0.$

3. The roots of the cubic

$$x^3 + qx + r = 0$$

are α, β, γ; form the equation whose roots are

$$\beta^2 + \beta\gamma + \gamma^2, \quad \gamma^2 + \gamma\alpha + \alpha^2, \quad \alpha^2 + \alpha\beta + \beta^2.$$

Ans. $(y + q)^3 = 0.$

4. The roots of the cubic

$$x^3 + px^2 + qx + r = 0$$

being α, β, γ; form the equation whose roots are

$$\beta^2 + \gamma^2 - \alpha^2, \quad \gamma^2 + \alpha^2 - \beta^2, \quad \alpha^2 + \beta^2 - \gamma^2.$$

Ans. $y^3 - (p^2 - 2q)\, y^2 - (p^4 - 4p^2 q + 8pr)\, y + p^6 - 6p^4 q + 8p^3 r$
$+ 8p^2 q^2 - 16pqr + 8r^2 = 0.$

5. If α, β, γ be the roots of the cubic

$$x^3 - 3(1 + a + a^2)\, x + 1 + 3a + 3a^2 + 2a^3 = 0;$$

prove that $\quad (\beta - \gamma)(\gamma - \alpha)(\alpha - \beta)$ is a rational function of a.

Ans. $\pm 9(1 + a + a^2).$

6. Find the relation between G and H of the cubic

$$a_0 x^3 + 3a_1 x^2 + 3a_2 x + a_3 = 0$$

when its roots are so related that $(\beta - \gamma)^2$, $(\gamma - \alpha)^2$, $(\alpha - \beta)^2$ are in arithmetical progression.

Ans. $G^2 + 2H^3 = 0.$

7. If α, β, γ, δ be the roots of

$$c^2 x^4 - 2c^2 x^3 + 2x - 1 = 0,$$

find the value of

$$(\beta^2 - \gamma^2)^2 (\alpha^2 - \delta^2)^2 + (\gamma^2 - \alpha^2)^2 (\beta^2 - \delta^2)^2 + (\alpha^2 - \beta^2)^2 (\gamma^2 - \delta^2)^2.$$

Ans. $0.$

8. Prove that, if

$$\beta\gamma + \gamma\alpha + \alpha\beta + \alpha\delta + \beta\delta + \gamma\delta = 0,$$

$$\{(\beta-\gamma)^2(\alpha-\delta)^2 + (\gamma-\alpha)^2(\beta-\delta)^2 + (\alpha-\beta)^2(\gamma-\delta)^2\}^2$$

$$= 18\{(\beta^2-\gamma^2)^2(\alpha^2-\delta^2)^2 + (\gamma^2-\alpha^2)^2(\beta^2-\delta^2)^2 + (\alpha^2-\beta^2)^2(\gamma^2-\delta^2)^2\}.$$

9. Solve the equation

$$x^5 - x^4 + 8x^2 - 9x - 15 = 0,$$

which has one root of the form $1 + \alpha\sqrt{-1}$.

Diminish the roots by 1; substitute $\alpha\sqrt{-1}$ for x; we find that α must satisfy $\alpha^4 - 3\alpha^2 - 4 = 0$, and $\alpha^4 - 6\alpha^2 + 8 = 0$; hence $\alpha = \pm 2$. Hence the factor $x^2 - 2x + 5$. The other factors are $(x+1)$ and (x^2-3), as is evident.

10. The roots of the cubic

$$a_0 x^3 + 3a_1 x^2 + 3a_2 x + a_3 = 0$$

are α, β, γ; form the equation whose roots are

$$\beta + \gamma, \quad \gamma + \alpha, \quad \alpha + \beta.$$

This question has been already solved in Art. 40. We give here another solution which, although in this particular instance it is not the simplest, will be found convenient in many examples. Let the roots of the given equation be diminished by h. The transformed equation is (Art. 36)

$$a_0 y^3 + 3A_1 y^2 + 3A_2 y + A_3 = 0,$$

whose roots are $\alpha - h$, $\beta - h$, $\gamma - h$. We express the condition that this equation should have two roots equal with opposite signs. This condition is (see Ex. 17, Art. 24)

$$9A_1 A_2 - a_0 A_3 = 0.$$

This equation is a cubic in h whose roots are

$$\tfrac{1}{2}(\beta + \gamma), \quad \tfrac{1}{2}(\gamma + \alpha), \quad \tfrac{1}{2}(\alpha + \beta);$$

for the above condition is

$$(\beta - h) + (\gamma - h) = 0,$$

or

$$2h = \beta + \gamma;$$

where β, γ represent indifferently any two of the roots.

11. The roots of the biquadratic

$$a_0 x^4 + 4a_1 x^3 + 6a_2 x^2 + 4a_3 x + a_4 = 0$$

are α, β, γ, δ; form the sextic whose roots are

$$\beta + \gamma, \quad \gamma + \alpha, \quad \alpha + \beta, \quad \alpha + \delta, \quad \beta + \delta, \quad \gamma + \delta.$$

Employing the method of Ex. 10, the required equation can be obtained from the condition of Ex. 20, Art. 24.

The condition is in this case

$$6A_1 A_2 A_3 - A_1^2 A_4 - a_0 A_3^2 = 0.$$

This is a sextic in h whose roots are $\tfrac{1}{2}(\beta + \gamma)$, &c., and the required equation can be obtained by forming the equation whose roots are double the roots of this equation.

12. Form for the cubic of Ex. 10, the equation whose roots are

$$\frac{\beta\gamma - \alpha^2}{\beta + \gamma - 2\alpha}, \quad \frac{\gamma\alpha - \beta^2}{\gamma + \alpha - 2\beta}, \quad \frac{\alpha\beta - \gamma^2}{\alpha + \beta - 2\gamma}.$$

Diminish the roots by h, and express the condition that the resulting cubic should have its roots in geometric progression (see Ex. 18, Art. 24). The condition is

$$A_1^3 A_3 - a_0 A_2^3 = 0.$$

This will be found to reduce to a cubic in h; whose roots are the values above written, since

$$(\alpha - h)^2 = (\beta - h)(\gamma - h), \text{ or } h = \frac{\beta\gamma - \alpha^2}{\beta + \gamma - 2\alpha}.$$

13. Form for the same cubic the equation whose roots are

$$\frac{2\beta\gamma - \alpha\beta - \alpha\gamma}{\beta + \gamma - 2\alpha}, \quad \frac{2\gamma\alpha - \beta\gamma - \beta\alpha}{\gamma + \alpha - 2\beta}, \quad \frac{2\alpha\beta - \gamma\alpha - \gamma\beta}{\alpha + \beta - 2\gamma}.$$

Diminish the roots by h, and express the condition that the transformed cubic should have its roots in harmonic progression (see Ex. 19, Art. 24). We have

$$\frac{2}{\alpha - h} = \frac{1}{\beta - h} + \frac{1}{\gamma - h},$$

or

$$h = \frac{2\beta\gamma - \alpha\beta - \alpha\gamma}{\beta + \gamma - 2\alpha}.$$

The equation in h is

$$a_0 A_3^2 - 3A_1 A_2 A_3 + 2A_2^3 = 0,$$

which will be found to reduce to a cubic.

14. The roots of the biquadratic

$$a_0 x^4 + 4a_1 x^3 + 6a_2 x^2 + 4a_3 x + a_4 = 0$$

are α, β, γ, δ; find the cubic whose roots are

$$\frac{\beta\gamma - \alpha\delta}{\beta + \gamma - \alpha - \delta}, \quad \frac{\gamma\alpha - \beta\delta}{\gamma + \alpha - \beta - \delta}, \quad \frac{\alpha\beta - \gamma\delta}{\alpha + \beta - \gamma - \delta}.$$

Diminish the roots by h, and employ the condition of Ex. 22, Art. 24. The condition is in this case

$$A_1^2 A_4 - a_0 A_3^2 = 0,$$

which reduces to a cubic in h whose roots are the values above written.

15. Find the equation whose roots are the ratios of the roots of the cubic

$$x^3 + qx + r = 0.$$

The general problem can be solved by elimination. Let $f(x) = 0$ be the given equation, and $\rho = \dfrac{\beta}{\alpha} = $ the ratio of two roots; then since $f(\beta) = 0$, we have $f(\rho\alpha) = 0$, also $f(\alpha) = 0$; and the required equation in ρ is obtained by eliminating

a between these two latter equations. For the cubic in the present example the result is

$$r^2(\rho^2 + \rho + 1)^3 + q^3\rho^2(\rho + 1)^2 = 0.$$

16. If a, β, γ be the roots of

$$x^3 + px^2 + qx + r = 0,$$

form the equation whose roots are

$$\beta^2 + \gamma^2, \quad \gamma^2 + a^2, \quad a^2 + \beta^2.$$

Ans. $x^3 - 2(p^2 - 2q)x^2 + (p^4 - 4p^2q + 5q^2 - 2pr)x - (p^2q^2 - 2p^3r + 4pqr - 2q^3 - r^2) = 0.$

17. Form for the same cubic the equation whose roots are

$$\frac{\beta}{\gamma} + \frac{\gamma}{\beta}, \quad \frac{\gamma}{a} + \frac{a}{\gamma}, \quad \frac{a}{\beta} + \frac{\beta}{a}.$$

Ans. $r^2x^3 - (pqr - 3r^2)x^2 + (p^3r - 5pqr + 3r^2 + q^3)x$
$$- (p^2q^2 - 2p^3r + 4pqr - 2q^3 - r^2) = 0.$$

18. If a, β, γ be the roots of the cubic

$$x^3 + qx + r = 0,$$

form the equation whose roots are

$$la + m\beta\gamma, \quad l\beta + m\gamma a, \quad l\gamma + ma\beta.$$

Ans. $y^3 - mqy^2 + (l^2q + 3lmr)y + l^3r - l^2mq^2 - 2lm^2qr - m^3r^2 = 0.$

19. If a, β, γ be the roots of the cubic

$$a_0x^3 + 3a_1x^2 + 3a_2x + a_3 = 0,$$

find the equation whose roots are

$$(a - \beta)(a - \gamma), \quad (\beta - \gamma)(\beta - a), \quad (\gamma - a)(\gamma - \beta).$$

$$\textit{Ans.} \quad y^3 + \frac{9H}{a_0^2}y^2 - \frac{27(G^2 + 4H^3)}{a_0^6} = 0.$$

20. Form, for the cubic of Ex. 19, the equation whose roots are

$$(\beta - \gamma)^2(2a - \beta - \gamma)^2, \quad (\gamma - a)^2(2\beta - \gamma - a)^2, \quad (a - \beta)^2(2\gamma - a - \beta)^2.$$

The required equation can be obtained by forming the equation of the squares of the differences of the roots of the cubic (4) of Art. 41, since

$$(\gamma - a)^2 - (a - \beta)^2 = (\beta - \gamma)(2a - \beta - \gamma).$$

21. Form, for the cubic, Ex. 16, the equation whose roots are

$$a(\beta - \gamma)^2, \quad \beta(\gamma - a)^2, \quad \gamma(a - \beta)^2.$$

Let the transformed equation be $x^3 + Px^2 + Qx + R = 0$.

$$\textit{Ans.} \quad P = pq - 9r, \quad Q = q^3 - 9pqr + 27r^2 + p^3r,$$
$$R = -r(4q^3 + 27r^2 + 4p^3r - p^2q^2 - 18pqr).$$

22. Form, for the cubic, Ex. 16, the equation whose roots are

$$a^2 + 2\beta\gamma, \quad \beta^2 + 2\gamma a, \quad \gamma^2 + 2a\beta.$$

$$\textit{Ans.} \quad P = -p^2, \quad Q = q(2p^2 - 3q),$$
$$-R = 4p^3r - 18pqr + 2q^3 + 27r^2.$$

CHAPTER V.

45. Reciprocal Equations.—It has been shown in Art. 33 that all reciprocal equations can be reduced to a standard form, in which the degree is even, and the coefficients counting from the beginning and end equal with the same sign. We now proceed to prove that *a reciprocal equation of the standard form can always be depressed to another of half the dimensions.*

Consider the equation

$$a_0 x^{2m} + a_1 x^{2m-1} + \ldots + a_m x^m + \ldots + a_1 x + a_0 = 0.$$

Dividing by x^m, and uniting terms equally distant from the extremes, we have

$$a_0\left(x^m + \frac{1}{x^m}\right) + a_1\left(x^{m-1} + \frac{1}{x^{m-1}}\right) + \ldots + a_{m-1}\left(x + \frac{1}{x}\right) + a_m = 0.$$

Assume $x + \dfrac{1}{x} = z$, and let $x^p + \dfrac{1}{x^p}$ be denoted for brevity by V_p. We have plainly the relation

$$V_{p+1} = V_p z - V_{p-1}.$$

Giving p in succession the values 1, 2, 3, &c., we have

$$V_2 = V_1 z - V_0 = z^2 - 2,$$
$$V_3 = V_2 z - V_1 = z^3 - 3z,$$
$$V_4 = V_3 z - V_2 = z^4 - 4z^2 + 2,$$
$$V_5 = V_4 z - V_3 = z^5 - 5z^3 + 5z;$$

and so on. Substituting these values in the above equation, we get an equation of the mth degree in z; and from the values of z those of x can be obtained.

EXAMPLES.

1. Find the roots of the equation

$$x^5 + x^4 + x^3 + x^2 + x + 1 = 0.$$

Dividing by $x + 1$ (see Art. 33), we have

$$x^4 + x^2 + 1 = 0.$$

This equation may be depressed to the form

$$z^2 - 1 = 0, \quad \text{giving } z = \pm 1 ;$$

whence

$$x + \frac{1}{x} = 1, \quad x + \frac{1}{x} = -1,$$

and the roots of these equations are

$$\frac{1 \pm \sqrt{-3}}{2}, \quad \frac{-1 \pm \sqrt{-3}}{2}.$$

2. Find the roots of the equation

$$x^{10} - 3x^8 + 5x^6 - 5x^4 + 3x^2 - 1 = 0.$$

Dividing by $x^2 - 1$, which may be done briefly as follows (see Art. 8),

$$
\begin{array}{rrrrrr}
1 & -3 & 5 & -5 & 3 & -1 \\
  & 1 & -2 & 3 & -2 & 1 \\
\hline
-2 & 3 & -2 & 1 & 0 &
\end{array}
$$

we have the reciprocal equation

$$x^8 - 2x^6 + 3x^4 - 2x^2 + 1 = 0, \tag{1}$$

or

$$\left(x^4 + \frac{1}{x^4}\right) - 2\left(x^2 + \frac{1}{x^2}\right) + 3 = 0.$$

Substituting for V_4, $z^4 - 4z^2 + 2$; and for V_2, $z^2 - 2$, we have the equation

$$z^4 - 6z^2 + 9 = 0, \quad \text{or } (z^2 - 3)^2 = 0,$$

whence

$$z^2 = 3, \quad \text{and} \quad z = \pm \sqrt{3},$$

giving

$$x + \frac{1}{x} = \sqrt{3}, \quad x + \frac{1}{x} = -\sqrt{3},$$

and the roots of these equations are

$$\frac{\sqrt{3} \pm \sqrt{-1}}{2}, \quad \frac{-\sqrt{3} \pm \sqrt{-1}}{2}.$$

These roots are double roots of the equation (1).

3. Solve the equation

$$x^5 - 1 = 0.$$

Dividing by $x - 1$ we have

$$x^4 + x^3 + x^2 + x + 1 = 0 ;$$

from which we obtain

$$z^2 + z - 1 = 0.$$

Solving this equation, we have the quadratics

$$x^2 + \tfrac{1}{2}(1 + \sqrt{5})x + 1 = 0,$$

$$x^2 + \tfrac{1}{2}(1 - \sqrt{5})x + 1 = 0,$$

from which we obtain

$$x = \tfrac{1}{4}\{-1 + \theta\sqrt{5} \pm (10 + 2\theta\sqrt{5})^{\frac{1}{2}}\sqrt{-1}\},$$

where $\theta^2 = 1$.

This expression gives the four values of x.

4. Find the quadratic factors of

$$x^6 + 1 = 0.$$

Transforming this, we have

$$z^3 - 3z = 0,$$

whence
$$z = 0, \text{ and } z = \pm\sqrt{3}.$$

The quadratic factors of the given equation are, therefore,

$$x^2 + 1 = 0, \quad x^2 \pm \sqrt{3}\,x + 1 = 0.$$

5. Solve the equations

$$(1). \ (1 + x)^4 = a(1 + x^4), \quad (2). \ (1 + x)^5 = a(1 + x^5).$$

6. Reduce to an equation of the fourth degree in z

$$\frac{(1 + x)^5}{1 + x^5} + \frac{(1 - x)^5}{1 - x^5} = 2a.$$

$$Ans. \ (1 - a)z^4 + (7 + 3a)z^2 - (4 + a) = 0.$$

46. **Binomial Equations. General Properties.**—
In this and the following Articles will be proved the leading general properties of Binomial Equations.

Prop. I.—*If a be an imaginary root of $x^n - 1 = 0$, then a^m also will be a root, m being any integer.*

Since a is a root,

$$a^n = 1, \text{ and therefore } (a^n)^m = 1, \text{ or } (a^m)^n = 1;$$

that is,
$$a^m \text{ is a root of } x^n - 1 = 0.$$

The same is true of the equation $x^n + 1 = 0$, except that in this case m must be an *odd* integer.

47. Prop. II.—*If m and n be prime to each other, the equations $x^m - 1 = 0$, $x^n - 1 = 0$ have no common root except unity.*

To prove this we make use of the following property of numbers :—

If m and n be integers prime to each other, integers a and b can be found such that $mb - na = \pm 1$. For, in fact, when $\dfrac{m}{n}$ is turned into a continued fraction, $\dfrac{a}{b}$ is the approximation preceding the final restoration of $\dfrac{m}{n}$.

Now, if possible, let a be any common root of the given equations ; then

$$a^m = 1, \text{ and } a^n = 1;$$

also

$$a^{mb} = 1, \text{ and } a^{na} = 1;$$

whence

$$a^{(mb-na)} = 1, \quad \text{or} \quad a^{\pm 1} = 1, \text{ or } a = 1;$$

that is, 1 is the only root common to the given equations.

48. PROP. III.—*If k be the greatest common measure of two integers m and n, the roots common to the equations $x^m - 1 = 0$, and $x^n - 1 = 0$, are roots of the equation $x^k - 1 = 0$.*

To prove this, let

$$m = km', \quad n = kn'.$$

Now, since m' and n' are prime to each other, integers b and a may be found such that $m'b - n'a = \pm 1$; hence

$$mb - na = \pm k.$$

If, therefore, a be a common root of $x^m - 1 = 0$, and $x^n - 1 = 0$,

$$a^{(mb-na)} = 1, \text{ or } a^k = 1;$$

which proves that a is a root of the equation $x^k - 1 = 0$.

49. PROP. IV.—*When n is a prime number, and a any imaginary root of $x^n - 1 = 0$, all the roots are included in the series*

$$1, \ a, \ a^2, \ \ldots \ a^{n-1}.$$

For, by Prop. (I.), these quantities are all roots of the equation ; and they are all different ; for, if possible, let any two be equal :

$$a^p = a^q, \text{ whence } a^{(p-q)} = 1;$$

but, by Prop. II., this equation is impossible, since n is necessarily prime to $(p - q)$, which is a number less than n.

50. PROP. V.—*When n is a composite number formed of the factors, p, q, r, &c., the roots of the equations $x^p - 1 = 0$, $x^q - 1 = 0$, $x^r - 1 = 0$, &c., all satisfy the equation $x^n - 1 = 0$.*

For, consider a root a of the equation $x^p - 1 = 0$; then $a^p = 1$; from which we derive

$$(a^p)^{qr\,\cdots} = 1 \; ; \; \text{ or } \; a^n - 1 = 0 \; ;$$

which proves the proposition.

51. PROP. VI.—*When n is a composite number formed of the prime factors p, q, r, &c., the roots of the equation $x^n - 1 = 0$ are the n terms of the product*

$$(1 + a + a^2 + \ldots + a^{p-1})(1 + \beta + \ldots + \beta^{q-1})(1 + \gamma + \ldots + \gamma^{r-1}) \ldots,$$

where a is a root of $x^p - 1 = 0$, β of $x^q - 1 = 0$, γ of $x^r - 1 = 0$, &c.

We prove this for the case of three factors p, q, r. A similar proof applies in general. Any term, *e.g.* $a^a \beta^b \gamma^c$, of the product is evidently a root of the equation $x^n - 1 = 0$, since $a^{an} = 1, \beta^{bn} = 1, \gamma^{cn} = 1$, and, therefore, $(a^a \beta^b \gamma^c)^n = 1$. And no two terms of the product can be equal; for, if possible let $a^a \beta^b \gamma^c$ be equal to another term $a^{a'} \beta^{b'} \gamma^{c'}$; then $a^{a'-a} = \beta^{b-b'} \gamma^{c-c'}$. The first member of this equation is a root of $x^p - 1 = 0$, and the second member is a root of $x^{qr} - 1 = 0$. Now these two equations cannot have a common root since p and qr are prime to each other (Prop. II.) ; hence $a^a \beta^b \gamma^c$ cannot be equal to $a^{a'} \beta^{b'} \gamma^{c'}$.

52. PROP. VII.—*The roots of the equation $x^n - 1 = 0$, where $n = p^a q^b r^c$, and p, q, r are the prime factors of n, are the n products of the form $a\beta\gamma$, where a is a root of $x^{p^a} = 1$, β of $x^{q^b} = 1$, and γ of $x^{r^c} = 1$.* This is an extension of Prop. VI. to the case where the prime factors occur more than once in n. The proof is exactly similar. Any such product $a\beta\gamma$ must be a root, since $a^n = 1$, $\beta^n = 1$, $\gamma^n = 1$; n being a multiple of p^a, q^b, r^c; and a proof similar to that of Art. 51 shows that no two such products can be equal, since p^a, q^b, r^c are prime to one another. We have, for convenience, stated this proposition for three factors only of n. A similar proof can be applied to the general case.

From this and the preceding propositions we are now able to derive the following general conclusion :—

The determination of the nth roots of unity is reduced to the case where n is a prime number, or a power of a prime number.

53. The Special Roots of the Equation $x^n - 1 = 0$.— Every equation $x^n - 1 = 0$ has certain roots which do not belong to any equation of similar form and lower degree. Such roots we call *special roots** of that equation, or *special n^{th} roots of unity*. If n be a prime number, all the imaginary roots are roots of this kind. If $n = p^a$, where p is a prime number, any n^{th} root of a lower degree than n must belong to the equation $x^{p^{a-1}} - 1 = 0$, since every divisor of p^a is a divisor of p^{a-1} (except n itself); hence there are $p^a\left(1 - \dfrac{1}{p}\right)$ roots which belong to no lower degree. If, again, $n = p^a q^b$, where p and q are prime to each other, there are $p^a\left(1 - \dfrac{1}{p}\right)$, and $q^b\left(1 - \dfrac{1}{q}\right)$ special roots of $x^{p^a} - 1 = 0$, and $x^{q^b} - 1 = 0$, respectively. Now, if a and β be any two special roots of these equations, $a\beta$ is a special root of $x^n - 1 = 0$; for if not, suppose $(a\beta)^m = 1$, where m is less than n; we have then $a^m = \beta^{-m}$; but a^m is a root of $x^{p^a} - 1 = 0$, and β^{-m} is a root of $x^{q^b} - 1 = 0$, and these equations cannot have a common root other than 1, as their degrees are prime to each other; consequently m cannot be less than n, and $a\beta$ is a special root of $x^n - 1 = 0$. Also as there are

$$p^a\left(1 - \frac{1}{p}\right) q^b\left(1 - \frac{1}{q}\right), \text{ or } n\left(1 - \frac{1}{p}\right)\left(1 - \frac{1}{q}\right),$$

such products, there are the same number of special n^{th} roots. This proof may be extended without difficulty to any form of n.

All the roots of $x^n - 1 = 0$ are given by the series $1, a, a^2, \ldots a^{n-1}$; where a is any special n^{th} root. For it is plain that a, a^2, &c., are all roots; and no two are equal; for, if $a^p = a^q$, $a^{(p-q)} = 1$; and therefore a is not a special n^{th} root, since $p - q$ is less than n.

When one special n^{th} root a is given, we may obtain all the other special n^{th} roots of unity.

---

* The term "special root" is here used in preference to the usual term "primitive root," since the latter has a different signification in the theory of numbers.

Since a is a special root, all the roots $1, a, a^2, \ldots a^{n-1}$ are different n^{th} roots, as we have just proved; and if we select a root a^p of this series, where p is prime to n, the roots

$$a^p, a^{2p}, \ldots a^{(n-1)p}, a^{np}(=1)$$

are all different, since the exponents of a when divided by n give different remainders in every case; that is, the series of numbers $0, 1, 2, 3, \ldots n-1$ in some order; whence this series of roots is the same as the former except that the terms occur in a different order. To each number p, prime to n and less than it (1 included), corresponds a special n^{th} root of unity; for a^{mp} cannot be equal to 1 when m is less than n, for if it were we should have two roots in the series equal to 1, and the series could not give all the roots in that case; therefore a^p is not a root of any binomial equation of a degree inferior to n: that is, a^p is a special n^{th} root of unity. What is here proved agrees with the result above established, since the number of integers less than n and prime to it is, by a known property of numbers, $n\left(1 - \dfrac{1}{p}\right)\left(1 - \dfrac{1}{q}\right)$ when $n = p^a q^b$, which is also, as above proved, the number of special roots of $x^n - 1 = 0$.

EXAMPLES.

1. To determine the special roots of $x^6 - 1 = 0$.

Here, $6 = 2 \times 3$. Consequently the roots of the equations $x^2 - 1 = 0$, and $x^3 - 1 = 0$ are roots of $x^6 - 1 = 0$. Now, dividing $x^6 - 1$ by $x^3 - 1$ we have $x^3 + 1$; and dividing $x^3 + 1$ by $\dfrac{x^2 - 1}{x - 1}$, or $x + 1$, we have $x^2 - x + 1 = 0$, which determines the special roots of $x^6 - 1 = 0$.

Solving this quadratic, the roots are

$$\alpha = \frac{1 + \sqrt{-3}}{2},$$

$$\alpha_1 = \frac{1 - \sqrt{-3}}{2};$$

also since

$$\alpha\alpha_1 = 1 = \alpha^6,$$

$$\alpha_1 = \alpha^5,$$

which may be easily verified.

The special roots are, therefore,

$$\alpha, \ \alpha^5; \ \text{or } \alpha_1{}^5, \ \alpha_1; \ \text{or } \alpha, \ \frac{1}{\alpha}.$$

2. On the special roots of $x^{12} - 1 = 0$.

Since 2 and 3 are the prime factors of 12, and $\dfrac{12}{2} = 6$, $\dfrac{12}{3} = 4$, the roots of $x^6 - 1 = 0$, and $x^4 - 1 = 0$, are roots of $x^{12} - 1 = 0$; now, dividing $x^{12} - 1$ by $x^4 - 1$, and $x^6 - 1$, and equating the quotients to zero, we have the two equations $x^8 + x^4 + 1 = 0$, and $x^6 + 1 = 0$, both of which must be satisfied by the special roots of $x^{12} - 1 = 0$; therefore, taking the greatest common measure of $x^8 + x^4 + 1$, and $x^6 + 1$, and equating it to zero, the special roots are the roots of the equation $x^4 - x^2 + 1 = 0$.

The same result would plainly have been arrived at by dividing $x^{12} - 1$ by the least common multiple of $x^4 - 1$ and $x^6 - 1$. Now, solving the reciprocal equation $x^4 - x^2 + 1 = 0$, we have $x + \dfrac{1}{x} = \pm \sqrt{3}$; whence, if a and a_1 be two special roots,

$$\left(a, \frac{1}{a} \right) = \frac{\sqrt{3} \pm \sqrt{-1}}{2}, \qquad \left(a_1, \frac{1}{a_1} \right) = \frac{-\sqrt{3} \pm \sqrt{-1}}{2}$$

are the four special roots of $x^{12} - 1 = 0$.

We proceed now to express the four special roots in terms of any one of them a.

Since
$$a + \frac{1}{a} + a_1 + \frac{1}{a_1} = 0, \text{ or } (a + a_1)\left(1 + \frac{1}{a a_1} \right) = 0,$$

we take $a a_1 = -1$ (as consistent with the values we have assigned to a and a_1); and since a and a_1 are roots of $x^6 + 1 = 0$, $a^6 = -1$, and $a^5 = -\dfrac{1}{a} = a_1$. The roots $a, a_1, \dfrac{1}{a_1}, \dfrac{1}{a}$ may therefore be expressed by the series a, a^5, a^7, a^{11}, since $a^{12} = 1$.

Further, replacing a by a^5, a^7, a^{11}, we have, including the series just determined, the four following series, by omitting multiples of 12 in the exponents of a:

$$\begin{array}{cccc}
a, & a^5, & a^7, & a^{11}, \\
a^5, & a, & a^{11}, & a^7, \\
a^7, & a^{11}, & a, & a^5, \\
a^{11}, & a^7, & a^5, & a;
\end{array}$$

where the same roots are reproduced in every row and column, their order only being changed. We have therefore proved that this property is not peculiar to any one root of the four special roots; and it will be noticed, in accordance with what is above proved in general, that 1, 5, 7, and 11 are all the numbers prime to 12, and less than it. We may obtain all the roots of $x^{12} - 1 = 0$ by the powers of any one of the four special roots a, a^5, a^7, a^{11}, as follows :—

$$\begin{array}{cccccccccccc}
a, & a^2, & a^3, & a^4, & a^5, & a^6, & a^7, & a^8, & a^9, & a^{10}, & a^{11}, & 1, \\
a^5, & a^{10}, & a^3, & a^8, & a, & a^6, & a^{11}, & a^4, & a^9, & a^2, & a^7, & 1, \\
a^7, & a^2, & a^9, & a^4, & a^{11}, & a^6, & a, & a^8, & a^3, & a^{10}, & a^5, & 1, \\
a^{11}, & a^{10}, & a^9, & a^8, & a^7, & a^6, & a^5, & a^4, & a^3, & a^2, & a, & 1.
\end{array}$$

3. Prove that the special roots of $x^{15} - 1 = 0$ are roots of the equation

$$x^8 - x^7 + x^5 - x^4 + x^3 - x + 1 = 0.$$

4. Show that the eight roots of the equation in the preceding example may be obtained by multiplying the two roots of $x^2 + x + 1 = 0$ by the four roots of

$$x^4 + x^3 + x^2 + x + 1 = 0.$$

54. Solution of Binomial Equations by Circular Functions.—We take the most general binomial equation

$$x^n = a + b \sqrt{-1},$$

where a and b are constants.

Let $\qquad a = R \cos a, \quad b = R \sin a$;

then $\qquad x^n = R (\cos a + \sqrt{-1} \sin a)$;

now, if $\qquad r (\cos \theta + \sqrt{-1} \sin \theta)$

be a root of this equation, we have, by De Moivre's Theorem,

$$r^n (\cos n\theta + \sqrt{-1} \sin n\theta) = R (\cos a + \sqrt{-1} \sin a);$$

and, therefore,

$$r^n \cos n\theta = R \cos a,$$

$$r^n \sin n\theta = R \sin a :$$

squaring these two equalities, and adding,

$$r^{2n} = R^2, \quad \text{giving} \quad r^n = R;$$

where we take R and r both positive, since in expressions of the kind here considered the factor containing the angle may always be taken to involve the sign.

We have then

$$\cos n\theta = \cos a, \quad \sin n\theta = \sin a ;$$

and consequently

$$n\theta = a + 2k\pi,$$

k being any integer; whence the assumed n^{th} root is of the general type

$$\sqrt[n]{R} \left(\cos \frac{a + 2k\pi}{n} + \sqrt{-1} \sin \frac{a + 2k\pi}{n} \right).$$

Giving to k in this expression any n consecutive values in the series of numbers between $-\infty$ and $+\infty$, we get all the nth roots; and no more than n, since the n values recur in periods.

We may write the expression for the nth root under the form

$$\left\{\sqrt[n]{R}\left(\cos\frac{a}{n} + \sqrt{-1}\sin\frac{a}{n}\right)\right\}\left(\cos\frac{2k\pi}{n} + \sqrt{-1}\sin\frac{2k\pi}{n}\right);$$

making $R = 1$, and $a = 0$, the equation $x^n = a + b\sqrt{-1}$ becomes $x^n = 1 + 0\sqrt{-1}$; the general type, therefore, of an nth root of $1 + 0\sqrt{-1}$, or 1, is

$$\cos\frac{2k\pi}{n} + \sqrt{-1}\sin\frac{2k\pi}{n}.$$

If we give k any definite value, for instance zero,

$$\sqrt[n]{R}\left(\cos\frac{a}{n} + \sqrt{-1}\sin\frac{a}{n}\right)$$

is one n^{th} root of $a + b\sqrt{-1}$.

The preceding formula shows, therefore, that *all the nth roots of any imaginary quantity may be obtained by multiplying any one of them by the nth roots of unity.*

Taking in conjunction the binomial equations

$$x^n = a + b\sqrt{-1}, \text{ and } x^n = a - b\sqrt{-1},$$

we see that the factors of the trinomial

$$x^{2n} - 2R\cos a \cdot x^n + R^2$$

are

$$\sqrt[n]{R}\left\{\cos\frac{a + 2k\pi}{n} \pm \sqrt{-1}\sin\frac{a + 2k\pi}{n}\right\},$$

where k has the values $0, 1, 2, 3 \ldots n - 1$.

Miscellaneous Examples.

1. Solve the equation $x^7 - 1 = 0$.

Dividing by $x - 1$, this is reduced to the standard form of reciprocal equation. Assuming $z = x + \dfrac{1}{x}$, we obtain the cubic

$$z^3 + z^2 - 2z - 1 = 0,$$

from whose solution that of the required equation is obtained.

2. Resolve $(x + 1)^7 - x^7 - 1$ into factors.

Ans. $7x(x + 1)(x^2 + x + 1)^2$.

3. Find the quintic on whose solution that of the binomial equation $x^{11} - 1 = 0$ depends.

Ans. $z^5 + z^4 - 4z^3 - 3z^2 + 3z + 1 = 0$.

4. When a binomial equation is reduced to the standard form of reciprocal equation (by division by $x - 1$, $x + 1$, or $x^2 - 1$), show that the reduced equation has all its roots imaginary. (Cf. Examples 15, 16, p. 33.)

5. When this reduced reciprocal equation is transformed by the substitution $z = x + \dfrac{1}{x}$; show that the equation in z has all its roots real, and situated between -2 and 2.

For the roots of the equation in x are of the form $\cos\alpha + \sqrt{-1}\,\sin\alpha$ (see Art. 54); hence $x + \dfrac{1}{x}$ is of the form $2\cos\alpha$, and the value of this is real and between -2 and 2.

6. Show that the following equation is reciprocal, and solve it:—

$$4(x^2 - x + 1)^3 - 27x^2(x - 1)^2 = 0.$$

Ans. Roots: $2, 2, \tfrac{1}{2}, \tfrac{1}{2}, -1, -1$.

7. Exhibit all the roots of the equation $x^9 - 1 = 0$.

The solution of this is reduced to the solution of the three cubics

$$x^3 - 1 = 0, \qquad x^3 - \omega = 0, \qquad x^3 - \omega^2 = 0;$$

where ω, ω^2 are the imaginary cube roots of unity. The nine roots may be represented as follows:—

$$1, \ \omega^{\frac{1}{3}}, \ \omega^{\frac{2}{3}}, \ \omega, \ \omega^{\frac{4}{3}}, \ \omega^{\frac{5}{3}}, \ \omega^2, \ \omega^{\frac{7}{3}}, \ \omega^{\frac{8}{3}}.$$

Excluding 1, ω, ω^2; the other six roots are special roots of the given equation; and are the roots of the sextic

$$x^6 + x^3 + 1 = 0.$$

8. Reducing the equation of the 8th degree in Ex. 3, Art. 53, by the substitution $z = x + \dfrac{1}{x}$, we obtain

$$z^4 - z^3 - 4z^2 + 4z + 1 = 0;$$

prove that the roots of this equation are

$$2 \cos \frac{2\pi}{15}, \quad 2 \cos \frac{4\pi}{15}, \quad 2 \cos \frac{8\pi}{15}, \quad 2 \cos \frac{14\pi}{15}.$$

9. Reduce the equation

$$4x^4 - 85x^3 + 357x^2 - 340x + 64 = 0$$

to a reciprocal equation, and solve it.

Assume $$z = \frac{x}{2} + \frac{2}{x}.$$ *Ans.* Roots: $\frac{1}{4}$, 1, 4, 16.

10. Solve the equation

$$x^4 + mp\,x^3 + m^2 q x^2 + m^3 px + m^4 = 0.$$

Dividing the roots by m, this reduces to a reciprocal equation.

11. If α be an imaginary root of the equation $x^n - 1 = 0$, where n is a prime number; prove the relation

$$(1 - \alpha)(1 - \alpha^2)(1 - \alpha^3) \ldots (1 - \alpha^{n-1}) = n.$$

12. Show that a cubic equation can be reduced immediately to the reciprocal form when the relation of Ex. 18, Art. 24, exists amongst its coefficients.

13. Show that a biquadratic can be reduced immediately to the reciprocal form when the relation of Ex. 22, Art. 24, exists amongst its coefficients.

14. Form the cubic whose roots are

$$\alpha + \alpha^6, \quad \alpha^3 + \alpha^4, \quad \alpha^2 + \alpha^5,$$

where α is an imaginary root of $x^7 - 1 = 0$.

Ans. $x^3 + x^2 - 2x - 1 = 0.$

15. Form the cubic whose roots are

$$\alpha + \alpha^8 + \alpha^{12} + \alpha^5, \quad \alpha^2 + \alpha^3 + \alpha^{11} + \alpha^{10}, \quad \alpha^4 + \alpha^6 + \alpha^9 + \alpha^7,$$

where α is an imaginary root of $x^{13} - 1 = 0$.

Ans. $x^3 + x^2 - 4x + 1 = 0.$

16. If $\alpha_1, \alpha_2, \alpha_3 \ldots \alpha_n$ be the roots of the equation

$$x^n + p_1 x^{n-1} + p_2 x^{n-2} + \ldots + p_{n-1} x + p_n = 0,$$

show how to form the equation whose roots are

$$\alpha_1 + \frac{1}{\alpha_1}, \quad \alpha_2 + \frac{1}{\alpha_2}, \ldots \alpha_n + \frac{1}{\alpha_n}.$$

We have here the identity

$$x^n + p_1 x^{n-1} + p_2 x^{n-2} + \ldots + p_{n-1} x + p_n \equiv (x - \alpha_1)(x - \alpha_2) \ldots (x - \alpha_n) ;$$

and changing x into $\frac{1}{x}$ (see Art. 32),

$$p_n x^n + p_{n-1} x^{n-1} + \ldots + p_2 x^2 + p_1 x + 1 \equiv p_n \left(x - \frac{1}{\alpha_1} \right) \left(x - \frac{1}{\alpha_2} \right) \ldots \left(x - \frac{1}{\alpha_n} \right).$$

H 2

Multiplying together these identities, and dividing by x^n, the factors on the right-hand side take the form $x + \dfrac{1}{x} - \left(a + \dfrac{1}{a} \right)$; and assuming $x + \dfrac{1}{x} = z$, the left-hand side can be expressed as a polynomial of the nth degree in z by means of the relations of Art. 45.

17. Find the value of the symmetric function $\Sigma \alpha^2 \beta^2 (\gamma - \delta)^2$ of the roots of the equation

$$a_0 x^4 + 4a_1 x^3 + 6a_2 x^2 + 4a_3 x + a_4 = 0.$$

This can be derived from the result of Ex. 19, p. 52, by changing the roots into their reciprocals, forming $\Sigma \left(\dfrac{1}{\alpha} - \dfrac{1}{\beta} \right)^2$ of the transformed equation, and multiplying by $\alpha^2 \beta^2 \gamma^2 \delta^2$, which is equal to $\dfrac{a_4^2}{a_0^2}$.

$$\textit{Ans.}\quad a_0^2 \, \Sigma \alpha^2 \beta^2 (\gamma - \delta)^2 = 48 \, (a_3^2 - a_2 a_4).$$

From the values of the symmetric functions given in Chapter III. several others can be obtained by the process here indicated.

18. Find the value of the symmetric function $\Sigma (\alpha_1 - \alpha_2)^2 \alpha_3^2 \alpha_4^2 \ldots \alpha_n^2$ of the roots of the equation

$$a_0 x^n + n a_1 x^{n-1} + \frac{n(n-1)}{1 \cdot 2} a_2 x^{n-2} + \ldots + n a_{n-1} x + a_n = 0.$$

We easily obtain $a_0^2 \, \Sigma (\alpha_1 - \alpha_2)^2 = n^2 (n - 1) (a_1^2 - a_0 a_2)$; and changing the roots into their reciprocals we have

$$a_0^2 \Sigma (\alpha_1 - \alpha_2)^2 \alpha_3^2 \alpha_4^2 \ldots \alpha_n^2 = n^2 (n - 1) (a_{n-1}^2 - a_{n-2} a_n).$$

19. Show that the five roots of the equation

$$x^5 + 5px^3 + 5p^2 x + q = 0$$

are

$$\sqrt[5]{a} + \sqrt[5]{b}, \quad \theta \sqrt[5]{a} + \theta^4 \sqrt[5]{b}, \quad \theta^2 \sqrt[5]{a} + \theta^3 \sqrt[5]{b},$$

$$\theta^4 \sqrt[5]{a} + \theta \sqrt[5]{b}, \quad \theta^3 \sqrt[5]{a} + \theta^2 \sqrt[5]{b},$$

where $\sqrt[5]{ab} = -p$, $a + b = -q$, and θ is an imaginary fifth root of unity.

Note.—A quintic reducible to this form can consequently be immediately solved.

CHAPTER VI.

55. On the Algebraic Solution of Equations.—Before proceeding with the solution of cubic and biquadratic equations we make some introductory remarks, with a view of putting clearly before the student the general principles on which the algebraic solution of these equations depends. With this object we give in the present Article three methods of solution of the quadratic, and state as we proceed how these methods may be extended to cubic and biquadratic equations, leaving to subsequent Articles the complete development of the principles involved.

(1). *First method of solution : by resolving into factors.* Let it be required to resolve the quadratic $x^2 + Px + Q$ into its simple factors. For this purpose we put it under the form

$$x^2 + Px + Q + \theta - \theta,$$

and determine θ so that

$$x^2 + Px + Q + \theta$$

may be a perfect square, *i.e.* we make

$$\theta + Q = \frac{P^2}{4}, \quad \text{or } \theta = \frac{P^2 - 4Q}{4};$$

whence, putting for θ its value, we have

$$x^2 + Px + Q = \left(x + \frac{P}{2}\right)^2 - \left(0x + \frac{\sqrt{P^2 - 4Q}}{2}\right)^2.$$

Thus we have reduced the quadratic to the form $u^2 - v^2$; and its simple factors are $u + v$, and $u - v$.

Subsequently we shall reduce the cubic to the form

$$(lx + m)^3 - (l'x + m')^3, \quad \text{or} \quad u^3 - v^3,$$

and obtain its solution from the simple equations

$$u - v = 0, \quad u - \omega v = 0, \quad u - \omega^2 v = 0.$$

It will be shown also that the biquadratic may be reduced to either of the forms

$$(lx^2 + mx + n)^2 - (l'x^2 + m'x + n')^2,$$
$$(x^2 + px + q)(x^2 + p'x + q'),$$

by solving a cubic equation; and, consequently, the solution of the biquadratic completed by solving two quadratics, viz., in the first case, $lx^2 + mx + n = \pm (l'x^2 + m'x + n')$; and in the second case, $x^2 + px + q = 0$, and $x^2 + p'x + q' = 0$.

(2). *Second method of solution : by assuming for a root a general form involving radicals.*

Assuming $x = p + \sqrt{q}$ to be a root of the equation $x^2 + Px + Q = 0$, and rationalizing the equation $x = p + \sqrt{q}$, we have

$$x^2 - 2px + p^2 - q = 0.$$

Now, if this equation be identical with $x^2 + Px + Q = 0$, we have

$$2p = - P, \quad p^2 - q = Q,$$

giving

$$x = p + \sqrt{q} = \frac{-P \pm \sqrt{P^2 - 4Q}}{2},$$

which is the solution of the quadratic equation.

In the case of the cubic equation we shall find that

$$x = \sqrt[3]{p} + \frac{A}{\sqrt[3]{p}}$$

is the proper form to represent a root; this formula giving precisely three values for x, in consequence of the manner in which the cube root enters into it.

In the case of the biquadratic equation we shall find that

$$\sqrt{p} + \sqrt{q} + \frac{A}{\sqrt{p}\,\sqrt{q}}, \quad \sqrt{q}\,\sqrt{r} + \sqrt{r}\,\sqrt{p} + \sqrt{p}\,\sqrt{q}$$

are forms which represent a root; these formulas each giving four, and only four, values of x when the square roots receive their double signs.

(3). *Third method of solution : by symmetric functions of the roots.*

Consider the quadratic equation $x^2 + Px + Q = 0$, of which the roots are a, β.

Then
$$a + \beta = - P,$$
$$a\beta = \quad Q.$$

If we attempt to determine a and β by these equations, we fall back on the original equation (see Art. 24); but if we could obtain a second equation between the roots and coefficients, of the form $la + m\beta = f(P, Q)$, we could easily find a and β by means of this equation and the equation $a + \beta = -P$.

Now in the case of the quadratic there is no difficulty in finding the required equation; for, obviously,

$$(a - \beta)^2 = P^2 - 4Q; \text{ and, therefore, } a - \beta = \sqrt{P^2 - 4Q}.$$

In the case of the cubic equation $x^3 + Px^2 + Qx + R = 0$, we require *two* simple equations of the form

$$la + m\beta + n\gamma = f(P, Q, R),$$

in addition to the equation $a + \beta + \gamma = -P$, to determine the roots a, β, γ. It will subsequently be proved that the functions

$$(a + \omega\beta + \omega^2\gamma)^3, \quad (a + \omega^2\beta + \omega\gamma)^3$$

may be expressed in terms of the coefficients by solving a *quadratic* equation; and when their values are known the roots of the cubic may be easily found.

In the case of the biquadratic equation

$$x^4 + Px^3 + Qx^2 + Rx + S = 0$$

we require *three* simple equations of the form

$$la + m\beta + n\gamma + r\delta = f(P, Q, R, S),$$

in addition to the equation

$$a + \beta + \gamma + \delta = - P,$$

to determine the roots a, β, γ, δ.   It will be proved in Art. 66, that the three functions

$$(\beta + \gamma - a - \delta)^2, \quad (\gamma + a - \beta - \delta)^2, \quad (a + \beta - \gamma - \delta)^2$$

may be expressed in terms of the coefficients by solving a *cubic* equation ; and when their values are known the roots of the biquadratic equation may be immediately obtained.

In applying the principles here explained to the solution of the cubic and biquadratic the order of the present Article is not followed.   The student will have no difficulty in perceiving under which of the methods here described any such solution should be included.

56. The Algebraic Solution of the Cubic Equation.—Let the general cubic equation

$$ax^3 + 3bx^2 + 3cx + d = 0$$

be put under the form

$$z^3 + 3Hz + G = 0,$$

where $z = ax + b$, $H = ac - b^2$, $G = a^2 d - 3abc + 2b^3$ (see Art. 37).

To solve this equation, assume*

$$z = \sqrt[3]{p} + \sqrt[3]{q} \, ;$$

hence, cubing,

$$z^3 = p + q + 3\sqrt[3]{p}\,\sqrt[3]{q}\,(\sqrt[3]{p} + \sqrt[3]{q}),$$

therefore

$$z^3 - 3\sqrt[3]{p}\,\sqrt[3]{q} \cdot z - (p + q) = 0.$$

Now, comparing coefficients, we have

$$\sqrt[3]{p} \cdot \sqrt[3]{q} = -H, \quad p + q = -G \, ;$$

from which equations we obtain

$$p = \tfrac{1}{2}\left(-G + \sqrt{G^2 + 4H^3}\right), \quad q = \tfrac{1}{2}\left(-G - \sqrt{G^2 + 4H^3}\right) ;$$

---

* This solution is usually called *Cardan's solution of the cubic.*   See Note A at the end of the volume.

and, substituting for $\sqrt[3]{q}$ its value $\dfrac{-H}{\sqrt[3]{p}}$, we have

$$z = \sqrt[3]{p} + \dfrac{-H}{\sqrt[3]{p}}$$

as the algebraic solution of the equation

$$z^3 + 3Hz + G = 0.$$

It should be noticed that if p be replaced by q this value of z is unchanged, as the terms are then simply interchanged; also, since $\sqrt[3]{p}$ has the three values $\sqrt[3]{p}$, $\omega\sqrt[3]{p}$, $\omega^2\sqrt[3]{p}$ obtained by multiplying any *one* of its values by the three cube roots of unity, we obtain three, and only three, values for z, namely,

$$\sqrt[3]{p} + \dfrac{-H}{\sqrt[3]{p}}, \quad \omega\sqrt[3]{p} + \omega^2\dfrac{-H}{\sqrt[3]{p}}, \quad \omega^2\sqrt[3]{p} + \omega\dfrac{-H}{\sqrt[3]{p}};$$

the order of these values only changing according to the cube root of p selected.

Now, if z be replaced by its value $ax + b$ we have, finally,

$$ax + b = \sqrt[3]{p} + \dfrac{-H}{\sqrt[3]{p}}$$

(where p has the value previously determined in terms of the coefficients) as the *complete algebraic solution of the cubic equation*

$$ax^3 + 3bx^2 + 3cx + d = 0,$$

the square root and cube root involved being taken in their entire generality.

57. Application to Numerical Equations.—The solution of the cubic which has been obtained, unlike the solution of the quadratic, is of little practical value when the coefficients of the equation are given numbers; although as an algebraic solution it is complete.

For, when the roots of the cubic are all real, $G^2 + 4H^3 = -K^2$, an essentially negative number (see Art. 42); and, substituting for p and q their values

$$\tfrac{1}{2}\left(-G \pm K\sqrt{-1}\right)$$

in the formula $\sqrt[3]{p} + \sqrt[3]{q}$, we have the following expression for a root of the cubic :—

$$\left(\frac{-G+K\sqrt{-1}}{2}\right)^{\frac{1}{3}} + \left(\frac{-G-K\sqrt{-1}}{2}\right)^{\frac{1}{3}}.$$

Now there is no general arithmetical process for extracting the cube root of such complex numbers, and consequently this formula is useless for purposes of arithmetical calculation.

But when the cubic has a pair of imaginary roots, an approximate numerical value may be obtained from the formula

$$\left(\frac{-G+\sqrt{G^2+4H^3}}{2}\right)^{\frac{1}{3}} + \left(\frac{-G-\sqrt{G^2+4H^3}}{2}\right)^{\frac{1}{3}},$$

since $G^2 + 4H^3$ is positive in this case. As a practical method, however, of obtaining the real root of a numerical cubic, this process is of little value.

In the first case; namely, where the roots are all real, we can make use of Trigonometry to obtain the numerical values of the roots in the following manner :—

Assuming $\quad 2R\cos\phi = -G$, and $2R\sin\phi = K$,

we have $\qquad p = Re^{\phi\sqrt{-1}}, \quad q = Re^{-\phi\sqrt{-1}}$;

also $\quad \tan\phi = -\dfrac{K}{G}$, and $R = \frac{1}{2}(G^2+K^2)^{\frac{1}{2}} = (-H)^{\frac{3}{2}}$;

and finally, since $\omega = \cos\dfrac{2\pi}{3} \pm \sqrt{-1}\sin\dfrac{2\pi}{3} = e^{\pm\frac{2\pi}{3}\sqrt{-1}}$,

the three roots of the cubic equation

$$z^3 + 3Hz + G = 0,$$

i. e. $\qquad \sqrt[3]{p}+\sqrt[3]{q},\quad \omega\sqrt[3]{p}+\omega^2\sqrt[3]{q},\quad \omega^2\sqrt[3]{p}+\omega\sqrt[3]{q},$

become

$$2(-H)^{\frac{1}{2}}\cos\frac{\phi}{3},\quad -2(-H)^{\frac{1}{2}}\cos\frac{\pi\pm\phi}{3};$$

from which formulas we obtain the numerical values of the roots of the cubic by aid of a table of sines and cosines. This process is not convenient in practice; and in general, for purposes of arithmetical calculation of real roots, the methods of solution of numerical equations to be hereafter explained (Chap. X.) should be employed.

58. **Expression of the Cubic as the Difference of two Cubes.**—Let the given cubic

$$ax^3 + 3bx^2 + 3cx + d \equiv \phi(x)$$

be put under the form

$$z^3 + 3Hz + G,$$

where $z \equiv ax + b$.

Now assume

$$z^3 + 3Hz + G \equiv \frac{1}{\mu - \nu}\left\{\mu(z+\nu)^3 - \nu(z+\mu)^3\right\}, \qquad (1)$$

where μ and ν are quantities to be determined; the second side of this identity becomes, when reduced,

$$z^3 - 3\mu\nu z - \mu\nu(\mu+\nu).$$

Comparing coefficients,

$$\mu\nu = -H, \quad \mu\nu(\mu+\nu) = -G;$$

therefore

$$\mu+\nu = \frac{G}{H}, \quad \mu-\nu = \frac{a\sqrt{\Delta}}{H};$$

where $a^2\Delta = G^2 + 4H^3$, as in Art. 41;

also

$$(z+\mu)(z+\nu) \equiv z^2 + \frac{G}{H}z - H. \qquad (2)$$

Whence, putting for z its value, $ax + b$, we have from (1)

$$a^3\phi(x) \equiv \left(\frac{G + a\Delta^{\frac{1}{2}}}{2\Delta^{\frac{1}{2}}}\right)\left(ax + b + \frac{G - a\Delta^{\frac{1}{2}}}{2H}\right)^3 - \left(\frac{G - a\Delta^{\frac{1}{2}}}{2\Delta^{\frac{1}{2}}}\right)\left(ax + b + \frac{G + a\Delta^{\frac{1}{2}}}{2H}\right)^3,$$

which is the required expression of $\phi(x)$ as the difference of two cubes.

The function (2), when transformed and reduced, becomes

$$\frac{a^2}{H}\left\{(ac - b^2)x^2 + (ad - bc)x + (bd - c^2)\right\},$$

which contains the two factors $ax + b + \mu$, $ax + b + \nu$.

The expression of the roots of this quadratic in terms of the roots of the given cubic may be seen on referring to Ex. 23, p. 57.

Thus the cubic may be resolved (as observed in Art. 55) into three factors. We add as examples some other instances of resolution into factors.

EXAMPLES.

1. Resolve into simple factors the expression

$$(\beta - \gamma)^2 (x - \alpha)^2 + (\gamma - \alpha)^2 (x - \beta)^2 + (\alpha - \beta)^2 (x - \gamma)^2.$$

Let $U = (\beta - \gamma)(x - \alpha)$, $V = (\gamma - \alpha)(x - \beta)$, $W = (\alpha - \beta)(x - \gamma)$.

$$\text{Ans. } \tfrac{2}{3}(U + \omega V + \omega^2 W)(U + \omega^2 V + \omega W).$$

2. Prove that the several equations of the system

$$(\beta - \gamma)^3 (x - \alpha)^3 = (\gamma - \alpha)^3 (x - \beta)^3 = (\alpha - \beta)^3 (x - \gamma)^3$$

have two factors common.

Making use of the notation in the last Example, we have

$$U^3 = V^3 = W^3 ;$$

whence

$$U^3 - V^3 = (U - V)(U^2 + UV + V^2) = \tfrac{1}{2}(U - V).(U^2 + V^2 + W^2),$$

since

$$U + V + W = 0 ;$$

therefore

$$(\beta - \gamma)^2 (x - \alpha)^2 + (\gamma - \alpha)^2 (x - \beta)^2 + (\alpha - \beta)^2 (x - \gamma)^2$$

is the common quadratic factor required.

3. Resolve into simple factors the expression

$$(\beta - \gamma)^3 (x - \alpha)^3 + (\gamma - \alpha)^3 (x - \beta)^3 + (\alpha - \beta)^3 (x - \gamma)^3.$$

$$\text{Ans. } 3(\beta - \gamma)(\gamma - \alpha)(\alpha - \beta)(x - \alpha)(x - \beta)(x - \gamma).$$

4. Resolve

$$(x - \alpha)(x - \beta)(x - \gamma)$$

into the difference of two cubes.

Assume

$$(x - \alpha)(x - \beta)(x - \gamma) = U_1^3 - V_1^3 ;$$

whence

$$U_1 - V_1 = \lambda(x - \alpha),$$

$$\omega U_1 - \omega^2 V_1 = \mu(x - \beta),$$

$$\omega^2 U_1 - \omega V_1 = \nu(x - \gamma) :$$

adding these we have

$$\lambda + \mu + \nu = 0, \quad \lambda\alpha + \mu\beta + \nu\gamma = 0 ;$$

and, therefore,

$$\lambda = \rho(\beta - \gamma), \quad \mu = \rho(\gamma - \alpha), \quad \nu = \rho(\alpha - \beta) ;$$

but $\lambda\mu\nu = 1$; whence

$$\frac{1}{\rho^3} = (\beta - \gamma)(\gamma - \alpha)(\alpha - \beta).$$

Substituting these values of λ, μ, ν; and using the notation of Ex. 1,

$$U_1 - V_1 = \rho U, \quad \omega U_1 - \omega^2 V_1 = \rho V, \quad \omega^2 U_1 - \omega V_1 = \rho W;$$

whence

$$3U_1 = \rho(U + \omega^2 V + \omega \ W),$$

$$-3V_1 = \rho(U + \omega \ V + \omega^2 W);$$

and U_1 and V_1 are completely determined.

59. Solution of the Cubic by Symmetric Functions of the Roots.—Since the three values of the expression

$$\tfrac{1}{3}\{a + \beta + \gamma + \theta(a \mp \omega\beta + \omega^2\gamma) + \theta^2(a + \omega^2\beta + \omega\gamma)\},$$

where θ has the values 1, ω, ω^2,   or $\theta^3 = 1$, are a, β, γ; it is plain that if the functions

$$\theta(a + \omega\beta + \omega^2\gamma), \quad \theta^2(a + \omega^2\beta + \omega\gamma)$$

were expressed in terms of the coefficients of the cubic, we could, by substituting their values in the formula given above, arrive at an algebraic solution of the cubic equation. Now this cannot be done directly by solving a quadratic equation; for, although the product of the two functions above written is a rational symmetric function of a, β, γ, their sum is not so. It will be found, however, that the sum of the cubes of the two functions in question is a symmetric function of the roots, and can, therefore, be expressed by the coefficients, as we proceed to show. For convenience we adopt the notation

$$L = a + \omega\beta + \omega^2\gamma, \quad M = a + \omega^2\beta + \omega\gamma.$$

We have then

$$(\theta L)^3 = A + B\omega + C\omega^2, \quad (\theta^2 M)^3 = A + B\omega^2 + C\omega,$$

where

$$A = a^3 + \beta^3 + \gamma^3 + 6a\beta\gamma, \quad B = 3(a^2\beta + \beta^2\gamma + \gamma^2 a), \quad C = 3(a\beta^2 + \beta\gamma^2 + \gamma a^2);$$

from which we obtain

$$L^3 + M^3 = 2\Sigma a^3 - 3\Sigma a^2\beta + 12a\beta\gamma = -27\frac{G}{a^3}.$$

(Cf. Ex. 5, p. 45; Ex. 15, p. 50.)

Again,

$$(\theta L)(\theta^2 M) = LM = a^2 + \beta^2 + \gamma^2 - \beta\gamma - \gamma a - a\beta = -9\frac{H}{a^2};$$

whence
$$(\alpha + \omega\beta + \omega^2\gamma)^3, \quad (\alpha + \omega^2\beta + \omega\gamma)^3$$

are the roots of the quadratic equation

$$t^2 + 3^3\frac{G}{a^3}t - 3^6\frac{H^3}{a^6} = 0.$$

These roots we denote by t_1 and t_2; their values are

$$\frac{3^3}{2a^3}\left(-G \pm \sqrt{G^2 + 4H^3}\right);$$

and therefore the original formula expressed in terms of the coefficients of the cubic gives for the three roots the expressions

$$\alpha = -\frac{b}{a} + \frac{1}{3}\left(\sqrt[3]{t_1} + \sqrt[3]{t_2}\right),$$

$$\beta = -\frac{b}{a} + \frac{1}{3}\left(\omega\sqrt[3]{t_1} + \omega^2\sqrt[3]{t_2}\right),$$

$$\gamma = -\frac{b}{a} + \frac{1}{3}\left(\omega^2\sqrt[3]{t_1} + \omega\sqrt[3]{t_2}\right).$$

These values of α, β, γ are the same as were obtained in Art. 56, by the former method of solution, if only

$$\frac{1}{3}\sqrt[3]{t_1} \quad \text{be replaced by} \quad \frac{\sqrt[3]{p}}{a}.$$

It is important to observe that the functions

$$(\alpha + \omega\beta + \omega^2\gamma)^3, \quad (\alpha + \omega^2\beta + \omega\gamma)^3$$

are remarkable as being the simplest functions of *three* variables which have but *two* values when the variables are interchanged in every way.   And it is owing to this property of these functions that the solution of a cubic equation can be reduced to that of a quadratic equation.

In the following Examples will be found other properties and applications of the functions L and M.

EXAMPLES.

1. The functions L and M are functions of the differences of the roots.

For, $$L = \alpha + \omega\beta + \omega^2\gamma = \alpha - h + \omega(\beta - h) + \omega^2(\gamma - h)$$

for all values of h, since $1 + \omega + \omega^2 = 0$; and giving to h the values α, β, γ, in succession, we obtain three forms for L in terms of the differences $\beta - \gamma$, $\gamma - \alpha$, $\alpha - \beta$. Similarly for M.

2. To express the product of the squares of the differences of the roots in terms of the coefficients.

We have
$$L + M = 2\alpha - \beta - \gamma, \quad L + \omega^2 M = (2\beta - \gamma - \alpha)\omega, \quad L + \omega M = (2\gamma - \alpha - \beta)\omega^2;$$
and, again,
$$L - M = (\beta - \gamma)(\omega - \omega^2), \quad \omega^2 L - \omega M = (\gamma - \alpha)(\omega - \omega^2), \quad \omega L - \omega^2 M = (\alpha - \beta)(\omega - \omega^2);$$

from which we obtain, as in Art. 26,

$$L^3 + M^3 = (2\alpha - \beta - \gamma)(2\beta - \gamma - \alpha)(2\gamma - \alpha - \beta),$$

$$L^3 - M^3 = -3\sqrt{-3}\,(\beta - \gamma)(\gamma - \alpha)(\alpha - \beta);$$

and since

$$(L^3 - M^3)^2 = (L^3 + M^3)^2 - 4L^3 M^3,$$

we have, substituting the previous results,

$$a_0^6(\beta - \gamma)^2(\gamma - \alpha)^2(\alpha - \beta)^2 = -27(G^2 + 4H^3).$$

(See Art. 41.)

3. Prove the following identities :—

$$L^3 + M^3 = \tfrac{1}{3}\{(2\alpha - \beta - \gamma)^3 + (2\beta - \gamma - \alpha)^3 + (2\gamma - \alpha - \beta)^3\},$$

$$L^3 - M^3 = \sqrt{-3}\,\{(\beta - \gamma)^3 + (\gamma - \alpha)^3 + (\alpha - \beta)^3\}.$$

These are easily obtained by cubing and adding the values of

$$L + M, \ \&c. \ ; \ L - M, \ \&c.$$

in the preceding example.

4. There are six functions of the type of L or M, viz.,

$$\alpha + \omega\beta + \omega^2\gamma, \quad \omega\alpha + \omega^2\beta + \gamma, \quad \omega^2\alpha + \beta + \omega\gamma,$$

$$\alpha + \omega^2\beta + \omega\gamma, \quad \omega\alpha + \beta + \omega^2\gamma, \quad \omega^2\alpha + \omega\beta + \gamma,$$

to form the equation whose roots are these six quantities.

These functions may be expressed as follows :—

$$L, \qquad \omega L, \qquad \omega^2 L,$$

$$M, \qquad \omega M, \qquad \omega^2 M;$$

hence they are the roots of the equation

$$(\phi - L)(\phi - \omega L)(\phi - \omega^2 L)(\phi - M)(\phi - \omega M)(\phi - \omega^2 M) = 0,$$

or
$$\phi^6 - (L^3 + M^3)\,\phi^3 + L^3 M^3 = 0.$$

Substituting for L and M from the equations

$$LM = -\frac{9H}{a^2}, \quad L^3 + M^3 = -27\,\frac{G}{a^3},$$

we have this equation expressed in terms of the coefficients as follows :—

$$\phi^6 + 3^3\,\frac{G}{a^3}\,\phi^3 - 3^6\,\frac{H^3}{a^6} = 0.$$

5. To obtain expressions for L^2, M^2, &c., in terms of α, β, γ.

The following forms for L^2 and M^2 are obtained by subtracting

$(\alpha^2 + \beta^2 + \gamma^2)(1 + \omega + \omega^2) \equiv 0$ from $(\alpha + \omega\beta + \omega^2\gamma)^2$, and $(\alpha + \omega^2\beta + \omega\gamma)^2$:

$$-L^2 = (\beta - \gamma)^2 + \omega^2(\gamma - \alpha)^2 + \omega\,(\alpha - \beta)^2,$$

$$-M^2 = (\beta - \gamma)^2 + \omega\,(\gamma - \alpha)^2 + \omega^2\,(\alpha - \beta)^2.$$

In a similar manner, we find from these formulas

$$-L^4 = (\beta - \gamma)^2(2\alpha - \beta - \gamma)^2 + \omega\,(\gamma - \alpha)^2(2\beta - \gamma - \alpha)^2 + \omega^2\,(\alpha - \beta)^2(2\gamma - \alpha - \beta)^2,$$

$$-M^4 = (\beta - \gamma)^2(2\alpha - \beta - \gamma)^2 + \omega^2(\gamma - \alpha)^2(2\beta - \gamma - \alpha)^2 + \omega\,(\alpha - \beta)^2(2\gamma - \alpha - \beta)^2.$$

Also, without difficulty, we have the following forms for LM, and $L^2 M^2$:—

$$2LM = (\beta - \gamma)^2 + (\gamma - \alpha)^2 + (\alpha - \beta)^2,$$

$$L^2 M^2 = (\alpha - \beta)^2(\alpha - \gamma)^2 + (\beta - \gamma)^2(\beta - \alpha)^2 + (\gamma - \alpha)^2(\gamma - \beta)^2.$$

6. To form the equation of the squares of the differences of the roots of the general cubic equation in terms of L and M.

Let
$$\phi = (\alpha - \beta)^2\,;$$

hence, by former results,

$$\sqrt{-3\phi} = \omega L - \omega^2 M.$$

Rationalizing this, we obtain

$$\phi\,(\phi - LM)^2 + \frac{(L^3 - M^3)^2}{27} = 0,$$

which is the required equation.

In a similar manner, by the aid of the results of Ex. 5, the equation of the squares of the differences of the roots of this equation, or the equation whose roots are

$$(\beta - \gamma)^2(2\alpha - \beta - \gamma)^2, \quad (\gamma - \alpha)^2(2\beta - \gamma - \alpha)^2, \quad (\alpha - \beta)^2(2\gamma - \alpha - \beta)^2,$$

is obtained by substituting $-L^2$ and $-M^2$ for M and L, respectively, in the last equation ; and this process may be repeated any number of times. Finally, all these equations may be easily expressed in terms of the coefficients of the cubic by means of the relations

$$LM = 9\,\frac{H}{a^2}, \quad \text{and} \quad L^3 + M^3 = -27\,\frac{G}{a^3}.$$

For instance, the first equation is

$$\phi\left(\phi + 9\,\frac{H}{a^2}\right)^2 + 27\,\frac{G^2 + 4H^3}{a^6} = 0.$$

(Cf. Art. 41.)

7. If α, β, γ and α', β', γ' be the roots of the cubic equations

$$ax^3 + 3bx^2 + 3cx + d = 0,$$

$$a'x^3 + 3b'x^2 + 3c'x + d' = 0\,;$$

to form the equation which has for roots the six values of the function

$$\phi \equiv \alpha\alpha' + \beta\beta' + \gamma\gamma'.$$

The easiest mode of procedure is first to form the corresponding equation for the cubics

$$z^3 + 3Hz + G = 0, \quad z^3 + 3H'z + G' = 0,$$

and thence deduce the equation in the general case ; for in this case the corresponding function

$$\phi_0 \equiv (a\alpha + b)(a'\alpha' + b') + (a\beta + b)(a'\beta' + b') + (a\gamma + b)(a'\gamma' + b')$$

$$\equiv aa'\phi - 3bb'.$$

Also, substituting for the roots of these latter their values expressed by radicals, we have

$$\phi_0 = \left(\sqrt[3]{p} + \sqrt[3]{q}\right)\left(\sqrt[3]{p'} + \sqrt[3]{q'}\right) + \left(\omega\sqrt[3]{p} + \omega^2\sqrt[3]{q}\right)\left(\omega\sqrt[3]{p'} + \omega^2\sqrt[3]{q'}\right)$$

$$+ \left(\omega^2\sqrt[3]{p} + \omega\sqrt[3]{q}\right)\left(\omega^2\sqrt[3]{p'} + \omega\sqrt[3]{q'}\right),$$

which reduces to

$$\phi_0 = 3\left(\sqrt[3]{pq'} + \sqrt[3]{p'q}\right).$$

Cubing this, we find

$$\phi_0^3 - 27\sqrt[3]{p\,q\,p'\,q'}\,\phi_0 - 27\,(pq' + p'q) = 0.$$

Now, substituting for p and q, p' and q', their values given by the equations

$$x^2 + Gx - H^3 = 0, \quad x^2 + G'x - H'^3 = 0,$$

we have the six values of ϕ_0 given by the two cubic equations

$$\phi_0^3 - 27\,HH'\,\phi_0 - \frac{27}{2}\,(GG' \pm aa'\sqrt{\Delta\Delta'}) = 0,$$

where

$$a^2\Delta = G^2 + 4H^3, \quad \text{and} \quad a'^2\Delta' = G'^2 + 4H'^3.$$

Finally, substituting for ϕ_0 its value $aa'\phi - 3bb'$, and multiplying these cubics together, we have the required equation. It may be noticed that if one of the cubics be $x^3 - 1 = 0$, $\phi = \alpha + \omega\beta + \omega^2\gamma$, &c., which case has been already considered in Ex. 4. Mr. M. Roberts, *Dublin Exam. Papers*, 1855.

I

8. Form the equation whose roots are the several values of ρ, where

$$\rho = \frac{\alpha - \beta}{\beta - \gamma}.$$

Since

$$\alpha - (1 + \rho)\beta + \rho\gamma = 0,$$

substituting for α, β, γ, their values in terms of p, q; and putting

$$\lambda = 1 - (1 + \rho)\omega + \rho\omega^2, \quad \mu = 1 - (1 + \rho)\omega^2 + \rho\omega,$$

we have

$$\lambda \sqrt[3]{\bar{p}} + \mu \sqrt[3]{\bar{q}} = 0.$$

Cubing, and substituting for p, q their values,

$$G(\lambda^3 + \mu^3) + a\sqrt{\bar{\Delta}}(\lambda^3 - \mu^3) = 0.$$

Squaring,

$$a^2 \Delta \lambda^3 \mu^3 = H^3 (\lambda^3 + \mu^3)^2,$$

and by previous results

$$\lambda\mu = 3(1 + \rho + \rho^2), \quad \lambda^3 + \mu^3 = -27\rho(1 + \rho) \; ;$$

substituting these values, we have the required equation

$$a^2 \Delta (1 + \rho + \rho^2)^3 - 27 H^3 (\rho + \rho^2)^2 = 0.$$

9. Find the relation between the coefficients of the cubics

$$ax^3 \; + 3bx^2 \; + 3cx \; + d = 0,$$
$$a'x'^3 + 3b'x'^2 + 3c'x' + d' = 0,$$

when the roots are connected by the equation

$$\alpha(\beta' - \gamma') + \beta(\gamma' - \alpha') + \gamma(\alpha' - \beta') = 0.$$

Multiplying by $\omega - \omega^2$, this equation becomes

$$LM' = L'M.$$

Cubing and introducing the coefficients, we find

$$G^2 H'^3 = G'^2 H^3,$$

the required relation.

10. Determine the condition in terms of the roots and coefficients that the cubics of Ex. 9 should become identical by the linear transformation

$$x' = px + q.$$

In this case

$$\alpha' = p\alpha + q, \quad \beta' = p\beta + q, \quad \gamma' = p\gamma + q.$$

Eliminating p and q, we have

$$\beta\gamma' - \beta'\gamma + \gamma\alpha' - \gamma'\alpha + \alpha\beta' - \alpha'\beta = 0,$$

which is the function of the roots considered in the last example. This relation, moreover, is unchanged if for α, β, γ; α', β', γ', we substitute

$$l\alpha \; + m, \quad l\beta \; + m, \quad l\gamma \; + m,$$
$$l'\alpha' + m', \quad l'\beta' + m', \quad l'\gamma' + m',$$

whence we may consider the cubics in the last example under the simple forms

$$z^3 + 3Hz + G = 0, \quad z'^3 + 3H'z' + G' = 0,$$

obtained by the linear transformations $z = ax + b$, $z' = a'x' + b'$; for if the condition holds for the roots of the former equations, it must hold for the roots of the latter. Now putting $z' = kz$, these equations become identical if

$$H' \equiv k^2 H, \quad G' \equiv k^3 G\,;$$

whence, eliminating k,

$$G^2 H'^3 = G'^2 H^3$$

is the required condition, the same as that obtained in Ex. 9. It may be observed that the reducing quadratics of the cubics necessarily become identical by the same transformation, viz.,

$$\frac{H'}{G'}\,(a'\,x' + b') = \frac{H}{G}\,(ax + b).$$

60. Homographic Relation between two Roots of a Cubic.—Before proceeding to the discussion of the biquadratic we prove the following important proposition relative to the cubic :—

The roots of the cubic are connected in pairs by a homographic relation in terms of the coefficients.

Referring to Example 13, Art. 27, we have the relations

$$a_0^2\{\ (\beta - \gamma)^2 + \ (\gamma - a)^2 + \ (a - \beta)^2\} = 18\,(a_1^2 - a_0 a_2),$$
$$a_0^2\{a\,(\beta - \gamma)^2 + \beta\,(\gamma - a)^2 + \gamma\,(a - \beta)^2\} = \ 9\,(a_0 a_3 - a_1 a_2),$$
$$a_0^2\{a^2(\beta - \gamma)^2 + \beta^2(\gamma - a)^2 + \gamma^2(a - \beta)^2\} = 18\,(a_2^2 - a_1 a_3).$$

We adopt the notation

$$a_0 a_2 - a_1^2 = H, \quad a_0 a_3 - a_1 a_2 = 2H_1, \quad a_1 a_3 - a_2^2 = H_2.$$

Now, multiplying the above equations by $a\beta$, $-(a + \beta)$, 1, respectively, and adding, since

$$a^2 - a(a + \beta) + a\beta \equiv 0, \quad \beta^2 - \beta(a + \beta) + a\beta \equiv 0,$$

we have

$$a_0^2(\beta - \gamma)(\gamma - a)(a - \beta)^2 = 18\{Ha\beta + H_1(a + \beta) + H_2\}\,;$$

but

$$a_0^4(\beta - \gamma)^2(\gamma - a)^2(a - \beta)^2 = -27\Delta \equiv 108\,(HH_2 - H_1^2)$$

(see Art. 41) ; whence

$$\pm\sqrt{-\frac{\Delta}{3}}\left(\frac{a - \beta}{2}\right) = Ha\beta + H_1(a + \beta) + H_2,$$

and, therefore,

$$Ha\beta + \left(H_1 + \frac{1}{2}\sqrt{-\frac{\Delta}{3}}\right)a + \left(H_1 - \frac{1}{2}\sqrt{-\frac{\Delta}{3}}\right)\beta + H_2 = 0,$$

which is the required homographic relation (see Art. 39).

61. First Solution by Radicals of the Biquadratic.

Euler's Assumption :—Let the biquadratic equation

$$ax^4 + 4bx^3 + 6cx^2 + 4dx + e = 0$$

be put under the form

$$z^4 + 6Hz^2 + 4Gz + a^2 I - 3H^2 = 0,$$

where $z = ax + b$,

$$H \equiv ac - b^2, \quad I \equiv ae - 4bd + 3c^2, \quad G \equiv a^2 d - 3abc + 2b^3.$$

(See Art. 38.)

To solve this equation (a biquadratic wanting the second term) Euler assumes as the general expression for a root

$$z = \sqrt{p} + \sqrt{q} + \sqrt{r}.$$

Squaring,

$$z^2 - p - q - r = 2(\sqrt{q}\,\sqrt{r} + \sqrt{r}\,\sqrt{p} + \sqrt{p}\,\sqrt{q}).$$

Squaring again, and reducing, we obtain the equation

$$z^4 - 2(p+q+r)z^2 - 8\sqrt{p} \cdot \sqrt{q} \cdot \sqrt{r} \cdot z + (p+q+r)^2 - 4(qr+rp+pq) = 0.$$

Comparing this equation with the former equation in z, we have

$$p + q + r = -3H, \quad qr + rp + pq = 3H^2 - \frac{a^2 I}{4}, \quad \sqrt{p} \cdot \sqrt{q} \cdot \sqrt{r} = -\frac{G}{2};$$

and consequently p, q, r are the roots of the equation

$$t^3 + 3Ht^2 + \left(3H^2 - \frac{a^2 I}{4}\right)t - \frac{G^2}{4} = 0;$$

or, since

$$-G^2 = 4H^3 - a^2 HI + a^3 J \qquad \text{(see Art. 38),}$$

where

$$J \equiv ace + 2bcd - ad^2 - eb^2 - c^3,$$

we may write this equation under the form

$$4(t + H)^3 - a^2 I(t + H) + a^3 J = 0;$$

and finally, putting $t + H = a^2 \theta$, we obtain the equation

$$4a^3 \theta^3 - Ia\theta + J = 0,$$

which we call *the reducing cubic* of the biquadratic equation.

Also, since $t = b^2 - ac + a^2\theta$; if θ_1, θ_2, θ_3 be the roots of the reducing cubic, we have

$$p = b^2 - ac + a^2\theta_1, \quad q = b^2 - ac + a^2\theta_2, \quad r = b^2 - ac + a^2\theta_3;$$

and, therefore,

$$z = \sqrt{b^2 - ac + a^2\theta_1} + \sqrt{b^2 - ac + a^2\theta_2} + \sqrt{b^2 - ac + a^2\theta_3}.$$

The radicals in this formula have not complete generality; for if they had, eight values of z in place of four would be given by the formula. This limitation is imposed by the relation

$$\sqrt{p} \cdot \sqrt{q} \cdot \sqrt{r} = -\frac{G}{2},$$

which (lost sight of in squaring to obtain the value of pqr) requires such signs to be attached to each of the quantities $\sqrt{p}$, $\sqrt{q}$, $\sqrt{r}$, that their product may maintain the sign determined by the above equation; thus,

$$\sqrt{p}\,\sqrt{q}\,\sqrt{r} = \sqrt{p}(-\sqrt{q})(-\sqrt{r}) = (-\sqrt{p})\sqrt{q}(-\sqrt{r})$$
$$= (-\sqrt{p})(-\sqrt{q})\sqrt{r}$$

are all the possible combinations of $\sqrt{p}$, $\sqrt{q}$, $\sqrt{r}$ fulfilling this condition, provided $\sqrt{p}$, $\sqrt{q}$, $\sqrt{r}$ retain the same signs throughout, whatever those signs may be. But we may avoid all ambiguity as regards sign, and express in a single algebraic formula the four values of z, by eliminating one of the quantities $\sqrt{p}$, $\sqrt{q}$, $\sqrt{r}$ from the formula

$$z = \sqrt{p} + \sqrt{q} + \sqrt{r}$$

by means of the relation given above, and leaving the other two quantities unrestricted in sign. We have then

$$z = \sqrt{p} + \sqrt{q} - \frac{G}{2\sqrt{p}\,\sqrt{q}},$$

a formula free from all ambiguity, since it gives four, and only four, values of z when $\sqrt{p}$ and $\sqrt{q}$ receive their double signs:

the sign given to each of these in the first two terms determining that which must be attached to it in the denominator of the third term. And finally, restoring p, q, and z their values given before, we have

$$ax + b = \sqrt{b^2 - ac + a^2\theta_1} + \sqrt{b^2 - ac + a^2\theta_2}$$

$$- \frac{G}{2\sqrt{(b^2 - ac + a^2\theta_1)} \cdot \sqrt{b^2 - ac + a^2\theta_2}}$$

as *the complete algebraic solution of the biquadratic equation;* θ_1 and θ_2 being roots of the equation

$$4a^3\theta^3 - Ia\theta + J = 0.$$

To assist the student in justifying Euler's apparently arbitrary assumption as to the form of solution of the biquadratic, we remark, that since the second term of the equation in z is absent, the sum of the four roots is zero, or $z_1 + z_2 + z_3 + z_4 = 0$; and consequently the functions $(z_1 + z_2)^2$, &c., of which there are in general *six* (the combinations of four quantities two and two), are in this case reduced to *three* only; so that we may assume

$$(z_2 + z_3)^2 = (z_1 + z_4)^2 = 4p,$$

$$(z_3 + z_1)^2 = (z_2 + z_4)^2 = 4q,$$

$$(z_1 + z_2)^2 = (z_3 + z_4)^2 = 4r;$$

from which we have z_1, z_2, z_3, z_4, included in the formula

$$\sqrt{p} + \sqrt{q} + \sqrt{r}.$$

Examples.

1. Show that the two biquadratic equations

$$A_0 x^4 + 6A_2 x^2 \pm 4A_3 x + A_4 = 0$$

have the same reducing cubic.

2. Find the reducing cubic of the two biquadratic equations

$$x^4 - 6lx^2 \pm 8\sqrt{(l^3 + m^3 + n^3 - 3lmn)} \cdot x + 3(4mn - l^2) = 0.$$

$$Ans. \quad \theta^3 - 3mn\,\theta - (m^3 + n^3) = 0.$$

3. Prove that the eight roots of the equation

$$\{x^4 - 6lx^2 + 3(4mn - l^2)\}^2 = 64(l^3 + m^3 + n^3 - 3lmn)x^2$$

are given by the formula

$$\sqrt{l + m + n} + \sqrt{l + \omega m + \omega^2 n} + \sqrt{l + \omega^2 m + \omega n}.$$

(Compare Ex. 20, p. 34.)

4. If the expression

$$\sqrt{l + m + n} + \sqrt{l + \omega m + \omega^2 n} + \sqrt{l + \omega^2 m + \omega n}$$

be a root of the equation

$$z^4 + 6Hz^2 + 4Gz + a^2 I - 3H^2 = 0,$$

determine H, I, J in terms of l, m, n.

$$Ans. \quad H = -l, \quad I = 12mn, \quad J = -4(m^3 + n^3).$$

5. Prove that J vanishes for the biquadratic

$$m(x - n)^4 - n(x - m)^4.$$

6. Write down the formulas expressing the root of a biquadratic in the particular cases when $I = 0$, and $J = 0$.

7. What is the quantity under the *final* square root in the formula expressing a root?

$$Ans. \quad 27J^2 - I^3.$$

8. Prove that the coefficients of the equation of the squares of the differences of the roots of the biquadratic equation

$$a_0 x^4 + 4a_1 x^3 + 6a_2 x^2 + 4a_3 x + a_4 = 0$$

may be expressed in terms a_0, H, I, and J.

Removing the second term from the equation, we obtain

$$y^4 + \frac{6H}{a_0^2} y^2 + \frac{4G}{a_0^3} y + \frac{a_0^2 I - 3H^2}{a_0^4} = 0 \, ;$$

and changing the signs of the roots, we have

$$y^4 + \frac{6H}{a_0^2} y^2 - \frac{4G}{a_0^3} y + \frac{a_0^2 I - 3H^2}{a_0^4} = 0.$$

These transformations leave the functions $(\alpha - \beta)^2$, &c., unaltered; but G becomes $-G$, the other coefficients of the latter equation remaining unchanged; therefore G can enter the coefficients of the equation of the squares of the differences in *even* powers only. And since

$$-G^2 = 4H^3 - a_0^2 HI + a_0^3 J,$$

G^2 may be eliminated, introducing a_0, H, I, J. In a similar manner we may prove that every even function of the differences of the roots α, β, γ, δ may be expressed in terms of a_0, H, I, J, the function G of odd degree not entering.

62. Second Solution by Radicals of the Biquadratic.—Let the biquadratic equation

$$ax^4 + 4bx^3 + 6cx^2 + 4dx + e = 0$$

be put, as before, under the form

$$z^4 + 6Hz^2 + 4Gz + a^2I - 3H^2 = 0,$$

where $z = ax + b$.

We now assume as the general expression for a root of this equation

$$z = \sqrt{q} \cdot \sqrt{r} + \sqrt{r} \cdot \sqrt{p} + \sqrt{p} \cdot \sqrt{q},$$

a formula involving *three* radicals.

Squaring twice, and reducing, we have

$$(z^2 - qr - rp - pq)^2 = 4pqr\,(2z + p + q + r),$$

or

$$z^4 - 2\,(qr + rp + pq)\,z^2 - 8pqrz + (qr + rp + pq)^2 - 4\,(p + q + r)\,pqr = 0.$$

Comparing this equation with the former equation in z, we easily find

$$qr + rp + pq = -3H, \quad pqr = -\frac{G}{2}, \quad p + q + r = \frac{a^2I - 12H^2}{2G};$$

whence p, q, r are the roots of the equation

$$2\,Gt^3 + (12H^2 - a^2I)\,t^2 - 6HGt + G^2 = 0.$$

Now, making the substitution

$$t = \frac{\tfrac{1}{2}G}{H - a^2\theta},$$

and putting for G^2 its value in terms of H, I, and J, we may reduce this equation to the standard form of the reducing cubic, viz.,

$$4a^3\theta^3 - Ia\theta + J = 0.$$

It is important that the student should clearly understand that the expression

$$\sqrt{q}\,\sqrt{r} + \sqrt{r}\,\sqrt{p} + \sqrt{p}\,\sqrt{q}$$

has only four values due to the double signs of $\sqrt{p}$, $\sqrt{q}$, $\sqrt{r}$, whilst $\sqrt{p} + \sqrt{q} + \sqrt{r}$ has eight values. This will appear from the following identical equation, viz.,

$$(\sqrt{p} + \sqrt{q} + \sqrt{r})^2 \equiv p + q + r + 2(\sqrt{q}\sqrt{r} + \sqrt{r}\sqrt{p} + \sqrt{p}\sqrt{q}),$$

which shows that $\sqrt{q}\sqrt{r} + \sqrt{r}\sqrt{p} + \sqrt{p}\sqrt{q}$ has just as many distinct values as $(\sqrt{p} + \sqrt{q} + \sqrt{r})^2$, namely four.

We now proceed to express p, q, r in terms of the roots α, β, γ, δ of the biquadratic. Since

$$z = ax + b = \sqrt{q}\sqrt{r} + \sqrt{r}\sqrt{p} + \sqrt{p}\sqrt{q},$$

we have, giving to x the four values α, β, γ, δ,

$$z_1 = a\alpha + b = \quad \sqrt{q}\sqrt{r} - \sqrt{r}\sqrt{p} - \sqrt{p}\sqrt{q},$$

$$z_2 = a\beta + b = -\sqrt{q}\sqrt{r} + \sqrt{r}\sqrt{p} - \sqrt{p}\sqrt{q},$$

$$z_3 = a\gamma + b = -\sqrt{q}\sqrt{r} - \sqrt{r}\sqrt{p} + \sqrt{p}\sqrt{q},$$

$$z_4 = a\delta + b = \quad \sqrt{q}\sqrt{r} + \sqrt{r}\sqrt{p} + \sqrt{p}\sqrt{q};$$

whence, from the values of $z_2 + z_3 - z_1 - z_4$, and $z_2 z_3 - z_1 z_4$, we obtain

$$a(\beta + \gamma - \alpha - \delta) = -4\sqrt{q}\sqrt{r},$$

$$a^2(\beta\gamma - \alpha\delta) + ab(\beta + \gamma - \alpha - \delta) = 4p\sqrt{q}\sqrt{r};$$

from these and similar equations we have, employing the relation $G = -2pqr$, the following modes of expressing p, q, r in terms of the roots α, β, γ, δ :—

$$-p = a\frac{\beta\gamma - \alpha\delta}{\beta + \gamma - \alpha - \delta} + b = \frac{8G}{a^2(\beta + \gamma - \alpha - \delta)^2},$$

$$-q = a\frac{\gamma\alpha - \beta\delta}{\gamma + \alpha - \beta - \delta} + b = \frac{8G}{a^2(\gamma + \alpha - \beta - \delta)^2},$$

$$-r = a\frac{\alpha\beta - \gamma\delta}{\alpha + \beta - \gamma - \delta} + b = \frac{8G}{a^2(\alpha + \beta - \gamma - \delta)^2}.$$

We have also from the above expressions for z_1, z_2, z_3, z_4,

$$a(\beta - \gamma) = -2\sqrt{p}\,(\sqrt{q} - \sqrt{r}), \quad a(\alpha - \delta) = -2\sqrt{p}\,(\sqrt{q} + \sqrt{r});$$

whence,

$$a^2(\beta - \gamma)(\alpha - \delta) = 4p(q - r) \equiv 4(R - Q),$$
$$a^2(\gamma - \alpha)(\beta - \delta) = 4q(r - p) \equiv 4(P - R),$$
$$a^2(\alpha - \beta)(\gamma - \delta) = 4r(p - q) \equiv 4(Q - P),$$

where

$$P \equiv qr, \quad Q \equiv rp, \quad R \equiv pq.$$

EXAMPLES.

1. Prove that

$$a^2 I = \frac{2}{3}\left\{ (Q - R)^2 + (R - P)^2 + (P - Q)^2 \right\},$$

$$-a^3 J = \frac{4}{27}(2P - Q - R)(2Q - R - P)(2R - P - Q),$$

by means of the equations

$$P + Q + R = -3H, \quad 2pqr = -G, \quad QR + RP + PQ = 3H^2 - \frac{a^2 I}{4},$$

and

$$G^2 + 4H^3 \equiv a^2(HI - aJ).$$

2. Express the values in Ex. 1 of

$$\frac{I}{a^2}, \quad \text{and} \quad \frac{J}{a^3},$$

in terms of

$$(\beta - \gamma)(\alpha - \delta), \quad (\gamma - \alpha)(\beta - \delta), \quad (\alpha - \beta)(\gamma - \delta);$$

and prove

$$a^6(\beta - \gamma)^2(\gamma - \alpha)^2(\alpha - \beta)^2(\alpha - \delta)^2(\beta - \delta)^2(\gamma - \delta)^2 = 256(I^3 - 27J^2).$$

3. Prove that p, q, r are in harmonic progression when α, β, γ, δ are in harmonic progression, and conversely.

63. The Resolution of the Quartic into its Quadratic Factors.—Let the given quartic be*

$$ax^4 + 4bx^3 + 6cx^2 + 4dx + e.$$

Multiplying this by a, and comparing it then with

$$(ax^2 + 2bx + c + 2a\theta)^2 - (2Mx + N)^2,$$

we have

$$M^2 = b^2 - ac + a^2\theta, \quad MN = bc - ad + 2ab\theta, \quad N^2 = (c + 2a\theta)^2 - ae.$$

---

* *Ferrari* was the first to solve the biquadratic, by expressing it as the difference of two squares. Some writers call this *Simpson's Solution*. The method of Art. 64 is due to *Descartes*. See note A.

Eliminating M and N from these equations, we find

$$4a^3\theta^3 - (ae - 4bd + 3c^2)\,a\theta + ace + 2bcd - ad^2 - eb^2 - c^3 = 0,$$

which is the reducing cubic before obtained.

From this equation we have three values of θ (θ_1, θ_2, θ_3), with three corresponding values of M^2, MN, N^2; and thus all the coefficients of the assumed form for the quartic are determined in three distinct ways; moreover, it should be noticed that to each value of M corresponds a *single* value of N, since

$$MN = bc - ad + 2ab\theta.$$

The quartic

$$(ax^2 + 2bx + c + 2a\theta)^2 - (2Mx + N)^2$$

may plainly be resolved into the two quadratic factors

$$ax^2 + 2(b - M)x + c + 2a\theta - N,$$

$$ax^2 + 2(b + M)x + c + 2a\theta + N;$$

when θ receives the three values θ_1, θ_2, θ_3, we have the three pairs of quadratic factors of the original quartic, and the problem is completely solved. If now, corresponding to these pairs of factors *in order*, the roots of the quartic be taken as

$$\beta,\ \gamma \text{ and } \alpha,\ \delta; \quad \gamma,\ \alpha \text{ and } \beta,\ \delta; \quad \alpha,\ \beta \text{ and } \gamma,\ \delta; \quad (1)$$

we have

$$\beta + \gamma = -\frac{2}{a}(b - M_1), \quad \gamma + \alpha = -\frac{2}{a}(b - M_2), \quad \alpha + \beta = -\frac{2}{a}(b - M_3),$$

$$\alpha + \delta = -\frac{2}{a}(b + M_1), \quad \beta + \delta = -\frac{2}{a}(b + M_2), \quad \gamma + \delta = -\frac{2}{a}(b + M_3),$$

where

$$M_1 = \sqrt{b^2 - ac + a^2\theta_1}, \quad M_2 = \sqrt{b^2 - ac + a^2\theta_2}, \quad M_3 = \sqrt{b^2 - ac + a^2\theta_3}.$$

Subtracting the last equations in pairs, we have

$$\beta + \gamma - \alpha - \delta = 4\frac{M_1}{a}, \quad \gamma + \alpha - \beta - \delta = 4\frac{M_2}{a}, \quad \alpha + \beta - \gamma - \delta = 4\frac{M_3}{a};$$

and since

$$\alpha + \beta + \gamma + \delta = -4\frac{b}{a},$$

we find also
$$
\begin{aligned}
a\alpha + b &= -M_1 + M_2 + M_3,\\
a\beta + b &= M_1 - M_2 + M_3,\\
a\gamma + b &= M_1 + M_2 - M_3,\\
a\delta + b &= -M_1 - M_2 - M_3.
\end{aligned}
\qquad (2)
$$

The signs of the radicals involved in M_1, M_2, M_3 were tacitly restricted when we arranged the *order* (1) of the pairs of quadratic factors.   The restriction imposed in this way is expressed by the equation

$$
(\beta + \gamma - \alpha - \delta)(\gamma + \alpha - \beta - \delta)(\alpha + \beta - \gamma - \delta) = 64\,\frac{M_1 M_2 M_3}{a^3},
$$

or
$$
M_1 M_2 M_3 = \frac{G}{2}
$$

(see Ex. 22, Art. 27).

Moreover, eliminating M_3 from equations (2), we find all the roots of the biquadratic given by the *single* formula

$$
ax + b = M_1 + M_2 - \frac{G}{2 M_1 M_2}
$$

when $M_1 = \sqrt{b^2 - ac + a^2\,\theta_1}$, and $M_2 = \sqrt{b^2 - ac + a^2\,\theta_2}$ take their double signs, thus bringing this solution into harmony with the solution of the biquadratic before given.

EXAMPLES.

1. Form the equation whose roots are λ, μ, ν, or
$$
\beta\gamma + \alpha\delta, \quad \gamma\alpha + \beta\delta, \quad \alpha\beta + \gamma\delta.
$$
Adding the last coefficients of the quadratic factors of the quartic, we have

$$
\beta\gamma + \alpha\delta = 4\theta_1 + 2\frac{c}{a},
$$

$$
\gamma\alpha + \beta\delta = 4\theta_2 + 2\frac{c}{a},
$$

$$
\alpha\beta + \gamma\delta = 4\theta_3 + 2\frac{c}{a},
$$

where θ_1, θ_2, θ_3 are the roots of the reducing cubic; therefore the required equation is

$$
\textit{Ans.}\quad (ax - 2c)^3 - 4I(ax - 2c) + 16J = 0.
$$

(Compare Examples 4, 5, Art. 43.)

2. Express the roots of the reducing cubic in terms of the roots of the biquadratic.

Substituting for $\dfrac{2c}{a}$ its value in terms of α, β, γ, δ, in the equations of the last Example, we find

$$12\,\theta_1 = (\gamma - \alpha)\,(\beta - \delta) - (\alpha - \beta)\,(\gamma - \delta) \equiv 2\lambda - \mu - \nu,$$
$$12\,\theta_2 = (\alpha - \beta)\,(\gamma - \delta) - (\beta - \gamma)\,(\alpha - \delta) \equiv 2\mu - \nu - \lambda,$$
$$12\,\theta_3 = (\beta - \gamma)\,(\alpha - \delta) - (\gamma - \alpha)\,(\beta - \delta) \equiv 2\nu - \lambda - \mu.$$

3. If two roots of the reducing cubic are imaginary, two roots of the biquadratic are real and two imaginary.

Let θ_1, θ_2 have the imaginary values $p \pm q\sqrt{-1}$; substituting these values in the formulas

$$M_1 = \sqrt{b^2 - ac + a^2\theta_1}, \qquad M_2 = \sqrt{b^2 - ac + a^2\theta_2},$$

we find

$$M_1 = \sqrt{P + Q\sqrt{-1}}, \qquad M_2 = \sqrt{P - Q\sqrt{-1}}.$$

Now, let

$$\tan 2\phi = \frac{P}{Q},$$

and we have

$$M_1 = \sqrt{P^2 + Q^2} \cdot e^{\phi\sqrt{-1}}, \quad M_2 = \sqrt{P^2 + Q^2} \cdot e^{-\phi\sqrt{-1}};$$

whence

$$M_1 + M_2 = 2\sqrt{P^2 + Q^2}\cos\phi,$$

$$M_1 - M_2 = 2i\sqrt{P^2 + Q^2}\sin\phi, \quad \text{where } i^2 = -1;$$

also, since the general solution of the biquadratic is

$$ax + b = M_1 + M_2 - \frac{G}{2M_1 M_2},$$

where M_1, M_2 have double signs, it is plain that the two roots of the biquadratic involving $M_1 + M_2$ are real, and the two roots involving $M_1 - M_2$ are imaginary.

4. If the roots of the biquadratic are all real or all imaginary, the roots of the reducing cubic are all real.

It is easily proved that

$$\beta\gamma + \alpha\delta, \quad \gamma\alpha + \beta\delta, \quad \alpha\beta + \gamma\delta$$

are real, and consequently θ_1, θ_2, θ_3.

5. Prove similarly that when two roots of the biquadratic are real and two imaginary, the reducing cubic has imaginary roots.

6. Form the equation whose roots are the functions

$$\tfrac{1}{8}(\beta\gamma - \alpha\delta)(\beta + \gamma - \alpha - \delta), \quad \tfrac{1}{8}(\gamma\alpha - \beta\delta)(\gamma + \alpha - \beta - \delta), \quad \tfrac{1}{8}(\alpha\beta - \gamma\delta)(\alpha + \beta - \gamma - \delta).$$

From the quadratic factors of the quartic we find

$$\frac{4M_1}{a} = \beta + \gamma - \alpha - \delta, \quad -\frac{2N_1}{a} = \beta\gamma - \alpha\delta\,;$$

also

$$M_1 N_1 = bc - ad + 2ab\,\theta_1 = -a^2\,\phi_1,$$

the roots of the required cubic being represented by ϕ_1, ϕ_2, ϕ_3.

We obtain, therefore, the required equation by a linear transformation of the reducing cubic.

$$\text{\textit{Ans.}}\quad (a^2\phi + bc - ad)^3 - b^2 I\,(a^2\phi + bc - ad) - 2b^3 J = 0.$$

7. Form the equation whose roots are

$$\frac{\beta\gamma - a\delta}{\beta + \gamma - a - \delta}, \quad \frac{\gamma a - \beta\delta}{\gamma + a - \beta - \delta}, \quad \frac{a\beta - \gamma\delta}{a + \beta - \gamma - \delta}.$$

If ϕ denote one of these functions generally, we have, employing former results,

$$-2\phi = \frac{MN}{M^2} = \frac{bc - ad + 2ab\theta}{b^2 - ac + a^2\theta};$$

and thus we obtain the required equation by a homographic transformation of the reducing cubic.   This formula may be put under the more convenient form

$$a\phi + b = \frac{\tfrac{1}{2}G}{a^2\theta - H},$$

by means of which we obtain the required cubic in the following form:—

$$2G\,(a\phi + b)^3 + (a^2 I - 12 H^2)(a\phi + b)^2 - 6HG\,(a\phi + b) - G^2 = 0,$$

which, expanded and divided by a^3, becomes

$$2G\phi^3 + (a^2 e + 6b^2 c - 9ac^2 + 2abd)\,\phi^2 + 2\,(abe + 2b^2 d - 3acd)\,\phi + b^2 e - ad^2 = 0.$$

(Compare Ex. 14, p. 86.)

64. **The Resolution of the Quartic into Quadratic Factors. Second Method.**—If the quartic

$$ax^4 + 4bx^3 + 6cx^2 + 4dx + e$$

be resolved into the quadratic factors

$$a\,(x^2 + 2px + q)(x^2 + 2p'x + q'),$$

we have, by comparing these two forms, the equations

$$p + p' = 2\frac{b}{a}, \quad q + q' + 4pp' = 6\frac{c}{a}, \quad pq' + p'q = 2\frac{d}{a}, \quad qq' = \frac{e}{a}. \qquad (1)$$

If now we had any fifth equation of the form

$$F(p,\, q,\, p',\, q') = \phi,$$

we could eliminate $p,\, p',\, q,\, q'$; and thus find an equation giving the several values of ϕ.

Thus, if (for reasons to be explained presently) we assume as the fifth equation

$$\phi = \frac{c}{a} - pp' = \frac{1}{4}\left(q + q' - \frac{2c}{a}\right),$$

these two functions of p, p', q, q', being equal by the second of equations (1), we easily find from the same equations

$$pq + p'q' = \frac{4abc - 2a^2d}{a^3} + \frac{8b\phi}{a};$$

and eliminating p, p', q, q', by means of the identical relation

$$(p^2 + p'^2)(q^2 + q'^2) \equiv (pq' - p'q)^2 + (pq + p'q')^2,$$

there results the equation

$$4a^3\phi^3 - Ia\phi + J = 0,$$

which is the reducing cubic obtained by the previous methods of solution.

Having thus found pp', or $q + q'$, we complete the resolution of the quartic by means of the equations (1).   We now explain why it might have been seen *à priori* that the form above assumed for ϕ would lead to a convenient equation.   If ϕ be expressed in terms of λ, μ, ν (see Ex. 1, Art. 63), we find

$$\phi = \frac{c}{a} - pp' = \frac{1}{4}\left(q + q' - \frac{2c}{a}\right) \equiv \frac{2\lambda - \mu - \nu}{12},$$

a function of the differences of λ, μ, ν; and every such function is an *even* function of the differences of a, β, γ, δ, in virtue of the equations

$$-\mu + \nu = (\beta - \gamma)(a - \delta), \quad -\nu + \lambda = (\gamma - a)(\beta - \delta),$$
$$-\lambda + \mu = (a - \beta)(\gamma - \delta);$$

hence the equation for ϕ cannot involve any function of the coefficients except a, H, I, and J; and in fact it is, as we have just proved, the reducing cubic involving only a, I, and J.

EXAMPLES.

1. Resolve into quadratic factors

$$z^4 + 6Hz^2 + 4Gz + a^2I - 3H^2.$$

Comparing this form with the product

$$(z^2 + 2pz + q)(z^2 - 2pz + q').$$

we find the following equation for p:—

$$4p^6 + 12Hp^4 + 12\left(H^2 - \frac{a^2I}{12}\right)p^2 - G^2 = 0 ;$$

and putting

$$a^2\phi = p^2 + H \equiv \tfrac{1}{4}(q + q' - 2H),$$

this equation, when divided by a^3, becomes

$$4a^3\phi^3 - Ia\phi + J = 0.$$

2. If a quartic be resolved into the two quadratic factors

$$x^2 + px + q, \quad x^2 + p'x + q',$$

prove that ϕ is determined by a cubic equation when it has all the values corresponding to each of the following types:—

$$q + q', \quad \frac{q - q'}{p - p'}, \quad \frac{pq' - p'q}{p - p'}, \quad \frac{pq' - p'q}{q - q'},$$

$$(p - p')^2, \quad (p - p')(q - q'), \quad (q - q')^2, \quad (pq' - p'q)^2 ;$$

and by an equation of the sixth degree when it has all the values corresponding to

$$p, \quad q, \quad p - p', \quad q - q', \quad pq' - p'q, \quad \text{or} \quad p^2 - 4q.$$

Express these functions in terms of the roots, and the number of values each function has becomes apparent.

65. **Transformation of the Biquadratic into the Reciprocal Form.**—To effect this transformation we make the linear substitution $x = ky + \rho$ in the equation

$$ax^4 + 4bx^3 + 6cx^2 + 4dx + e = 0,$$

which then assumes the form

$$ak^4y^4 + 4U_1k^3y^3 + 6U_2k^2y^2 + 4U_3ky + U_4 = 0,$$

where

$$U_1 \equiv a\rho + b, \quad U_2 \equiv a\rho^2 + 2b\rho + c, \quad U_3 \equiv a\rho^3 + 3b\rho^2 + 3c\rho + d, \quad \&c.$$

(Cf. Art. 36.) If this equation be reciprocal, we have two equations to determine k and ρ, viz.,

$$ak^4 = U_4, \quad k^3U_1 = kU_3 ;$$

eliminating k, we have the following equation to determine ρ :—

$$a U_3^2 - U_1^2 U_4 = 0 ;$$

and since

$$k^2 = \frac{U_3}{U_1} = \frac{a\rho^3 + 3b\rho^2 + 3c\rho + d}{a\rho + b},$$

there are two values of k, equal with opposite signs, corresponding to each value of ρ.

The equation

$$a U_3^2 - U_1^2 U_4 = 0,$$

when reduced by the substitutions (see Arts. 37, 38)

$$a^2 U_3 = U_1^3 + 3 H U_1 + G,$$

$$a^3 U_4 = U_1^4 + 6 H U_1^2 + 4 G U_1 + a^2 I - 3 H^2,$$

becomes

$$2 G U_1^3 + (a^2 I - 12 H^2) U_1^2 - 6 G H U_1 - G^2 = 0, \tag{1}$$

which is a cubic equation determining $U_1 = a\rho + b$; and if we put

$$a\rho + b = \frac{\frac{1}{2} G}{a^2 \theta - H},$$

θ is determined by the standard reducing cubic

$$4 a^3 \theta^3 - I a \theta + J = 0.$$

This transformation* may be employed to solve the biquadratic ; and it is important to observe that the cubic (1) which here presents itself differs from that which results in the mode of solution of Art. 62 only in having roots with contrary signs.

We proceed now to express k and ρ in terms of a, β, γ, δ, the roots of the biquadratic equation. Since the equation in y, obtained by putting $x = ky + \rho$, is reciprocal, its roots are of the form $y_1,\ y_2,\ \dfrac{1}{y_2},\ \dfrac{1}{y_1}$; hence we may write

$$a = ky_1 + \rho, \quad \beta = ky_2 + \rho, \quad \delta = k\frac{1}{y_1} + \rho, \quad \gamma = k\frac{1}{y_2} + \rho ;$$

---

* This method of solving the biquadratic by transforming it to the reciprocal form was given by Mr. S. S. Greatheed in the *Camb. Math. Journal*, vol. i.

K

and, therefore,

$$(a - \rho)(\delta - \rho) = (\beta - \rho)(\gamma - \rho) = k^2,$$

from which we find

$$\rho = \frac{\beta\gamma - a\delta}{\beta + \gamma - a - \delta},$$

and

$$-k^2 = \frac{(\gamma - a)(\beta - \delta)(a - \beta)(\gamma - \delta)}{(\beta + \gamma - a - \delta)^2}.$$

An important geometrical interpretation may be given to the quantities k and ρ which enter into this transformation. Let the distances OA, OB, OC, OD, of four points A, B, C, D, on a right line from a fixed origin O on the line be determined by the roots a, β, γ, δ, of the equation

$$ax^4 + 4bx^3 + 6cx^2 + 4dx + e = 0 ;$$

also let O_1, O_2, O_3 be the centres; and F_1, F_1' ; F_2, F_2'; F_3, F_3' the foci of the three systems of involution determined by the sets of quadratics arranged horizontally in pairs

$$(x - \beta)(x - \gamma) = 0, \quad (x - a)(x - \delta) = 0,$$

$$(x - \gamma)(x - a) = 0, \quad (x - \beta)(x - \delta) = 0,$$

$$(x - a)(x - \beta) = 0, \quad (x - \gamma)(x - \delta) = 0.$$

We have then the equations

$$O_1B \cdot O_1C = O_1A \cdot O_1D = O_1F_1^2, \ \&c.$$

which, transformed and compared with the equations

$$(\beta - \rho)(\gamma - \rho) = (a - \rho)(\delta - \rho) = k^2, \ \&c.,$$

prove that the three values of ρ are OO_1, OO_2, OO_3, the distances of the three centres of involution from the fixed origin O. Also since $O_1F_1^2 = k^2$, k has six values represented geometrically by the distances

$$O_1F_1, \ O_1F_1' ; \quad O_2F_2, \ O_2F_2' ; \quad O_3F_3, \ O_3F_3',$$

where $O_1F_1 + O_1F_1' = 0$, &c., as the distances are measured in opposite directions.

We can from geometrical considerations alone find the positions of the centres and foci of involution in terms of α, β, γ, δ, and thus confirm the results just established, as follows:—

Since the systems $\{F_1 B F_1' C\}$ and $\{F_1 A F_1' D\}$ are harmonic,

$$\frac{2}{F_1 F_1'} = \frac{1}{F_1 B} + \frac{1}{F_1 C} = \frac{1}{F_1 A} + \frac{1}{F_1 D};$$

and if x represent the distance of F_1 or F_1' from the fixed origin O, we have

$$\frac{1}{x - \beta} + \frac{1}{x - \gamma} = \frac{1}{x - \alpha} + \frac{1}{x - \delta}.$$

Solving this equation we find

$$x = \frac{\beta\gamma - \alpha\delta}{\beta + \gamma - \alpha - \delta} \pm \frac{\sqrt{-(\gamma - \alpha)(\beta - \delta)(\alpha - \beta)(\gamma - \delta)}}{\beta + \gamma - \alpha - \delta},$$

or $\qquad x = \rho \pm k,$

whence $\qquad \rho = \dfrac{OF_1 + OF_1'}{2},$

$$k = \pm \frac{OF_1 - OF_1'}{2} = \pm\, O_1 F_1.$$

EXAMPLE.

Transform the cubic

$$ax^3 + 3bx^2 + 3cx + d$$

to the reciprocal form.

The assumption $x = ky + \rho$ leads here to the equation

$$- G U_1^3 + 3H^2 U_1^2 + H^3 = 0, \text{ where } U_1 \equiv a\rho + b.$$

The values of ρ are easily seen to be

$$\frac{\beta\gamma - \alpha^2}{\beta + \gamma - 2\alpha}, \quad \frac{\gamma\alpha - \beta^2}{\gamma + \alpha - 2\beta}, \quad \frac{\alpha\beta - \gamma^2}{\alpha + \beta - 2\gamma}.$$

The geometrical interpretation in this case is, that if three points A', B', C' be taken on the axis such that A' is the harmonic conjugate of A with respect to B and C, B' of B with respect to C and A, and C' of C with respect to A and B; then we have the following values of ρ and k:—

$$\rho = \frac{OA + OA'}{2}, \quad k = \frac{OA - OA'}{2}.$$

For the values of OA', OB', OC' in terms of α, β, γ, see Ex. 13, p. 86.

K 2

66. Solution of the Biquadratic by Symmetric Functions of the Roots.—The possibility of reducing the solution of the biquadratic to that of a cubic by the present method depends on the possibility of forming functions of the four roots a, β, γ, δ, which admit of only three values when these roots are interchanged in every way. That there are several functions of this kind will be seen on referring to Ex. 2, Art. 64. We employ in the present Article the functions already referred to in Art. 55, since they lead in the most direct manner to the expressions for the roots of the biquadratic in terms of the coefficients.

We form the equation whose roots are the three values of the function

$$t = \left(\frac{a + \theta\beta + \theta^2\gamma + \theta^3\delta}{4}\right)^2,$$

when the roots are interchanged in every way, and $\theta = -1$.

These values are

$$t_1 = \left(\frac{\beta + \gamma - a - \delta}{4}\right)^2, \quad t_2 = \left(\frac{\gamma + a - \beta - \delta}{4}\right)^2, \quad t_3 = \left(\frac{a + \beta - \gamma - \delta}{4}\right)^2;$$

and since

$$(\beta + \gamma - a - \delta)^2 = \Sigma a^2 + 2\lambda - 2\mu - 2\nu,$$

$$\Sigma(a - \beta)^2 = 3\Sigma a^2 - 2\lambda - 2\mu - 2\nu = -48\frac{H}{a^2},$$

we find the following values of t_1, t_2, t_3:—

$$\frac{2\lambda - \mu - \nu}{12} - \frac{H}{a^2}, \qquad \frac{2\mu - \nu - \lambda}{12} - \frac{H}{a^2}, \qquad \frac{2\nu - \lambda - \mu}{12} - \frac{H}{a^2};$$

whence

$$t_1 + t_2 + t_3 = -3\frac{H}{a^2}.$$

Again, since

$$\Sigma(2\mu - \nu - \lambda)(2\nu - \lambda - \mu) = -3(\lambda^2 + \mu^2 + \nu^2 - \mu\nu - \nu\lambda - \lambda\mu) = -\frac{3}{2}\Sigma(\mu - \nu)^2,$$

and

$$\Sigma(\mu - \nu)^2 = 24\frac{I}{a^2},$$

we have

$$t_2 t_3 + t_3 t_1 + t_1 t_2 = 3\frac{H^2}{a^4} - \frac{1}{96}\Sigma(\mu - \nu)^2 = \frac{3H^2}{a^4} - \frac{I}{4a^2}.$$

We have also

$$t_1 t_2 t_3 = \left(\frac{32\,G}{64\,a^3}\right)^2 = \frac{G^2}{4a^6}.$$

Hence the equation whose roots are t_1, t_2, t_3 becomes

$$(a^2 t)^3 + 3H(a^2 t)^2 + \left(3H^2 - \frac{a^2 I}{4} \right)(a^2 t) - \frac{G^2}{4} = 0 \, ;$$

or, substituting for G^2 its value from Art. 38,

$$4\,(a^2 t + H)^3 - a^2 I\,(a^2 t + H) + a^3 J = 0,$$

which is transformed into the standard reducing cubic by the substitution $a^2 t + H = a^2 \theta$.

To determine a, β, γ, δ we have the following equations :—

$$-a + \beta + \gamma - \delta = 4\sqrt{t_1}, \quad a - \beta + \gamma - \delta = 4\sqrt{t_2}, \quad a + \beta - \gamma - \delta = 4\sqrt{t_3},$$

along with $\qquad\qquad a + \beta + \gamma + \delta = - 4\,\dfrac{b}{a} \, ;$

from which we find

$$a = -\frac{b}{a} - \sqrt{t_1} + \sqrt{t_2} + \sqrt{t_3},$$

$$\beta = -\frac{b}{a} + \sqrt{t_1} - \sqrt{t_2} + \sqrt{t_3},$$

$$\gamma = -\frac{b}{a} + \sqrt{t_1} + \sqrt{t_2} - \sqrt{t_3},$$

$$\delta = -\frac{b}{a} - \sqrt{t_1} - \sqrt{t_2} - \sqrt{t_3}.$$

We have also from the above values of $\sqrt{t_1}$, $\sqrt{t_2}$, $\sqrt{t_3}$ the equation

$$\sqrt{t_1} \cdot \sqrt{t_2} \cdot \sqrt{t_3} = \frac{G}{2a^3},$$

by means of which one radical can be expressed in terms of the other two, and the general formula for a root shown to be the same as those previously given.

The cubic of Ex. 4, Art. 43, whose roots are λ, μ, ν, might have been employed in a similar manner to obtain the values of a, β, γ, δ ; but it does not lead so directly to these values as the equation above given. We wish here to call attention to two functions of λ, μ, ν, which possess properties analogous to those established in Art. 59 for corresponding functions of the roots of a cubic. We use a notation similar to that of the Article referred to, and write

$$L \equiv (\beta\gamma + a\delta) + \omega\,(\gamma a + \beta\delta) + \omega^2\,(a\beta + \gamma\delta),$$

$$M \equiv (\beta\gamma + a\delta) + \omega^2\,(\gamma a + \beta\delta) + \omega\,(a\beta + \gamma\delta).$$

The following relations between λ, μ, ν and the roots of the reducing cubic

$$4a^3\,\theta^3 - Ia\theta + J = 0$$

are easily established (see Ex. 1, Art. 63) :—

$$\lambda = 4\theta_1 + 2\frac{c}{a}, \quad \mu = 4\theta_2 + 2\frac{c}{a}, \quad \nu = 4\theta_3 + 2\frac{c}{a};$$

hence the values of L and M may be written as follows :—

$$\frac{1}{4}L = \theta_1 + \omega\theta_2 + \omega^2\theta_3, \quad \frac{1}{4}M = \theta_1 + \omega^2\theta_2 + \omega\theta_3.$$

These functions are just as important in the theory of the biquadratic as the functions of Art. 59 in the theory of the cubic ; for they are the simplest functions of four variables which have but *two* values when the variables are interchanged in every way. They are the roots of the reducing quadratic of the reducing cubic above written ; and underlie every solution of the biquadratic which has been given.

We add some applications of these functions.

EXAMPLES.

1. Show that L and M are functions of the differences of a, β, γ, δ. Increasing a, β, γ, δ by h, these functions remain unaltered, since $1 + \omega + \omega^2 = 0$.

2. To find in terms of the coefficients the product of the squares of the differences of the roots a, β, γ, δ.

From the values of L and M in terms of θ_1, θ_2, θ_3, we find easily

$$12\theta_1 = \ \ L + M, \qquad L - M \ = (\beta - \gamma)(a - \delta)(\omega^2 - \omega),$$

$$12\theta_2 = \omega^2 L + \omega M, \quad \omega^2 L - \omega M = (\gamma - a)(\beta - \delta)(\omega^2 - \omega),$$

$$12\theta_3 = \ \omega L + \omega^2 M, \quad \omega L - \omega^2 M = (a - \beta)(\gamma - \delta)(\omega^2 - \omega).$$

Again, from these equations, multiplying the terms on both sides together, and remembering that θ_1, θ_2, θ_3 are the roots of

$$4a^3\theta^3 - Ia\theta + J = 0,$$

we find

$$L^3 + M^3 = -432\,\frac{J}{a^3},$$

$$L^3 - M^3 = 3\sqrt{-3}\,(\beta-\gamma)(\gamma-\alpha)(\alpha-\beta)(\alpha-\delta)(\beta-\delta)(\gamma-\delta)\,;$$

also, adding the squares of the same terms, we have

$$2LM = 24\,\frac{I}{a^2} = (\beta-\gamma)^2(\alpha-\delta)^2 + (\gamma-\alpha)^2(\beta-\delta)^2 + (\alpha-\beta)^2(\gamma-\delta)^2\,;$$

and, since

$$(L^3 - M^3)^2 \equiv (L^3 + M^3)^2 - 4L^3M^3,$$

substituting for these quantities their values derived from former equations, we have finally

$$a^6(\beta-\gamma)^2(\gamma-\alpha)^2(\alpha-\beta)^2(\alpha-\delta)^2(\beta-\delta)^2(\gamma-\delta)^2 = 256\,(I^3 - 27J^2).$$

3. Show by a comparison of the equations of the present Article and Art. 59 that the results of the previous Article may be extended to the biquadratic by changing

$$\beta-\gamma,\ \gamma-\alpha,\ \alpha-\beta \text{ into } -(\beta-\gamma)(\alpha-\delta),\ -(\gamma-\alpha)(\beta-\delta),\ -(\alpha-\beta)(\gamma-\delta),$$

respectively ; and, consequently, H into $-\dfrac{4}{3}I$, and G into $16J$.

67. Equation of the Squares of the Differences of the Roots of a Biquadratic.—The general problem of the formation of the equation whose roots are the squares of the differences of the roots of a given equation may be treated as follows :—Let the proposed equation be

$$f(x) \equiv (x - a_1)(x - a_2)(x - a_3)\ \ldots\ (x - a_n) = 0.$$

Substituting $x + a_r$ for x, and giving r the values 1, 2, 3, $\ldots n$, we have

$$\left.\begin{aligned}
f(x + a_1) &\equiv x\,(x + a_1 - a_2)(x + a_1 - a_3)\,\ldots\,(x + a_1 - a_n),\\
f(x + a_2) &\equiv x\,(x + a_2 - a_1)(x + a_2 - a_3)\,\ldots\,(x + a_2 - a_n),\\
&\quad\cdot\quad\cdot\quad\cdot\quad\cdot\quad\cdot\quad\cdot\quad\cdot\quad\cdot\quad\cdot\quad\cdot\\
f(x + a_n) &\equiv x\,(x + a_n - a_1)(x + a_n - a_2)\,\ldots\,(x + a_n - a_{n-1})\,;
\end{aligned}\right\} \quad (1)$$

also, since

$$f(a_r + x) = f(a_r) + xf'(a_r) + \frac{x^2}{1\cdot 2}f''(a_r) + \ldots + x^n,$$

and $f(a_r) = 0$, we have

$$\frac{1}{x} f(x + a_r) = f'(a_r) + \frac{x}{1 \cdot 2} f''(a_r) + \ldots + x^{n-1}.$$

Denoting the second side of this equation by $\phi(x, a_r)$, and multiplying both sides of equations (1), we find

$$\phi(x, a_1) \, \phi(x, a_2) \ldots \ldots \phi(x, a_n) = \{x^2 - (a_1 - a_2)^2\} \{x^2 - (a_1 - a_3)^2\}$$
$$\ldots \ldots \{x^2 - (a_{n-1} - a_n)^2\}.$$

Thus to form the equation of the squares of the differences, we can multiply together the n factors $\phi(x, a_1)$, $\phi(x, a_2)$, &c., and substitute for the symmetric functions of the roots which occur in the product their values in terms of the coefficients. The resulting equation will be of the $n(n-1)^{th}$ degree in x; but as it will contain only even powers of x it must be reduced to the $\frac{1}{2} n(n-1)^{th}$ degree by the substitution of x for x^2. Or we may form directly the product of the $\frac{1}{2} n(n-1)$ factors on the right-hand side of the above equation; and express the symmetric functions involved in terms of the coefficients. We adopt the latter of these methods in the following application to the biquadratic :—

The problem is equivalent to expressing the following product in terms of the coefficients of the biquadratic

$$\{\phi - (\beta - \gamma)^2\} \{\phi - (\gamma - \alpha)^2\} \{\phi - (\alpha - \beta)^2\} \{\phi - (\alpha - \delta)^2\} \{\phi - (\beta - \delta)^2\} \{\phi - (\gamma - \delta)^2\}.$$

The most convenient mode of procedure is to group these six factors in pairs involving all the roots, and to express the three products, which we denote by Π_1, Π_2, Π_3, separately in terms of the roots of the reducing cubic, and finally to express the product $\Pi_1 \Pi_2 \Pi_3$ in terms of a, H, I, J.

$$\Pi_1 \equiv \phi^2 - \{(\beta - \gamma)^2 + (\alpha - \delta)^2\} \phi + (\beta - \gamma)^2 (\alpha - \delta)^2 \, ;$$

and, by previous results, we have the following expressions for $(\beta - \gamma)^2$, $(\alpha - \delta)^2$:—

$$4 \left(\sqrt{\theta_2 - \frac{H}{a^2}} - \sqrt{\theta_3 - \frac{H}{a^2}} \right)^2, \qquad 4 \left(\sqrt{\theta_2 - \frac{H}{a^2}} + \sqrt{\theta_3 - \frac{H}{a^2}} \right)^2 ;$$

also,

$$\theta_2 \theta_3 + \theta_3 \theta_1 + \theta_1 \theta_2 = \theta_2 \theta_3 - (\theta_2 + \theta_3)^2 = -3\theta_2 \theta_3 - (\theta_2 - \theta_3)^2 = -\frac{I}{4a^2} ;$$

whence

$$\Pi_1 \equiv \phi^2 + \left(8\theta_1 + 16 \frac{H}{a^2} \right) \phi + 4 \frac{I}{a^2} - 48\theta_2 \theta_3.$$

For convenience in the calculation we now put

$$16 H \equiv a^2 P, \quad 4 I \equiv a^2 Q, \quad 16 J \equiv a^3 R,$$

and

$$\phi^2 + P\phi + Q \equiv \Psi ;$$

whence $\Pi_1\Pi_2\Pi_3$, divided by 8^3, becomes

$$\left\{\frac{\Psi}{8} + \theta_1\phi - 6\theta_2\theta_3\right\}\left\{\frac{\Psi}{8} + \theta_2\phi - 6\theta_3\theta_1\right\}\left\{\frac{\Psi}{8} + \theta_3\phi - 6\theta_1\theta_2\right\}.$$

Reducing this product by the result of Example 18, page 87, we obtain

$$\Psi^3 + 3Q\Psi^2 - (4Q\phi^2 + 18R\phi)\,\Psi - (8R\phi^3 + 12Q^2\phi^2 + 36QR\phi + 27R^2) = 0.$$

Finally, restoring the value of Ψ, we have the equation of the squares of the differences expressed in terms of P, Q, R, as follows :—

$$\phi^6 + 3P\phi^5 + (3P^2 + 2Q)\,\phi^4 + (P^3 + 8PQ - 26R)\,\phi^3$$
$$+ (6P^2Q - 7Q^2 - 18PR)\,\phi^2 + 9Q\,(PQ - 6R)\,\phi + 4Q^3 - 27R^2 = 0.$$

We give for convenience of reference the result also in terms of a, H, I, J*:—

$$a^6\phi^6 + 48a^4 H\phi^5 + 8a^2\,(96H^2 + a^2 I)\,\phi^4 + 32\,(128H^3 + 16a^2 HI - 13a^3 J)\,\phi^3$$
$$+ 16\,(384H^2 I - 7a^2 I^2 - 288aHJ)\,\phi^2 + 1152\,(2HI - 3aJ)I\phi + 256\,(I^3 - 27J^2) = 0.$$

68. Criterion of the Nature of the Roots of the Biquadratic.—The quantity $I^3 - 27J^2$ is the *discriminant* of the biquadratic, and is denoted by Δ. The sign of Δ does not enable us to determine completely, as in the case of the cubic (see Art. 42), the character of the roots; but certain conclusions can be drawn from it, as we proceed to show.

(1). *When Δ is negative, the biquadratic has two real and two imaginary roots.* For, forming the product of the squares $(\beta - \gamma)^2$, $(\gamma - a)^2$, &c., it readily appears that this product is positive if the roots be either all real or all imaginary.

(2). *When Δ is positive, the biquadratic has its roots either all real, or all imaginary.* For, forming the product as before, it appears that this product is negative when two of the roots are real and two imaginary.

(3). *When $\Delta = 0$, the biquadratic has two roots equal.*

(4). *When $I = 0$, and $J = 0$, the biquadratic has three roots equal.* This is easily inferred from the expressions of the roots in terms of the roots of the reducing cubic.

The complete discrimination of the roots in the case where Δ is positive requires the consideration of the function H as well as I and J, and will be most conveniently discussed after the proof of Sturm's theorem in Chap. IX.

---

* The equation of the squares of the differences was first given in this form by Mr. M. Roberts in the *Nouvelles Annales de Mathématiques*, vol. xvi.

MISCELLANEOUS EXAMPLES.

1. Show that if H be positive, or if $H = 0$ (and G not $= 0$), the cubic will have a pair of imaginary roots.

2. Show that if H be negative, the cubic will have its roots (1) all real and unequal, (2) two equal, or (3) two imaginary, according as G^2 is (1) less than, (2) equal to, or (3) greater than $-4H^3$.

3. If the cubic equation

$$a_0 x^3 + 3a_1 x^2 + 3a_2 x + a_3 = 0$$

have two roots equal to α; prove

$$-\alpha = \frac{H_2}{H_1} = \frac{H_1}{H},$$

where $\quad a_0 a_2 - a_1^2 = H, \quad a_0 a_3 - a_1 a_2 = 2H_1, \quad a_1 a_3 - a_2^2 = H_2.$

4. If $\qquad ax^3 + 3bx^2 + 3cx + d + k(x - r)^3$

be a perfect cube, prove

$$(ac - b^2) r^2 + (ad - bc) r + (bd - c^2) = 0.$$

5. Find the condition that the cubic

$$ax^3 + 3bx^2 + 3cx + d$$

may be written under the form

$$l(x - \alpha_1)^3 + m(x - \beta_1)^3 + n(x - \gamma_1)^3,$$

where α_1, β_1, γ_1 are the roots of the cubic

$$a_1 x^3 + 3b_1 x^2 + 3c_1 x + d_1 = 0.$$

Comparing the forms, we have

$$
\begin{aligned}
a &= l + m + n, \\
-b &= l\alpha_1 + m\beta_1 + n\gamma_1, \\
c &= l\alpha_1^2 + m\beta_1^2 + n\gamma_1^2, \\
-d &= l\alpha_1^3 + m\beta_1^3 + n\gamma_1^3.
\end{aligned}
$$

Also $\qquad a_1 \alpha_1^3 + 3b_1 \alpha_1^2 + 3c_1 \alpha_1 + d_1 = 0, \quad$ &c.

Whence, multiplying these equations by d_1, $3c_1$, $3b_1$, a_1, respectively, and adding, we find

$$(ad_1 - a_1 d) - 3(bc_1 - b_1 c) = 0.$$

6. If α, β, γ be the roots of the cubic equation

$$a_0 x^3 + 3a_1 x^2 + 3a_2 x + a_3 = 0 ;$$

form the equation

$$\sqrt[4]{x - \alpha} + \sqrt[4]{x - \beta} + \sqrt[4]{x - \gamma} = 0$$

in terms of the coefficients.

$$\textit{Ans.} \quad 125 U_1^4 + 360 H U_1^2 + 128 G U_1 - 48 H^2 = 0.$$

7. Show that if H be positive, the biquadratic must have imaginary roots.

8. Show that if I be negative, the biquadratic must have two real and two imaginary roots.

9. Show that when the biquadratic has a pair of equal roots, the reducing cubic also has a pair of equal roots, and conversely.

10. Show that when the biquadratic has a double root, the cubic whose roots are the values of ρ (Art. 65), has the *same* double root.

11. If H and J are both positive, all the roots of the biquadratic are imaginary.

Since H is positive, there must be at least one pair of imaginary roots, $\alpha \pm \beta \sqrt{-1}$. Now diminishing all the roots by α, and dividing them by β (which transformations will not alter the character of the other pair of roots γ, δ, nor the signs of H and J), the biquadratic may be put under the form

$$(x^2 + 4px + q)(x^2 + 1),$$

or $\qquad x^4 + 4px^3 + 6cx^2 + 4px + q$, where $6c = q + 1$;

whence $\qquad H = c - p^2, \quad I = q - 4p^2 + 3c^2,$

$$J = qc + 2p^2c - p^2(q + 1) - c^3 = c(q - 4p^2 - c^2),$$

and, therefore,

$$q - 4p^2 = c^2 + \frac{J}{c} = (H + p^2)^2 + \frac{J}{H + p^2},$$

or $\qquad -\left(\frac{\gamma - \delta}{2\beta}\right)^2 = \left(H + p^2\right)^2 + \frac{J}{H + p^2},$

proving that γ and δ are imaginary if H and J are both positive.

12. If the biquadratic has two distinct pairs of equal roots, prove the relations

$$a_0^2 I = 12 H^2, \quad a_0^3 J = 8 H^3.$$

In this case the biquadratic divided by a_0 assumes the form

$$(x - \alpha)^2(x - \beta)^2 \equiv \left\{\left(x - \frac{\alpha + \beta}{2}\right)^2 - \left(\frac{\alpha - \beta}{2}\right)^2\right\}^2 = \left(\frac{z^2 - k^2}{a_0^2}\right)^2$$

where $\qquad z = a_0 x + a_1, \quad \text{and} \quad \frac{k}{a_0} = \frac{\alpha - \beta}{2};$

whence, comparing the forms

$$z^4 - 2k^2 z^2 + k^4$$

and $\qquad z^4 + 6Hz^2 + 4Gz + a_0^2 I - 3H^2,$

we find $\qquad 3H = -k^2, \quad G = 0, \quad a_0^2 I - 3H^2 = k^4,$

from which the above relations immediately follow. Also it should be noticed that in this case only one square root is involved in the solution of the biquadratic (coming from the solution of the quadratic $(x - \alpha)(x - \beta)$). Two roots of the equation in t (Art. 61) are zero.

13. Find the condition that the biquadratic may be put under the form

$$l(x^2 + 2px + q)^2 + m(x^2 + 2px + q) + n.$$

In this case the second and fourth coefficients are removed by the same transformation, and the general solution involves only two square roots.

Ans. $G = 0$.

14. Form the equation whose roots are the six anharmonic functions of four points in a right line determined by the equation

$$a_0 x^4 + 4a_1 x^3 + 6a_2 x + 4a_3 x + a_4 = 0.$$

The six anharmonic ratios are

$$\phi_1, \ \frac{1}{\phi_1}, \ \phi_2, \ \frac{1}{\phi_2}, \ \phi_3, \ \frac{1}{\phi_3},$$

where

$$\phi_1 = -\frac{(\alpha - \beta)(\gamma - \delta)}{(\gamma - \alpha)(\beta - \delta)} \equiv \frac{\lambda - \mu}{\lambda - \nu} = \frac{\theta_1 - \theta_2}{\theta_1 - \theta_3},$$

$$\phi_2 = -\frac{(\beta - \gamma)(\alpha - \delta)}{(\alpha - \beta)(\gamma - \delta)} \equiv \frac{\mu - \nu}{\mu - \lambda} = \frac{\theta_2 - \theta_3}{\theta_2 - \theta_1},$$

$$\phi_3 = -\frac{(\gamma - \alpha)(\beta - \delta)}{(\beta - \gamma)(\alpha - \delta)} \equiv \frac{\nu - \lambda}{\nu - \mu} = \frac{\theta_3 - \theta_1}{\theta_3 - \theta_2};$$

also the equation whose roots are

$$(\beta - \gamma)(\alpha - \delta), \quad (\gamma - \alpha)(\beta - \delta), \quad (\alpha - \beta)(\gamma - \delta)$$

is one of the cubics

$$a_0^3 t^3 - 12 a_0 I t^2 \pm 16 \sqrt{I^3 - 27 J^2} = 0.$$

The equation whose roots are the ratios, with sign changed, of the roots of *either* of these cubics is

$$4\Delta(\phi^2 - \phi + 1)^3 - 27 I^3 \phi^2 (\phi - 1)^2 = 0 \qquad \text{(see Ex. 15, p. 86)},$$

where

$$\Delta \equiv I^3 - 27 J^2.$$

The roots of the equation in ϕ are the six anharmonic ratios. This equation can be written in a more expressive form, as will appear from the following propositions :—

(*a*). The six anharmonic ratios may be expressed in terms of any one of them, as follows :—

$$\phi, \ \frac{1}{\phi}, \ 1 - \phi, \ \frac{1}{1 - \phi}, \ \frac{\phi - 1}{\phi}, \ \frac{\phi}{\phi - 1}.$$

From the identical equation

$$(\beta - \gamma)(\alpha - \delta) + (\gamma - \alpha)(\beta - \delta) + (\alpha - \beta)(\gamma - \delta) \equiv 0$$

we have the relations

$$\phi_1 + \frac{1}{\phi_3} = 1, \quad \phi_2 + \frac{1}{\phi_1} = 1, \quad \phi_3 + \frac{1}{\phi_2} = 1,$$

which determine all the anharmonic ratios in terms of any one of them.

(b). If two of the anharmonic ratios become equal, the six values of ϕ are $-\omega$ and $-\omega^2$ *three* times repeated; and in this case $I = 0$.

For suppose $\phi_1 = \phi_2$; we have then from the second of the above relations

$$\phi_1{}^2 - \phi_1 + 1 = 0,$$

whence $\qquad\qquad\qquad \phi_1 = -\omega, \text{ or } -\omega^2 ;$

and substituting either of these values for ϕ in (a), we find all the anharmonic ratios.

Also, since

$$\frac{\lambda - \mu}{\lambda - \nu} + \frac{\mu - \nu}{\lambda - \mu} = 0, \text{ or } \Sigma(\mu - \nu)^2 = 0,$$

we have

$$I \equiv a_0 a_4 - 4 a_1 a_3 + 3 a_2{}^2 = 0.$$

(c). If one of the ratios is harmonic, the six values of ϕ are $-1,\ 2,\ \dfrac{1}{2},$ *twice* repeated; and in this case $J = 0$; for if

$$\phi_1 = -1, \quad \frac{\lambda - \mu}{\lambda - \nu} = -1, \text{ or } 2\lambda - \mu - \nu = 0,$$

one of the factors of J (see Ex. 18, p. 51).

(d). These results, as well as the converse propositions, may be proved by writing the sextic in ϕ under the following form:—

$$I^3 \{(\phi + 1)(\phi - 2)(\phi - \tfrac{1}{2})\}^2 = 27 J^2 \{(\phi + \omega)(\phi + \omega^2)\}^3.$$

15. Solve the equation

$$\left(\frac{x^2 + 14x + 1}{\rho^4 + 14\rho^2 + 1}\right)^3 = \frac{x(x-1)^4}{\rho^2(\rho^2 - 1)^4}.$$

$$\textit{Ans. } \rho^2,\ \frac{1}{\rho^2},\ \left(\frac{1 + \theta\sqrt{\rho}}{1 - \theta\sqrt{\rho}}\right)^4, \text{ where } \theta^4 = 1.$$

16. Express $\Sigma(\alpha - \beta)^4(\gamma - \delta)^2$ as a rational function of $\theta_1,\ \theta_2,\ \theta_3$; and ultimately in terms of the coefficients of the quartic.

$$\textit{Ans. } -128\, \Sigma(\theta_2 - \theta_3)^2 \left(\theta_1 + \frac{2H}{a^2}\right) = -\frac{96}{a^4}(4HI + 3aJ).$$

17. Express

$$(\beta^2 - \gamma^2)^2(\alpha^2 - \delta^2)^2 + (\gamma^2 - \alpha^2)^2(\beta^2 - \delta^2)^2 + (\alpha^2 - \beta^2)^2(\gamma^2 - \delta^2)^2$$

as a rational function of $\theta_1,\ \theta_2,\ \theta_3$.

This symmetric function is equivalent to

$$(\mu^2 - \nu^2)^2 + (\nu^2 - \lambda^2)^2 + (\lambda^2 - \mu^2)^2 = 256\, \Sigma(\theta_2 - \theta_0)^2 \left(\theta_1 - \frac{c}{a}\right)^2.$$

18. Form the equation whose roots are the several products in pairs of the roots of a biquadratic.

The required equation is the product of three factors of the type

$$(\phi - \beta\gamma)(\phi - \alpha\delta) = \phi^2 - \lambda\phi + \frac{e}{a} = \phi^2 - 2\frac{c}{a}\phi + \frac{e}{a} - 4\phi\theta_1.$$

$$\textit{Ans. } (a\phi^2 - 2c\phi + e)^3 - 4I\phi^2(a\phi^2 - 2c\phi + e) + 16J\phi^3 = 0.$$

19. Form the equation whose roots are the several values of $\dfrac{\alpha + \beta}{2}$, where α, β, γ, δ are the roots of a biquadratic.

The required equation is the product of three factors of the type

$$\left(\phi - \frac{\beta + \gamma}{2}\right)\left(\phi - \frac{\alpha + \delta}{2}\right) = \phi^2 + 2\frac{b}{a}\phi + \frac{\mu + \nu}{4} = \phi^2 + 2\frac{b}{a}\phi + \frac{c}{a} - \theta_1.$$

Ans. $4\,(a\phi^2 + 2b\phi + c)^3 - I(a\phi^2 + 2b\phi + c) + J = 0.$

20. If α_1, β_1, and α_2, β_2 be the roots of the quadratic equations

$$a_1 x^2 + 2b_1 x + c_1 = 0, \quad a_2 x^2 + 2b_2 x + c_2 = 0 \ ;$$

find the equation whose roots are the four values of $\alpha_1 \alpha_2$.

Let $H_1 = a_1 c_1 - b_1^2, \quad H_2 = a_2 c_2 - b_2^2.$

Ans. $(a_1 a_2 \phi^2 - 2b_1 b_2 \phi + c_1 c_2)^2 - 4H_1 H_2 \phi^2 = 0.$

N. B.—This and the two following Examples may be solved by expressing ϕ in terms of radicals involving the coefficients.

21. Employing the notation of Ex. 20, form the equation whose roots are the four values of $\dfrac{\alpha_1 + \alpha_2}{2}$.

Let $2K_{12} = a_1 c_2 + a_2 c_1 + 2b_1 b_2.$

Ans. $(2a_1 a_2 \phi^2 + 2\,(a_1 b_2 + a_2 b_1)\,\phi + K_{12})^2 - H_1 H_2 = 0.$

In this Example the resulting biquadratic is such that $G = 0$.

22. In the same case, if $\phi = \frac{1}{2}\,(\alpha_1 - \alpha_2)^2$, form the equation whose roots are the several values of ϕ.

Let $M = a_1 b_2 - a_2 b_1, \quad 2H_{12} = a_1 c_2 + a_2 c_1 - 2b_1 b_2.$

Ans. $\{(a_1 a_2 \phi + H_{12})^2 - 2M^2 \phi + H_1 H_2\}^2 = 4H_1 H_2\,(a_1 a_2 \phi + H_{12})^2.$

23. Prove

$$\Sigma \frac{1}{(\alpha - \beta)^2} = \frac{9I}{2}\left(\frac{3aJ - 2HI}{I^3 - 27J^2}\right).$$

From the expressions for α, β, γ, δ in terms of θ_1, θ_2, θ_3, we have

$$\Sigma \frac{1}{(\alpha - \beta)^2} = -\frac{1}{2a^2}\left\{\frac{a^2\theta_1 + 2H}{(\theta_2 - \theta_3)^2} + \frac{a^2\theta_2 + 2H}{(\theta_3 - \theta_1)^2} + \frac{a^2\theta_3 + 2H}{(\theta_1 - \theta_2)^2}\right\},$$

which may be expressed in terms of a, H, I, J as above.

24. Prove

$$\Sigma \frac{\theta_1{}^m}{(\theta_2 - \theta_3)^2} = 0$$

if $I = 0$, and m of the form $3p$ or $3p + 1$.

25. Prove that

$$U \equiv ax^2 + cy^2 + ez^2 + 2dyz + 2czx + 2bxy$$

can be resolved into the sum or difference of two squares if

$$J \equiv ace + 2bcd - ad^2 - eb^2 - c^3 = 0.$$

Here $\qquad aU \equiv (ax + by + cz)^2 + (ac - b^2) y^2 + 2 (ad - bc) yz + (ae - c^2) z^2,$

and $\qquad (ac - b^2) y^2 + 2 (ad - bc) yz + (ae - c^2) z^2$

is a perfect square if

$$(ac - b^2)(ae - c^2) = (ad - bc)^2,$$

or $J = 0$.

26. If α, β, γ, δ be the roots of the equation

$$a_0 x^4 + 4a_1 x^3 + 6a_2 x^2 + 4a_3 x + a_4 = 0,$$

solve, in terms of the coefficients a_0, a_1, &c., the equation

$$\sqrt{x - \alpha} + \sqrt{x - \beta} + \sqrt{x - \gamma} + \sqrt{x - \delta} = 0.$$

When $\qquad \sqrt{\alpha} + \sqrt{\beta} + \sqrt{\gamma} + \sqrt{\delta} = 0$

is rationalized, and the coefficients substituted for α, β, γ, δ, we have

$$(3a_0 a_2 - 2a_1^2)^2 = 4a_0^3 a_4.$$

Now, substituting U_0, U_1, U_2, U_3, U_4 for a_0, a_1, a_2, a_3, a_4, and reducing, we find

$$a_0 x + a_1 = \frac{1}{G} \left(3H^2 - \frac{a_0^2 I}{4} \right).$$

27. To express the solution of the biquadratic in terms of a single root of the reducing cubic.

Substituting $x' + \rho$ for x in the equation

$$ax^4 + 4bx^3 + 6cx^2 + 4dx + e = 0,$$

we have

$$ax'^4 + 4 U_1 x'^3 + 6 U_2 x'^2 + 4 U_3 x' + U_4 = 0.$$

Now let this equation separate into the two equations

$$ax'^4 + 6 U_2 x'^2 + U_4 = 0, \quad U_1 x'^2 + U_3 = 0. \qquad (1)$$

Eliminating x'^2, and reducing as in Art. 65, we have

$$4 U_2^3 - I U_2 + J = 0 ;$$

whence $U_2 = a\theta$, where θ is a root of the reducing cubic, and therefore

$$U_1 = a\rho + b = \sqrt{a^2 \theta - H}.$$

Again, from (1)

$$x'^2 = - \frac{U_3}{U_1} = - \frac{1}{a^2} \left(U_1^2 + 3H + \frac{G}{U_1} \right);$$

whence, finally, since $x = x' + \rho$, or $ax + b = U_1 + ax'$, we have

$$ax + b = \sqrt{a^2\theta - H} + \sqrt{-a^2\theta - 2H - \frac{G}{\sqrt{a^2\theta - H}}},$$

an expression which has only four values.

This expression might of course be obtained from the resulting formula of Art. 61, or Art. 63. The method of arriving at it in the present Example is a distinct method of solving the biquadratic.

28. Prove in general that the solution of the biquadratic does not involve the extraction of a cube root when any relation among the roots α, β, γ, δ exists which can be expressed rationally in terms of a root θ of the reducing cubic.

Any rational function of θ can always be depressed to the second degree by the aid of the reducing cubic, which expresses θ^3 in terms of θ. Hence the determination of θ will not involve the extraction of a cube root; and the formula of the preceding Example shows that the expression for the root of the biquadratic will not then involve any cube root.

29. Find the relation which connects the roots of the biquadratic when the equation

$$4\rho^3 - I\rho + J = 0$$

is satisfied by each of the following values of ρ :—

(1) $\dfrac{H}{a}$, (2) c, (3) 0, (4) $\dfrac{\sqrt{ae - c}}{2}$, (5) $\sqrt[3]{\dfrac{-J}{4}}$, (6) $\sqrt{\dfrac{I}{12}}$, (7) $\sqrt{\dfrac{3J}{2I}}$ (8) $\dfrac{ad - bc}{2b}$.

Ans. (1) $\beta + \gamma - \alpha - \delta = 0$.

(2) (4) and (8) $\beta\gamma - \alpha\delta = 0$.

(3) $(\gamma - \alpha)(\beta - \delta) - (\alpha - \beta)(\gamma - \delta) = 0$.

(5) $(\gamma - \alpha)(\beta - \delta) - \omega(\alpha - \beta)(\gamma - \delta) = 0$.

(6) and (7) $\beta - \gamma = 0$.

30. Prove the following identity ; and by means of it verify the result of Ex. 11, p. 139 :—

$$a_0^6(I^3 - 27J^2) \equiv (a_0^2 I - 3H^2)(a_0^2 I - 12H^2)^2 + 27G^2(G^2 + 2a_0^3 J).$$

This may be proved by putting $a_1 = 0$, $a_2 = A_2$, &c., in the expanded value of Δ, and then substituting for A_2, A_3, A_4 the values of Art. 38.—Mr. M. Roberts.

CHAPTER VII.

69. Graphic Representation of the Derived Func-tion.—Let APB be the curve representing the polynomial $f(x)$, and P the point on it corresponding to any value of the variable $x = OM$. We proceed to determine the mode of representing the value of $f'(x)$ at the point P. Take a second point Q on the curve, corresponding to a value of

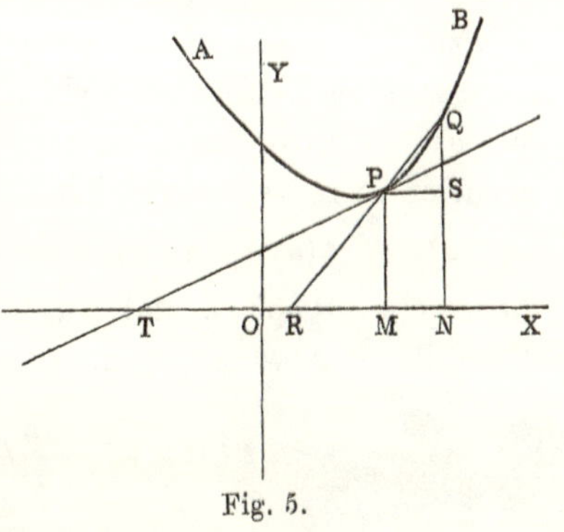

Fig. 5.

x which exceeds OM by a small quantity h. Thus

$$OM = x, \quad MN = h, \quad ON = x + h;$$

also
$$PM = f(x), \quad QN = f(x + h).$$

The expansion of Art. 6 gives

$$f(x + h) = f(x) + f'(x)\, h + \frac{f''(x)}{1.2}\, h^2 + \ldots \ldots ,$$

or
$$\frac{f(x + h) - f(x)}{h} = f'(x) + \frac{f''(x)}{1.2}\, h + \ldots \ldots \tag{1}$$

But
$$\frac{f(x + h) - f(x)}{h} = \frac{QS}{MN} = \frac{QS}{PS} = \tan QPS = \tan PRN.$$

Now, when h is indefinitely diminished, the point Q approaches, and ultimately coincides with, P; the chord PQ becomes the

"

tangent PT to the curve at P; the angle PRN becomes PTM. Also all terms of the right-hand member of equation (1) except the first diminish indefinitely, and ultimately vanish when $h = 0$. The equation (1) becomes

$$\tan PTM = f'(x);$$

from which we conclude that *the value assumed by the derived function $f'(x)$ on the substitution of any value of x is represented by the tangent of the angle made with the axis OX by the tangent at the corresponding point to the curve representing the function $f(x)$.*

70. Maxima and Minima Values of a Polynomial. Theorem.—*Any value of x which renders $f(x)$ a maximum or minimum is a root of the derived equation $f'(x) = 0$.*

Let a be a value of x which renders $f(x)$ a *minimum*. We proceed to prove that $f'(x) = 0$. Let h represent a small increment or decrement of x. Then

$$f(a) < f(a+h), \text{ also } f(a) < f(a-h);$$

hence $f(a+h) - f(a)$, and $f(a-h) - f(a)$ are both positive, *i. e.* the following two expressions are positive :—

$$f'(a)\, h + \frac{f''(a)}{1 \cdot 2}\, h^2 + \ldots \ldots ,$$

$$-f'(a)\, h + \frac{f''(a)}{1 \cdot 2}\, h^2 - \ldots \ldots$$

Now, when h is very small, we know (see Art. 5) that the signs of these expressions are the same as the signs of their first terms; hence, in order that both should be positive, $f'(a)$ must vanish; and, moreover, $f''(a)$ *must be positive.* An exactly similar proof shows that when $f(a)$ is a *maximum* $f'(a) = 0$, *and $f''(a)$ is negative.* Thus, in order to find the maximum and minimum values of a polynomial $f(x)$, we must solve the equation $f'(x) = 0$, and substitute the roots in $f(x)$. Each root will furnish a maximum or minimum value, the criterion to decide between these being the sign of $f''(x)$ when the root is substituted in it—*when $f''(x)$ is negative, the value is a maximum; and when $f''(x)$ is positive, the value is a minimum.*

The theorem of this Article follows at once from the construction of Art. 69; for it is plain that when the value of $f(x)$ is a maximum, as at P, P' (Fig. 6), or a minimum, as at p, p', the tangent to the curve will be parallel to the axis OX, and, consequently,

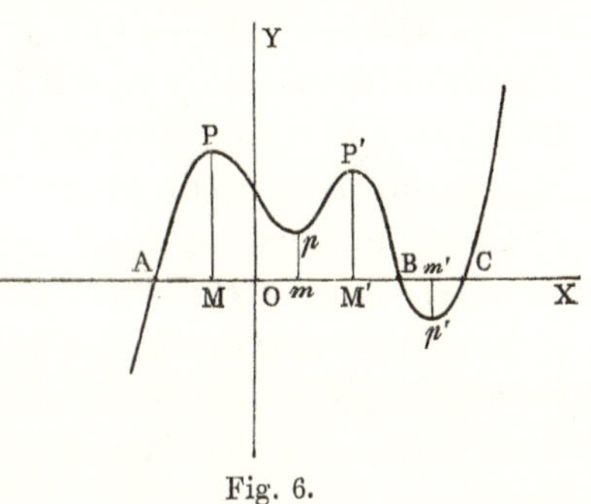

Fig. 6.

$$\tan PTM = f'(x) = 0.$$

Fig. 6 represents a polynomial of the 5th degree. Corresponding to the four roots of $f'(x) = 0$ (supposed all real in this case), i. e. OM, Om, OM', Om', there are two maxima values MP, $M'P'$, and two minima values mp, $m'p'$, of the function.

EXAMPLES.

1. Find the max. or min. value of

$$f(x) = 2x^2 + x - 6.$$

$$f'(x) = 4x + 1, \quad f''(x) = 4.$$

$$x = -\frac{1}{4} \text{ makes } f(x) = \frac{-49}{8}, \text{ a minimum.}$$

(See fig. 2, p. 15.)

2. Find the max. and min. values of

$$f(x) = 2x^3 - 3x^2 - 36x + 14.$$

$$f'(x) = 6(x^2 - x - 6), \quad f''(x) = 6(2x - 1).$$

$$x = -2 \text{ makes } f(x) = 68, \text{ a maximum.}$$

$$x = 3 \text{ makes } f(x) = -67, \text{ a minimum.}$$

3. Find the max. and min. values of

$$f(x) = 3x^4 - 16x^3 + 6x^2 - 48x + 7.$$

Here $f'(x)$ has only one real root $x = 4$; and it gives a minimum value $f(x) = -345$.

4. Find the max. and min. values of

$$f(x) = 10x^3 - 17x^2 + x + 6.$$

The roots of $f'(x)$ are, approximately, ·0302, 1·1031. The former gives a maximum value, the latter a minimum. (See fig. 3, p. 17.)

L 2

71. Rolle's Theorem.—*Between two consecutive real roots a and b of the equation $f(x) = 0$ there lies at least one real root of the equation $f'(x) = 0$.* For as x increases from a to b, $f(x)$, varying continuously from $f(a)$ to $f(b)$, must begin by increasing and then diminish, or must begin by diminishing and then increase. It must, therefore, pass through at least one maximum or minimum value during the passage from $f(a)$ to $f(b)$. This value ($f(a)$, suppose) corresponds to some value a of x between a and b, which by the Theorem of Art. 70 is a root of the equation $f'(x) = 0$.

The figure in the preceding Article illustrates this theorem. We observe that between the two points of section A and B there are *three* maximum or minimum values, and between the two points B and C there is one such value. The figure makes it plain also that the number of such values between two consecutive points of section of the axis is always odd.

Corollary.—*Two consecutive roots of the derived equation may not comprise between them any root of the original equation, and never can comprise more than one.* The first part of this proposition is merely a statement of the fact that between two adjacent zero values of a polynomial there may be several maxima and minima values; and the second part follows at once from the above theorem; for if two consecutive roots of $f'(x) = 0$ comprised between them more than one root of $f(x) = 0$, we should then have two consecutive roots of this latter equation comprising between them no root of $f'(x) = 0$, which is contradictory to the theorem.

72. Constitution of the Derived Functions.—Let the roots of the equation $f(x) = 0$ be $a_1, a_2, a_3, \ldots a_n$.

$$f(x) \equiv (x - a_1)(x - a_2)(x - a_3) \ldots (x - a_n).$$

In this identical equation substitute $y + x$ for x;

$$f(y + x) = (y + x - a_1)(y + x - a_2) \ldots (y + x - a_n)$$

$$= y^n + q_1 y^{n-1} + q_2 y^{n-2} + \ldots + q_{n-1} y + q_n,$$

where

$$q_1 \;=\; x - a_1 + x - a_2 + x - a_3 + \ldots + x - a_n,$$

$$q_2 \;=\; (x - a_1)(x - a_2) + (x - a_1)(x - a_3) + \ldots + (x - a_{n-1})(x - a_n),$$

$$\cdot \quad \cdot \quad \cdot \quad \cdot \quad \cdot \quad \cdot \quad \cdot \quad \cdot \quad \cdot \quad \cdot$$

$$q_{n-1} = (x - a_2)(x - a_3) \ldots (x - a_n) + (x - a_1)(x - a_3) \ldots (x - a_n) + \ldots$$

$$+ (x - a_1)(x - a_2) \ldots (x - a_{n-1}),$$

$$q_n \;=\; (x - a_1)(x - a_2)(x - a_3) \ldots (x - a_n).$$

We have, again,

$$f(y + x) = f(x) + f'(x)\, y + \frac{f''(x)}{1 \cdot 2}\, y^2 + \ldots + y^n.$$

Equating the two expressions for $f(y + x)$, we obtain

$$f(x) \;=\; (x - a_1)(x - a_2) \ldots (x - a_n),$$

$$f'(x) \;=\; (x - a_2)(x - a_3) \ldots (x - a_n) + \ldots, \text{ as above written,}$$

$$\frac{f''(x)}{1 \cdot 2} = \text{the similar value of } q_{n-2} \text{ in terms of } x \text{ and the roots,}$$

$$\cdot \quad \cdot \quad \cdot \quad \cdot \quad \cdot \quad \cdot \quad \cdot \quad \cdot \quad \cdot \quad \cdot$$

The value of $f'(x)$ may be conveniently written as follows:—

$$f'(x) = \frac{f(x)}{x - a_1} + \frac{f(x)}{x - a_2} + \ldots + \frac{f(x)}{x - a_n}.$$

73. Multiple Roots. Theorem.—*A multiple root of the order m of the equation $f(x) = 0$ is a multiple root of the order $m - 1$ of the first derived equation $f'(x) = 0$.*

This follows immediately from the expression given for $f'(x)$ in the preceding Article; for if the factor $(x - a_1)^m$ occurs in $f(x)$, *i.e.* if $a_1 = a_2 = \ldots = a_m$; we have

$$f'(x) = \frac{m f(x)}{x - a_1} + \frac{f(x)}{x - a_{m+1}} + \ldots + \frac{f(x)}{x - a_n}.$$

Each term in this will still have $(x - a_1)^m$ as a factor, except the first, which will have $(x - a_1)^{m-1}$ as a factor; hence $(x - a_1)^{m-1}$ is a factor in $f'(x)$.

Cor. 1.—Since $f''(x)$ is derived from $f'(x)$ in the same manner as $f'(x)$ is from $f(x)$, it is evident that $f''(x)$ will have $(x - a_1)^{m-2}$ for a factor. Similarly $f'''(x)$ will have $(x - a_1)^{m-3}$ for a factor; and so on. *Thus a root occurring m times in the equation $f(x) = 0$ occurs in degrees of multiplicity diminishing by unity in the first $m - 1$ derived equations.*

Cor. 2.—The converse of the preceding corollary—viz.: *If $f(x)$ and its first $m - 1$ derived functions $f_1(x)$, $f_2(x)$, . . . $f_{m-1}(x)$, all vanish for a value a of x, then $(x - a)^m$ is a factor in $f(x)$*—is most easily established as follows. For x substitute $a + x - a$. We have then

$$f(x) = f(a) + f_1(a)(x - a) + \frac{f_2(a)}{1 \cdot 2}(x - a)^2 + \ldots + \frac{f_{m-1}(a)}{1 \cdot 2 \cdot m - 1}(x - a)^{m-1}$$

$$+ \frac{f_m(a)}{1 \cdot 2 \ldots m}(x - a)^m + \ldots + \frac{f_n(a)}{1 \cdot 2 \ldots n}(x - a)^n,$$

from which the proposition in question is manifest.

74. Determination of Multiple Roots.—It is easily inferred from the preceding Article that if $f(x)$ and $f'(x)$ have a common factor $(x - a)^{m-1}$, $(x - a)^m$ will be a factor in $f(x)$; for, by Cor. 1, the $m - 2$ next succeeding derived functions vanish as well as $f(x)$ and $f'(x)$ when $x = a$; hence, by Cor 2, a is a root of $f(x)$ of multiplicity m. In the same way we see that if $f(x)$ and $f'(x)$ have other common factors

$$(x - \beta)^{p-1}, \quad (x - \gamma)^{q-1}, \text{ &c.,}$$

the equation $f(x) = 0$ will have p roots equal to β, q roots equal to γ, &c.

In order, then, to find whether any proposed equation has equal roots, and to determine those roots if it has, we must find the greatest common measure of $f(x)$ and $f'(x)$. Let this be $\phi(x)$. The determination of the equal roots will depend on the solution of the equation $\phi(x) = 0$.

EXAMPLES.

1. Find the multiple roots of the equation

$$x^3 + x^2 - 16x + 20 = 0.$$

The G. C. M. of $f(x)$ and $f'(x)$ is easily found to be $x - 2$; hence $(x - 2)^2$ is a factor in $f(x)$. The other factor is $x + 5$.

Whenever, after determining the multiple factors of $f(x)$, we wish to obtain the remaining factors, it will be found convenient to apply by repeated operations the method of division of Art. 8. Here, for example, we divide twice by $x - 2$, the operation being represented as follows:—

1	1	-16	20
	2	6	-20
1	3	-10	0
	2	10	
1	5	0	

Thus 1 and 5 being the two coefficients left, the third factor is $x + 5$. This operation verifies the previous result, the remainders after each division vanishing as they ought.

2. Find the multiple roots, and the remaining factor, of the equation

$$x^5 - 10x^2 + 15x - 6 = 0.$$

The G. C. M. of $f(x)$ and $f'(x)$ is found to be $x^2 - 2x + 1$. Hence $(x - 1)^3$ is a factor in $f(x)$. Dividing three times in succession by $x - 1$, we obtain

$$f(x) \equiv (x - 1)^3 (x^2 + 3x + 6).$$

3. Find the multiple roots of the equation

$$x^4 - 2x^3 - 11x^2 + 12x + 36 = 0.$$

The G. C. M. of $f(x)$ and $f'(x)$ is $x^2 - x - 6$. The factors of this are $x + 2$ and $x - 3$. Hence

$$f(x) \equiv (x + 2)^2 (x - 3)^2.$$

4. Find all the factors of the polynomial

$$f(x) \equiv x^6 - 5x^5 + 5x^4 + 9x^3 - 14x^2 - 4x + 8.$$

$$Ans. \ f(x) \equiv (x - 1) (x + 1)^2 (x - 2)^3.$$

Remark.—The process of finding the greatest common measure of two polynomials becomes laborious as the degree of the function increases. It is wrong, therefore, to speak, as writers on the Theory of Equations often do, of the determination in this way of the multiple roots of numerical equations as a simple process, and one preliminary to further investigations relative to the roots. It is chiefly in connexion with Sturm's theorem that the operation is of any practical value. The further consideration of multiple roots is deferred to Chap. IX., where the theorem in question is discussed. It will be shown also in Chap. X., that when any particular equation has equal roots, they can, in those cases which are of most usual occurrence, be determined from simple considerations not involving the process of finding the greatest common measure.

75. This and the succeeding Article will be occupied with theorems which will be found of considerable importance when we come to the discussion of the methods of separating the roots of equations.

Theorem.—*In passing continuously from a value $a - h$ of x a little less than a real root a of the equation $f(x) = 0$ to a value $a + h$ a little greater, the polynomials $f(x)$ and $f'(x)$ have unlike signs immediately before the passage through the root, and like signs immediately after.*

We have

$$f(a - h) = f(a) - f'(a)\, h + \frac{f''(a)}{1 \cdot 2}\, h^2 - \ldots ,$$

$$f'(a - h) = \qquad f'(a) - f''(a)\, h + \ldots ;$$

since $f(a) = 0$, the signs of these expressions, depending on those of their first terms, are unlike. When the sign of h is changed, the signs of the expressions become the same. The theorem is thus proved.

Corollary.—*This theorem is true no matter how many times the root a is repeated in $f(x) = 0$.*

Let the root be repeated r times. The following functions (using suffixes in place of the accents) all vanish :—

$$f(a),\ f_1(a),\ f_2(a),\ \ldots \ldots f_{r-1}(a).$$

In the series for $f(a - h)$ and $f'(a - h)$ the first terms which do not vanish are, respectively,

$$\frac{f_r(a)}{1 \cdot 2 \dots r}(-h)^r, \qquad \frac{f_r(a)}{1 \cdot 2 \dots r-1}(-h)^{r-1}.$$

These have plainly unlike signs; but after the sign of h is changed they will have like signs. Hence the proposition is established.

76. Extending the reasoning of the last Article to every consecutive pair of the series

$$f(x), \; f_1(x), \; f_2(x), \; \dots f_{r-1}(x),$$

we may state the proposition generally as follows:—

Theorem.—*When any equation $f(x) = 0$ has an r-multiple root a, a value a little inferior to a gives to this series of r functions signs alternately positive and negative, or negative and positive; and a value a little superior to it gives to all these functions the same sign; and this sign is, moreover, the same sign as the sign of $f_r(a)$, the first derived function which does not vanish when a is substituted for x.*

In order to give a precise idea of the use of this theorem, let us suppose that $f_5(a)$ is the first function which does not vanish when a is substituted, and let its sign be negative; what we are able to conclude is, that for a value $a - h$ of x the signs of the series of functions $f, f_1, f_2, f_3, f_4, f_5$, are

$$+ \quad - \quad + \quad - \quad + \quad - \; ;$$

and for a value $a + h$ of x they are

$$- \quad - \quad - \quad - \quad - \quad - \; ;$$

for before the passage through the root the sign of f_4 must be different from that of f_5; the sign of f_3 must be different from that of f_4, and so on; and after the passage the signs must be all the same. We of course assume that h is so small that no root of $f_5(x) = 0$ is included within the interval through which x travels.

Examples.

$$f(x) = x^4 - 7x^3 + 15x^2 - 13x + 4,$$

$$f_1(x) = 4x^3 - 21x^2 + 30x - 13,$$

$$f_2(x) = 2(6x^2 - 21x + 15),$$

$$f_3(x) = 2(12x - 21),$$

$$f_4(x) = 24 ;$$

Here $f_3(x)$ is the first function which does not vanish when $x = 1$; and $f_3(1)$ is negative. What we can conclude from the theorem is, that for a value a little less than 1 the signs of f, f_1, f_2, f_3 are $+ - + -$, and for a value a little greater than 1 they are all negative. We are able from this series of signs to trace the functions f, f_1, &c., in the neighbourhood of the point $x = 1$. Thus the curve representing $f(x)$ is above the axis before reaching the multiple point $x = 1$, and is below the axis immediately after reaching that point, and the axis must be regarded as cutting the curve in three coincident points, since $(x - 1)^3$ is a factor in $f(x)$. Again, the curve corresponding to $f_1(x)$ is below the axis both before and after the passage through the point $x = 1$. It touches the axis at that point. The curve representing $f_2(x)$ is above the axis before and below the axis after the passage, and cuts the axis at the point.

2. $f(x) \equiv x^6 - 10x^5 + 35x^4 - 60x^3 + 55x^2 - 26x + 5.$

$f_5(x)$ is the first function which does not vanish for $x = 1$, and $f_5(1)$ is negative.

Miscellaneous Examples.

1. Find the multiple roots of the equation

$$f(x) \equiv x^4 + 12x^3 + 32x^2 - 24x + 4 = 0.$$

                                       Ans. $f(x) \equiv (x^2 + 6x - 2)^2.$

2. Show that the binomial equation

$$x^n - a^n = 0$$

cannot have equal roots.

3. Show that the equation

$$x^n - nqx + (n - 1)r = 0$$

will have a pair of equal roots if $q^n = r^{n-1}$.

4. Prove that the equation

$$x^5 + 5px^3 + 5p^2x + q = 0$$

will have a pair of equal roots when $q^2 + 4p^5 = 0$; and that if it have one pair of equal roots it must have a second pair.

5. Apply the method of Art. 74, to determine the condition that the cubic

$$z^3 + 3Hz + G = 0$$

should have a pair of equal roots.

The last remainder in the process of finding the greatest common measure must vanish.

$$\text{\textit{Ans.}} \quad G^2 + 4H^3 = 0.$$

6. Apply the same method to show that both G and H vanish when the cubic has its three roots equal.

7. If α, β, γ, δ be the roots of the biquadratic $f(x) = 0$, prove that

$$f'(\alpha) + f'(\beta) + f'(\gamma) + f'(\delta)$$

can be expressed as a product of three factors.

$$\text{\textit{Ans.}} \quad (\alpha + \beta - \gamma - \delta)(\alpha + \gamma - \beta - \delta)(\alpha + \delta - \beta - \gamma).$$

8. If α, β, γ, δ, &c. be the roots of $f(x) = 0$, and α', β', γ', δ', &c. of $f'(x) = 0$; prove

$$f'(\alpha) f'(\beta) f'(\gamma) \dots = n^n f(\alpha') f(\beta') f(\gamma') \dots ,$$

and that each is equal to the absolute term in the equation of the squares of the differences.

9. If the equation

$$x^n + p_1 x^{n-1} + p_2 x^{n-2} + \dots + p_{n-1} x + p_n = 0$$

have a double root α; prove that α is a root of the equation

$$p_1 x^{n-1} + 2p_2 x^{n-2} + 3p_3 x^{n-3} + \dots + np_n = 0.$$

10. Show that the max. and min. values of the cubic

$$ax^3 + 3bx^2 + 3cx + d$$

are the roots of the equation

$$a^2 \rho^2 - 2G\rho + \Delta = 0,$$

where Δ is the discriminant.

If the curve representing the polynomial $f(x)$ be moved parallel to the axis of y (see Art. 10), through a distance equal to a max. or min. value ρ, the axis of x will become a tangent to it, *i.e.* the equation $f(x) - \rho = 0$ will have equal roots. Hence the max. and min. values are obtained by forming the discriminant of $f(x) - \rho$, or by putting $d - \rho$ for ρ in $G^2 + 4H^3 = 0$.

11. Prove similarly that the max. and min. values of

$$ax^4 + 4bx^3 + 6cx^2 + 4dx + e$$

are the roots of the equation

$$a^3 \rho^3 - 3(a^2 I - 9H^2)\rho^2 + 3(aI^2 - 18HJ)\rho - \Delta = 0,$$

where Δ is the discriminant of the quartic.

CHAPTER VIII.

77. Definition of Limits.—In attempting to discover the real roots of numerical equations, it is in the first place advantageous to narrow the region within which they must be sought. We here take up the inquiry referred to in the remark at the end of Art. 4, and proceed to prove certain propositions relative to the limits of the real roots of equations.

A *superior limit* of the positive roots is any greater positive number than the greatest of them; an *inferior limit* of the positive roots is any smaller positive number than the smallest of them. A superior limit of the negative roots is any greater negative number than the greatest of them; an inferior limit of the negative roots is any smaller negative number than the smallest of them; the greatest negative number meaning that nearest to $-\infty$.

When we have found limits within which all the real roots of an equation lie, the next step towards the solution of the equation is to discover the intervals in which the separate roots are situated. The methods which have been advanced for this latter purpose will form the subject of the next Chapter.

The following Propositions all relate to the superior limits of the positive roots; to which, as will be subsequently proved, the determination of inferior limits and limits of the negative roots can be immediately reduced.

78. Proposition I.—*In any equation*

$$x^n + p_1 x^{n-1} + p_2 x^{n-2} + \ldots + p_{n-1} x + p_n = 0,$$

if the first negative term be $-p_r x^{n-r}$, *and if the greatest negative coefficient be* $-p_k$, *then* $\sqrt[r]{p_k} + 1$ *is a superior limit of the positive roots.*

Any value of x which makes

$$x^n > p_k \left(x^{n-r} + x^{n-r-1} + \ldots + x + 1\right) > p_k \frac{x^{n-r+1} - 1}{x - 1}$$

will, *à fortiori*, make $f(x)$ positive.

Now, taking x greater than unity, this inequality is satisfied by the following :—

$$x^n > p_k \frac{x^{n-r+1}}{x - 1},$$

or
$$x^{n+1} - x^n > p_k x^{n-r+1},$$

or
$$x^{r-1} (x - 1) > p_k,$$

which inequality again is satisfied by the following :—

$$(x - 1)^{r-1} (x - 1) = \text{or} > p_k,$$

since plainly
$$x^{r-1} > (x - 1)^{r-1}.$$

We have, then, finally
$$(x - 1)^r = \text{or} > p_k,$$

or
$$x = \text{or} > 1 + \sqrt[r]{p_k}.$$

79. **Proposition II.**—*If in any equation each negative coefficient be taken positively, and divided by the sum of all the positive coefficients which precede it, the greatest quotient thus formed increased by unity is a superior limit of the positive roots.*

Let the equation be

$$a_0 x^n + a_1 x^{n-1} + a_2 x^{n-2} - a_3 x^{n-3} + \ldots - a_r x^{n-r} + \ldots + a_n = 0,$$

in which, in order to fix our ideas, we regard the fourth coefficient as negative, and we consider also a negative coefficient in general, i. e. $-a_r$.

Let each positive term in this equation be transformed by means of the formula

$$a_m x^m = a_m (x - 1)(x^{m-1} + x^{m-2} + \ldots + x + 1) + a_m,$$

which is derived at once from

$$\frac{x^m - 1}{x - 1} = x^{m-1} + x^{m-2} + \ldots + x + 1\,;$$

and let each negative term remain as it is.

The polynomial $f(x)$ becomes then, each horizontal line of the following corresponding to each term of $f(x)$ in succession—

$$a_0(x-1)\,x^{n-1} + a_0(x-1)\,x^{n-2} + a_0(x-1)\,x^{n-3} + \ldots + a_0(x-1)\,x^{n-r} + \ldots + a_0,$$
$$+\, a_1(x-1)\,x^{n-2} + a_1(x-1)\,x^{n-3} + \ldots + a_1(x-1)\,x^{n-r} + \ldots + a_1,$$
$$+\, a_2(x-1)\,x^{n-3} + \ldots + a_2(x-1)\,x^{n-r} + \ldots + a_2,$$
$$-\, a_3\,x^{n-3},$$
$$+\,\ldots\ldots\ldots\ldots\ldots\ldots$$
$$-\, a_r\,x^{n-r},$$
$$+\,\ldots\ldots\ldots\ldots$$
$$+\, a_n.$$

We now regard the vertical columns of this expression as successive terms in the polynomial; the successive coefficients of x^{n-1}, x^{n-2}, &c., being

$$a_0(x-1), \quad (a_0 + a_1)(x-1), \quad (a_0 + a_1 + a_2)(x-1) - a_3, \quad \text{&c.}$$

Any value of x greater than unity is sufficient to make positive every term in which no negative coefficient a_3, a_r, &c. occurs. To make the latter terms positive, we must have

$$(a_0 + a_1 + a_2)(x - 1) > a_3,$$
$$\ldots\ldots\ldots\ldots\ldots$$
$$(a_0 + a_1 + a_2 + \ldots + a_{r-1})(x - 1) > a_r, \quad \text{&c.}$$

Hence

$$x > \frac{a_3}{a_0 + a_1 + a_2} + 1, \;\ldots\; x > \frac{a_r}{a_0 + a_1 + a_2 + a_4 + \ldots + a_{r-1}} + 1, \quad \text{&c.}$$

And to ensure every term being made positive, we must take the value of the greatest of the quantities found in this way. Such a value of x, therefore, is a superior limit of the positive roots.

80. Practical Applications.—The propositions in the two preceding Articles furnish the most convenient *general* methods of finding in practice tolerably close limits of the roots. Sometimes one of the propositions will give the closer limit : sometimes the other. It is advantageous, therefore, to apply both methods, and take the smaller limit. Prop. I. will usually be found the more advantageous when the first negative coefficient is preceded by several positive coefficients, so that r is large ; and Prop. II. when large positive coefficients occur before the first large negative coefficient. In general, Prop. II. will give the closer limit. We speak of the integer next above the number given by either proposition as the limit.

EXAMPLES.

1. Find a superior limit of the positive roots of the equation

$$x^4 - 5x^3 + 40x^2 - 8x + 23 = 0.$$

Prop. I. gives $8 + 1$, or 9, as limit.

Prop. II. gives $\dfrac{5}{1} + 1$, or 6. Hence 6 is a superior limit.

2. Find a superior limit of the positive roots of the equation

$$x^5 + 3x^4 + x^3 - 8x^2 - 51x + 18 = 0.$$

Prop. I. gives $\sqrt[3]{51} + 1$; and 5 is, therefore, a limit.

Prop. II. gives $\dfrac{51}{1 + 3 + 1} + 1$, and 12 is a limit.

In this case Prop. I. gives the closer limit, *i.e.* 5.

3. Find a superior limit of the positive roots of

$$x^7 + 4x^6 - 3x^5 + 5x^4 - 9x^3 - 11x^2 + 6x - 8 = 0.$$

Of the fractions

$$\frac{3}{1 + 4}, \quad \frac{9}{1 + 4 + 5}, \quad \frac{11}{1 + 4 + 5}, \quad \frac{8}{1 + 4 + 5 + 6},$$

the third is the greatest, and Prop. II. gives the limit 3. Prop. I. gives 5.

4. Find a superior limit of the positive roots of

$$x^8 + 20x^7 + 4x^6 - 11x^5 - 120x^4 + 13x - 25 = 0.$$

Ans. Both Propositions give the limit 6.

5. Find a superior limit of the positive roots of

$$4x^5 - 8x^4 + 22x^3 + 98x^2 - 73x + 5 = 0.$$

Ans. Prop. I. gives 20. Prop. II. gives 3.

It is usually possible to determine by inspection a limit closer than that given by either of the preceding propositions. The method consists in general in arranging the terms of an equation in groups having a positive term first, and then observing what is the lowest integral value of x which will have the effect of rendering each group positive. The form of the equation will suggest the arrangement in any particular case.

6. The equation of Ex. 2 can be arranged as follows :—

$$x^2 (x^3 - 8) + x(3x^3 - 51) + x^3 + 18 = 0.$$

$x = 3$, or any greater number, renders each group positive ; hence 3 is a superior limit.

7. The equation of Ex. 4 may be arranged thus :

$$x^5 (x^3 - 11) + 20x^4 (x^3 - 6) + 4x^6 + 13x - 25 = 0.$$

$x = 3$, or any greater number, renders each group positive ; hence 3 is a limit.

8. Find a superior limit of the roots of the equation

$$x^4 - 4x^3 + 33x^2 - 2x + 18 = 0.$$

This can be arranged in the form

$$x^2 (x^2 - 4x + 5) + 28x (x - \tfrac{1}{14}) + 18 = 0.$$

Now the trinomial $x^2 - 4x + 5$, having imaginary roots, is positive for all values of x (Art. 12). Hence $x = 1$ is a superior limit.

The introduction in this way of a quadratic whose roots are imaginary, or of one with equal roots, will often be found useful.

9. Find a superior limit of the roots of the equation

$$5x^5 - 7x^4 - 10x^3 - 23x^2 - 90x - 317 = 0.$$

In examples of this kind it is convenient to distribute the highest power of x among the negative terms. Here the equation may be written

$$x^4 (x - 7) + x^3 (x^2 - 10) + x^2 (x^3 - 23) + x (x^4 - 90) + x^5 - 317 = 0,$$

so that 7 is evidently a superior limit of the roots. In this case the general methods give a very high limit.

10. Find a superior limit of the roots of the equation

$$x^4 - x^3 - 2x^2 - 4x - 24 = 0.$$

When there are several negative terms, and the coefficient of the highest term unity, it is convenient to multiply the whole equation by such a number as will enable us to distribute the highest term among the negative terms. Here, multiplying by 4, we can write the equation as follows :—

$$x^3 (x - 4) + x^2 (x^2 - 8) + x (x^3 - 16) + x^4 - 96 = 0,$$

and 4 is a superior limit. The general methods give 25.

81. **Proposition III.**—*Any number which renders positive the polynomial $f(x)$ and all its derived functions $f_1(x), f_2(x), \ldots f_n(x)$ is a superior limit of the positive roots of the equation $f(x) = 0$.*

This method of finding limits is due to Newton. It is much more laborious in its application than either of the preceding methods; but it has the advantage of giving always very close limits; and in the case of an equation all whose roots are real the limit found in this way is, as will be subsequently proved, the next integer above the greatest positive root.

To prove the proposition, let the roots of the equation $f(x) = 0$ be diminished by h; then $x - h = y$, and

$$f(y + h) = f(h) + f_1(h)\,y + \frac{f_2(h)}{1 \cdot 2}\,y^2 + \ldots + \frac{f_n(h)}{1 \cdot 2 \ldots n}\,y^n.$$

If now h be such as to make all the coefficients

$$f(h), \; f_1(h), \; f_2(h), \; \ldots f_n(h)$$

positive, the equation in y cannot have a positive root; that is to say, the equation in x has no root greater than h; hence h is a superior limit of the positive roots.

EXAMPLE.

$$f(x) \equiv x^4 - 2x^3 - 3x^2 - 15x - 3.$$

In applying Newton's method of finding limits to any example the general mode of procedure is as follows :—Take the smallest integral number which renders $f_{n-1}(x)$ positive; and proceeding upwards in order to $f_1(x)$, try the effect of substituting this number for x in the other functions of the series. When any function is reached which becomes negative for the integer in question, increase the integer successively by units till it makes that function positive; and then proceed with the new integer as before, increasing it again if another function in the series should become negative, and so on, till that integer is reached which renders all the functions in the series positive. In the present example the series of functions is

$$f(x) = x^4 - 2x^3 - 3x^2 - 15x - 3,$$
$$f_1(x) = 4x^3 - 6x^2 - 6x - 15,$$
$$\tfrac{1}{2} f_2(x) = 6x^2 - 6x - 3,$$
$$\tfrac{1}{6} f_3(x) = 4x - 2,$$
$$\tfrac{1}{24} f_4(x) = 1.$$

M

Here $x = 1$ makes $f_3(x)$ positive. We try then the effect of the substitution $x = 1$ in $f_2(x)$. It makes $f_2(x)$ negative. Increase by 1; and $x = 2$ makes $f_2(x)$ positive. Try the effect of $x = 2$ in $f_1(x)$; it gives a negative result. Increase by 1; and $x = 3$ makes $f_1(x)$ positive. Proceeding upwards, the substitution $x = 3$ makes $f(x)$ negative; and increasing again by unity, we find that $x = 4$ makes $f(x)$ positive. Hence 4 is the superior limit required.

It is assumed in this mode of applying Newton's rule, that when any number makes all the derived functions up to a certain stage positive, any higher number will also make them positive; so that there is no occasion to try the effect of that higher number on the functions in the series below that one where our upward progress is arrested. This is evident from the equation

$$\phi(a + h) = \phi(a) + \phi'(a)h + \phi''(a)\frac{h^2}{1 \cdot 2} + \cdots$$

(taking $\phi(x)$ to represent any function in the series, and using the common notation for derived functions), which shows that if $\phi(a)$, $\phi'(a)$, $\phi''(a)$, $\ldots$ are all positive, and h also positive, $\phi(a + h)$ must be positive.

It may be observed that one advantage of Newton's method is that often, as in the present instance, it gives us a knowledge of the two successive integers between which the highest root lies. Thus in the present example, since $f(x)$ is negative for $x = 3$, and positive for $x = 4$, we know that the greatest root of the equation lies between 3 and 4.

82. Inferior Limits, and Limits of the Negative Roots.—To find an inferior limit of the positive roots, the equation must be first transformed by the substitution $x = \dfrac{1}{y}$. Find then a superior limit h of the positive roots of the equation in y. The reciprocal of this, *i.e.* $\dfrac{1}{h}$, will be the required inferior limit; for since

$$y < h, \quad \frac{1}{y} > \frac{1}{h}, \quad \text{i.e. } x > \frac{1}{h}.$$

To find limits of the negative roots, we have only to transform the equation by the substitution $x = -y$. This changes the negative into positive roots. Let the superior and inferior limits of the positive roots of the equation in y be h and h'. Then $-h$ and $-h'$ are the limits of the negative roots of the proposed equation.

83. Limiting Equations.—*If all the real roots of the equation $f'(x) = 0$ could be found, it would be possible to determine the number of real roots of the equation $f(x) = 0$.* For, let the real roots of $f'(x) = 0$ be, in ascending order of magnitude, a', β', γ', ... λ'; and let the following series of values be substituted for x in $f(x)$:—

$$-\infty, \; a', \; \beta', \; \gamma', \; \ldots \; \lambda', \; +\infty.$$

When any successive two of these give results with different signs there is a root of $f(x) = 0$ between them; and by the Cor., Art. 71, there is only one; and when they give results with the same sign there is, by the same Cor., no root between them. Thus each change of sign in the results of the successive substitutions proves the existence of one real root of the proposed equation.

If all the roots of $f(x) = 0$ are real, it is evident, by the theorem of Art. 71, that all the roots of $f'(x) = 0$ are also real, and that they lie one by one between each adjacent pair of the roots of $f(x) = 0$. In the same case, and by the same theorem, it follows that the roots of $f''(x) = 0$, and of all the successive derived functions, are real also; and the roots of any function lie severally between each adjacent pair of the roots of the function from which it is immediately derived.

Equations of this kind, which are one degree below the degree of any proposed equation, and whose roots lie severally between each adjacent pair of the roots of the proposed, are called *limiting equations.*

It is evident that in the application of Newton's method of finding limits of the roots, when the roots of $f'(x) = 0$ are all real, in proceeding according to the method explained in Art. 81, the function $f(x)$ is itself the last which will be rendered positive, and therefore the superior limit arrived at is the integer next above the greatest root.

EXAMPLES.

1. Prove that any derived equation $f_m(x) = 0$ cannot have more imaginary roots, but may have more real roots, than the equation $f(x) = 0$ from which it is derived.

From this it follows that if any of the derived functions be found to have imaginary roots, the same number at least of imaginary roots must enter the primitive equation.

2. Apply the method of Art. 83, to determine the conditions that the equation

$$x^3 - qx + r = 0$$

should have all its roots real.

3. Determine by the same method the nature of the roots of the equation

$$x^n - nqx + (n - 1)r = 0.$$

Ans. When n is even, the equation has two real roots or none, according as $q^n >$ or $< r^{n-1}$.

When n is odd, the equation has three real roots or one, according as $q^n >$ or $< r^{n-1}$.

4. The equation $x^n (x - 1)^n = 0$ has all its roots real; hence show that the following equation has all its roots real, and situated between 0 and 1 :—

$$x^n - n\,\frac{n}{2n}\,x^{n-1} + \frac{n(n-1)}{1 \cdot 2}\,\frac{n(n-1)}{2n(2n-1)}\,x^{n-2} - \&c. = 0.$$

5. If any two of the quantities l, m, n in the following equation be put equal to zero, show that the quadratic to which the equation then reduces is a limiting equation; and hence prove that the roots of the proposed are all real :—

$$(x - a)(x - b)(x - c) - l^2(x - a) - m^2(x - b) - n^2(x - c) - 2lmn = 0.$$

CHAPTER IX.

84. By the methods of the preceding Chapter we are enabled to find limits between which all the real roots of any numerical equation lie. Before proceeding to the actual approximation to any particular root, it is necessary to separate the interval in which it is situated from the intervals which contain the remaining roots. The present Chapter will be occupied with certain theorems whose object is to determine the number of real roots between any two arbitrarily assumed values of the variable. It is plain that if this object can be effected, it will then be possible to tell not only the total number of real roots, but also the limits within which the roots separately lie.

The theorems given for this purpose by Fourier and Budan, although different in statement, are identical in principle. For purposes of exposition Fourier's statement is the more convenient, while with a view to practical application the statement of Budan will be found superior. The theorem of Sturm, although more laborious in practice, has the advantage over the preceding that it is unfailing in its application, giving always the exact number of real roots situated between any two proposed quantities; whereas the theorem of Fourier and Budan gives only a certain limit which the number of real roots in the proposed interval cannot exceed.

85. **Theorem of Fourier and Budan.**—*Let two numbers a and b, of which a is the less, be substituted in the series formed by $f(x)$ and its successive derived functions, viz.,*

$$f(x),\; f_1(x),\; f_2(x),\; \ldots,\; f_n(x)\,;$$

the number of real roots which lie between a and b cannot be greater than the excess of the number of changes of sign in the series when a is substituted for x, over the number of changes when b is substituted for x; and when the number of real roots in the interval falls short of that difference, it will be by an even number.

This is the form in which Fourier states the theorem.

It is to be understood here, as elsewhere, that, when we speak of two numbers a and b, of which a is the less, one or both of them may be negative, and all we mean is that a is nearer than b to $-\infty$.

The value of x is supposed to increase continuously from a to b; and we proceed to examine the changes which may occur among the signs of the functions in the above series. The following different cases can arise :—

(1). The value of x may pass through a single root of $f(x) = 0$.

(2). It may pass through a root occurring r times in $f(x) = 0$.

(3). It may pass through a root of one of the auxiliary functions $f_m(x) = 0$, this root not occurring in either $f_{m-1}(x) = 0$ or $f_{m+1}(x) = 0$.

(4). It may pass through a root occurring r times in $f_m(x) = 0$, and not occurring in $f_{m-1}(x) = 0$.

In what follows the symbol x is omitted after f for convenience.

(1). In the first case it is evident, from Art. 75, that in passing through a root of the equation $f(x) = 0$ one change of sign is lost; for f and f_1 have unlike signs immediately before the passage, and like signs immediately after the passage.

(2). In the second case, in passing through an r-multiple root of $f(x) = 0$, it is evident that r changes of sign are lost; for, by Art. 76, immediately before the passage the series of functions

$$f, \; f_1, \; f_2, \; \ldots \; f_{r-1}, \; f_r$$

have signs alternately $+$ and $-$, or $-$ and $+$, and immediately after the passage have all the same sign as f_r.

(3). In the third case, the root of $f_m(x) = 0$ must give to f_{m-1} and f_{m+1} either like signs or unlike signs. Suppose it to give like signs; then in passing through the root two changes of sign are lost, for before the passage the sign of f_m is different from these like signs, and after the passage it is the same (Art. 76). Suppose it to give unlike signs; then no change of sign is lost, for before the passage the signs of f_{m-1}, f_m, f_{m+1} must be either $+\ +\ -$, or $-\ -\ +$, and after the passage these become $+\ -\ -$, and $-\ +\ +$. On the whole, therefore, we conclude that no variation of sign can be gained, but two variations may be lost, on the passage through a root of $f_m(x) = 0$.

(4). In the fourth case x passes through a value (let us say a) which causes not only f_m but also $f_{m+1}, f_{m+2}, \ldots, f_{m+r-1}$ to vanish. It is evident from the theorem of Art. 76 that during the passage a number of changes of sign will always be lost. The definite number lost may be collected by considering the series of functions

$$f_{m-1},\ f_m,\ f_{m+1},\ \ldots\ldots,\ f_{m+r-1},\ f_{m+r}.$$

We easily obtain the following results :—

(a). When $f_{m-1}(a)$ and $f_{m+r}(a)$ have like signs.

> If r be even, r changes are lost.
> If r be odd, $r + 1$ changes are lost.

(b). When $f_{m-1}(a)$ and $f_{m+r}(a)$ have unlike signs.

> If r be even, r changes are lost.
> If r be odd, $r - 1$ changes are lost.

We conclude, therefore, on the whole, that an even number of changes is lost during the passage through an r-multiple root of $f_m(x)$.

It will bo observed that (1) is a particular case of (2), and (3) of (4), *i.e.* when $r = 1$. Since, however, these are the cases of ordinary occurrence, it is well to give them a separate classification.

Reviewing the above proof, we conclude that as x increases from a to b no change of sign can be gained; that for each passage through a root of $f(x) = 0$ one change is lost; and that under no circumstances except a passage through a root of $f(x) = 0$ can an odd number of changes be lost. Hence the number of changes lost during the whole variation of x from a to b must be either equal to the number of real roots of $f(x) = 0$ in the interval, or must exceed it by an even number. The theorem is thus proved.

86. **Application of the Theorem.**—The form in which the theorem has been stated by Budan is, as has been already observed, more convenient for practical purposes than that just given. It is as follows :—*Let the roots of an equation $f(x) = 0$ be diminished, first by a and then by b, where a and b are any two numbers of which a is the less ; then the number of real roots between a and b cannot be greater than the excess of the number of changes of sign in the first transformed equation over the number in the second.*

This is evidently included in Fourier's statement, for the two transformed equations are (see Art. 34)

$$f(a) + f_1(a)\,y + \frac{f_2(a)}{1\cdot 2}\,y^2 + \ldots + \frac{f_n(a)}{1\cdot 2\ldots n}\,y^n = 0,$$

$$f(b) + f_1(b)\,y + \frac{f_2(b)}{1\cdot 2}\,y^2 + \ldots + \frac{f_n(b)}{1\cdot 2\ldots n}\,y^n = 0 ;$$

from which, assuming the results of the last Article, the above proposition is manifest.

The reason why the theorem is in this form convenient in practice is, that we can apply the expeditious method of diminishing the roots given in Art. 34.

EXAMPLES.

1. Find the situations of the roots of the equation

$$x^5 - 3x^4 - 24x^3 + 95x^2 - 46x - 101 = 0.$$

We shall examine this function for values of x between the intervals

$$-10, \quad -1, \quad 0, \quad 1, \quad 10 ;$$

these numbers being assumed on account of the facility of calculation. Diminution of the roots by 1 gives the following series of coefficients of the transformed equation :—

$$1, \qquad 2, \qquad -26, \qquad 15, \qquad 65, \qquad -78.$$

In diminishing the roots by 10, it is apparent at the very outset of the calculation that the signs of the coefficients of the transformed equation will be all positive ; so that there is no occasion to complete the calculation in this case.

In diminishing the roots by -10 and -1, it is convenient to change the alternate signs of the equation, and diminish the roots by $+10$ and $+1$; and then in the result change the alternate signs again. The coefficients of the transformed equation when the roots are diminished by -1 are

$$1, \; -8, \; -2, \; 139, \; -291, \; 60.$$

In diminishing by -10 we observe in the course of the operation, as before, that the signs will be all positive in the result, *i. e.* when the alternate signs are changed, they will be alternately positive and negative.

Hence we have the following scheme :—

(-10)	$+$	$-$	$+$	$-$	$+$	$-$
(-1)	$+$	$-$	$-$	$+$	$-$	$+$
(0)	$+$	$-$	$-$	$+$	$-$	$-$, the equation itself.
(1)	$+$	$+$	$-$	$-$	$+$	$-$
(10)	$+$	$+$	$+$	$+$	$+$	$+$

These signs are the signs taken by $f(x)$ and the several derived functions f_1, f_2, f_3, f_4, f_5 on the substitution of the proposed numbers ; but it is to be observed that they are here written, not in the order of Art. 85, but in the reverse order, viz., $f_5, f_4, f_3, f_2, f_1, f$.

From these we draw the following conclusions :—All the real roots must lie between -10 and $+10$; one real root lies between -10 and -1, since one change of sign is lost ; one real root lies between -1 and 0, since one change of sign is lost ; no real root lies between 0 and 1 ; and between 1 and 10, since three changes of sign are lost, there is at least one real root ; but we are left in doubt as to the nature of the other two roots : whether they are imaginary, or whether there are three real roots between 1 and 10.

We might proceed to examine, by further transformations, the interval between 1 and 10 more closely, in order to determine the nature of the two doubtful roots ; but it is evident that the operations for this purpose might, if the roots were nearly equal, become very laborious. This is the weak side of the theorem of Fourier and Budan. Both writers have attempted to supply this defect, and have given methods of determining the nature of the roots in doubtful intervals ; but as these methods are complicated, we do not stop to explain them ; the more especially as the theorem of Sturm effects fully the purposes for which the supplementary methods of Fourier and Budan were invented.

2. Analyse the equation of Ex. 1, p. 98, viz.,

$$x^3 + x^2 - 2x - 1 = 0.$$

The roots of this are all real, and lie between -2 and 2 (see Ex. 5, p. 98). Whenever the roots of an equation are all real, the signs of Fourier's functions determine the exact number of real roots between any two proposed integers. We obtain the following results :—The roots lie in the intervals

$$(-2, -1); \quad (-1, 0); \quad (1, 2).$$

3. Analyse the equation of Ex. 3, p. 98, viz.,

$$x^5 + x^4 - 4x^3 - 3x^2 + 3x + 1 = 0.$$

Ans. Two roots in the interval $(-2, -1)$, and one root in each of the intervals $(-1, 0)$; $(0, 1)$; $(1, 2)$.

4. Analyse the equation

$$x^4 - 80x^3 + 1998x^2 - 14937x + 5000 = 0.$$

The equation can have no negative roots. Diminish the roots successively by 10 till the signs of the coefficients become all positive. We obtain the following result :—

(0)	+	−	+	−	+
(10)	+	−	+	+	−
(20)	+	0	−	+	+
(30)	+	+	+	−	+
(40)	+	+	+	+	+

Thus, there is one root between 0 and 10, and one between 10 and 20 ; no root between 20 and 30. Between 30 and 40 either there are two real roots, or there is an indication of a pair of imaginary roots. That the former is the case will appear by diminishing the roots of the third transformed equation by units. This process will separate the roots, which will be found to lie between $(2, 3)$ and $(4, 5)$; so that the proposed equation has a third real root in the interval $(32, 33)$, and a fourth in the interval $(34, 35)$.

87. Application of the Theorem to the Detection of Imaginary Roots.

—Since there exist only n changes of sign to be lost in the passage of x from $-\infty$ to $+\infty$, if we have any reason for knowing that a pair of changes is lost during the passage of x through an interval which includes no real root of the equation, we may be assured of the existence of a pair of imaginary roots. Circumstances of this nature will arise in the application of Fourier's theorem when any of the transformed equations contain vanishing coefficients. For we can assign by the principle of Art. 76 the proper sign to this coefficient, corresponding to values of x immediately before and immediately after that value which causes the coefficient to vanish; the whole interval being so small that it may be supposed not to include any root of the equation $f(x) = 0$. An example will make this plain.

Examples.

1. Analyse the equation

$$f(x) = x^4 - 4x^3 - 3x + 23 = 0.$$

We shall examine this function between the intervals 0, 1, 10. The transformed equations are

$$\tfrac{1}{24} f_4 \ (0)x^4 + \tfrac{1}{6}f_3 \ (0)x^3 + \tfrac{1}{2}f_2 \ (0)x^2 + f_1 \ (0) x + f \ (0) = 0,$$

$$\tfrac{1}{24} f_4 \ (1)x^4 + \tfrac{1}{6}f_3 \ (1)x^3 + \tfrac{1}{2}f_2 \ (1)x^2 + f_1 \ (1) x + f \ (1) = 0,$$

$$\tfrac{1}{24} f_4 (10)x^4 + \tfrac{1}{6}f_3 (10)x^3 + \tfrac{1}{2}f_2 (10) x^2 + f_1 (10) x + f (10) = 0,$$

the first of these being the proposed equation itself.

Making the calculations by the method of the preceding Article, we find that the coefficient $f_3(1) = 0$, and we get the following scheme :—

$$
\begin{array}{lccccc}
(0) & + & - & 0 & - & + \\
(1) & + & 0 & - & - & + \\
(10) & + & + & + & + & + \\
\end{array}
$$

We may now replace each of the rows containing a zero coefficient by two, the first corresponding to a value a little less, and the second to a value a little greater, than that which gives the zero coefficients; the signs being determined by the principle established in Art. 76. It must be remembered that in the above scheme the signs representing the derived functions are written in the reverse order to that of the Article referred to. The scheme will then stand as follows, using h to represent a very small quantity :—

$$
(0) \begin{cases} -h & + & - & + & - & + \\ +h & + & - & - & - & + \end{cases}
$$

$$
(1) \begin{cases} 1-h & + & - & - & - & + \\ 1+h & + & + & - & - & | \end{cases}
$$

$$
(10) \qquad + \ + \ + \ + \ +
$$

In this scheme the signs corresponding to $-h$ and $+h$ are determined by the condition that the sign of the coefficient which is zero when $x = 0$ must, when $x = -h$, be different from that next to it on the left-hand side; and when $x = +h$ it must be the same. Similarly the signs corresponding to $1 - h$ and $1 + h$ are determined.

Now since a pair of changes is lost in the interval $(-h, +h)$, and since the equation has no real root between $-h$ and $+h$, we have proved the existence of a pair of imaginary roots. Two changes of sign are lost between $1 + h$ and 10, so that this interval either includes a pair of real roots, or presents an indication of a pair of imaginary roots. Which of these is the case remains still doubtful.

2. If several coefficients vanish, we may be able to establish the existence of several pairs of imaginary roots. This will appear from the following example :—

$$x^6 - 1 = 0.$$

The signs corresponding to $-h$ and $+h$, are, applying Art. 76,

$$(-h) \qquad + \ - \ + \ - \ + \ - \ -$$
$$(+h) \qquad + \ + \ + \ + \ + \ + \ -$$

Hence, since no root exists between $-h$ and $+h$, and since 4 changes of sign are lost in passing from a value very little less than 0 to one very little greater, we are assured of the existence of two pairs of imaginary roots. The other two roots are in this case plainly real (see Art. 14). They are in fact 1 and -1.

The number of imaginary roots in any binomial equation can be determined in this way.

3. Find the character of the roots of the equation

$$x^8 + 10x^3 + x - 4 = 0.$$

In passing from a small negative to a small positive value of x we obtain the following series of signs :—

$$(-h) \qquad + \ - \ + \ - \ + \ + \ - \ + \ -$$
$$(0) \qquad + \ 0 \ 0 \ 0 \ 0 \ + \ 0 \ + \ -$$
$$(+h) \qquad + \ + \ + \ + \ + \ + \ + \ + \ -$$

Since six changes of sign are here lost, there are six imaginary roots. The remaining two roots are, by Art. 14, real ; one positive, and the other negative. The negative root lies between -2 and -1, and the positive between 0 and 1.

4. Analyse completely the equation

$$x^6 - 3x^2 - x + 1 = 0.$$

There are two imaginary roots. Whenever, as in the present instance, the roots are comprised within small limits, it is convenient to diminish by successive units. In this way we find here a root between 0 and 1, and another between 1 and 2. Proceeding to negative roots, we find on diminishing by -1 that -1 is itself a root, and writing down the signs corresponding to a value a little greater than -1, we observe an indication of a second negative root between -1 and 0.

5. Analyse the equation

$$x^5 + x^4 + x^2 - 25x - 36 = 0.$$

There are two imaginary roots ; one real positive root between 2 and 3 ; and two real negative roots in the intervals $(-3, -2)$, $(-2, -1)$.

88. Corollaries from the Theorem of Fourier and Budan.—The method of detecting the existence of imaginary roots explained in the preceding Article is called *The Rule of the Double Sign.* A similar rule, due to *De Gua*, was in use before the discovery of Fourier's theorem. This rule and Descartes' *Rule of Signs* are immediate corollaries from the theorem, as we proceed to show.

Cor. 1.—De Gua's Rule for finding Imaginary Roots.

The rule may be stated generally as follows :—*When $2m$ successive terms of an equation are absent, the equation has $2m$ imaginary roots; and when $2m + 1$ successive terms are absent, the equation has $2m + 2$, or $2m$ imaginary roots, according as the two terms between which the deficiency occurs have like or unlike signs.* This follows, as in case (4), Art. 85, by examining the number of changes of sign lost during the passage of x from a small negative value $- h$ to a small positive value h.

Cor. 2.—Descartes' Rule of Signs.

When 0 is substituted for x in the series of functions $f_n(x), f_{n-1}(x), \ldots f_2(x), f_1(x), f(x)$, the signs are the same as the signs of the coefficients $a_0, a_1, a_2, \ldots a_{n-1}, a_n$, of the proposed equation ; and when $+ \infty$ is substituted the signs are all positive. Fourier's theorem asserts that the number of roots between these limits, viz., the number of positive roots, cannot exceed the number of variations lost during the passage from 0 to $+ \infty$, that is the number of changes of sign in the series $a_0, a_1, a_2 \ldots a_n$. This is Descartes' rule for positive roots ; and the similar rule for negative roots follows in the usual way by changing the negative into positive roots.

Cor. 3.—Newton's Method of finding Limits.

When a number h has been found which renders positive each of the functions $f_n(x), f_{n-1}(x), \ldots f_2(x), f_1(x), f(x)$; since $+ \infty$ also renders each of them positive, it follows from Fourier's theorem that there can be no root between h and $+ \infty$, i.e. h is a superior limit of the positive roots ; and this is Newton's proposition (see Art. 81).

89. Sturm's Theorem.—We have already shown (Art. 74), that it is possible by performing the common algebraical operation of finding the greatest common measure of a polynomial $f(x)$ and its first derived polynomial to find the equal roots of the equation $f(x) = 0$. Sturm has employed the same operation for the formation of the auxiliary functions which enter into his method of separating the roots of an equation.

Let the process of finding the greatest common measure of $f(x)$ and its first derived be performed. The successive remainders will go on diminishing in degree till we come finally either to one which divides that immediately preceding without remainder, or to a remainder which does not contain the variable at all, *i. e.* which is numerical. The former is, as we have already seen, the case of equal roots. The latter is the case where no equal roots exist. It is convenient to divide the discussion of Sturm's theorem into these two cases. We shall in the present Article consider the case where no equal roots exist; and proceed in the next Article to the case of equal roots. The performance of the operation itself will of course disclose the class to which any particular example is to be referred.

The auxiliary functions employed by Sturm are not the remainders as they present themselves in the operation, but the remainders *with their signs changed*. In finding the greatest common measure of two expressions it is indifferent whether the signs of the remainders are changed or not. In the formation of Sturm's auxiliary functions it is essential that the sign of each remainder should be changed before it is made the next divisor.

Confining our attention for the present, therefore, to the case where no equal roots exist, Sturm's theorem may be stated as follows :—

Theorem.—*Let any two real quantities a and b be substituted for x in the series of $n + 1$ functions*

$$f(x),\ f_1(x),\ f_2(x),\ f_3(x),\ \ldots,\ f_{n-1}(x),\ f_n(x),$$

consisting of the given polynomial $f(x)$, its first derived $f_1(x)$, and *the successive remainders (with their signs changed) in the process of finding the greatest common measure of $f(x)$ and $f_1(x)$; then the difference between the number of changes of sign in the series when a is substituted for x, and the number when b is substituted for x expresses exactly the number of real roots of the equation $f(x) = 0$ between a and b.*

The mode of formation of Sturm's functions supplies the following series of equations, taking $q_1, q_2, \ldots q_{n-1}$ to represent the successive quotients in the operation :—

$$\left.\begin{aligned}
f(x) \ &= q_1 f_1(x) - f_2(x),\\
f_1(x) \ &= q_2 f_2(x) - f_3(x),\\
&\cdot \quad \cdot \quad \cdot \quad \cdot \quad \cdot \quad \cdot \quad \cdot\\
f_{r-1}(x) &= q_r f_r(x) - f_{r+1}(x),\\
&\cdot \quad \cdot \quad \cdot \quad \cdot \quad \cdot \quad \cdot \quad \cdot\\
f_{n-2}(x) &= q_{n-1} f_{n-1}(x) - f_n(x).
\end{aligned}\right\} \qquad (1)$$

These equations involve the theory of the method of finding the greatest common measure; for it follows from the first equation that if $f(x)$ and $f_1(x)$ have a common factor, this must be a factor in $f_2(x)$; and then from the second equation it follows, by the same reasoning, that it must be a factor in $f_3(x)$, and so on, till we come finally to the last remainder, which, when $f(x)$ and $f_1(x)$ have common factors, will be a polynomial consisting of these factors. In the present Article, where we suppose the given polynomial and its first derived to have no common factor, the last remainder $f_n(x)$ is numerical. It is essential for the proof of the theorem to observe also, that in this case no two consecutive functions in the series can have a common factor; for if they had we could, reasoning backwards, show from the equations that it must be a common factor in $f(x)$ and $f_1(x)$; and such, according to our hypothesis, is not here the case. In examining, then, what changes of sign can take place in the series during the passage of x from a to b, we may exclude the case of two consecutive functions vanishing for the same value of x; and the different cases in which any change of sign can take place are the following :—

(1). When x passes through a root of the proposed equation $f(x) = 0$.

(2). When x passes through a value which causes one of the auxiliary functions $f_1, f_2, \ldots f_{n-1}$ to vanish.

(3). When x passes through a value which causes two or more of the series $f, f_1, \ldots f_{n-1}$ to vanish together; no two of the vanishing functions, however, being consecutive :

(1). When x passes through a root of $f(x) = 0$, it follows from Art. 75 that one change of sign is lost, since immediately before the passage $f(x)$ and $f_1(x)$ have unlike signs, and immediately after the passage they have like signs.

(2). Suppose x to take a value a which satisfies the equation $f_r(x) = 0$. From the equation

$$f_{r-1}(x) = q_r f_r(x) - f_{r+1}(x)$$

we get $\qquad\qquad f_{r-1}(a) = -f_{r+1}(a),$

which proves that this value of x gives to $f_{r-1}(x)$ and $f_{r+1}(x)$ the same numerical value with different signs. In passing from a value a little less than a to one a little greater, we can suppose the interval so small that it contains no root of $f_{r-1}(x)$ or $f_{r+1}(x)$; hence, throughout the interval under consideration, these two functions retain their signs. The sign of $f_r(x)$ changes; but no variation of sign is either lost or gained thereby in the group of three; because, on account of the difference of signs of the two extremes $f_{r-1}(x)$ and $f_{r+1}(x)$, there will exist both before and after the passage one variation and one permanency of sign, whatever be the sign of the middle function. If, for example, before the passage the signs were $+ - -$; after the passage they are $+ + -$, *i. e.* a variation and a permanency are changed into a permanency and a variation; but no variation of sign is lost or gained on the whole.

(3). Since our arguments in the two preceding cases are founded on the relation of the function to those adjacent to it only; and since those relations remain unaltered in the present case, because no two adjacent functions can vanish together; we conclude that if $f(x)$ is one of the vanishing functions, one change of sign is lost, and if $f(x)$ is not one of them, no change is either lost or gained.

We have proved, therefore, that when x passes through a root of $f(x) = 0$ one change of sign is lost, and under no other circumstances is a change of sign either lost or gained. Hence the number of changes of sign lost during the variation of x from a to b is equal to the number of roots of the equation between a and b.[*]

Before proceeding to the case of equal roots, we add a few simple examples to illustrate the application of Sturm's theorem. It is convenient in practice to substitute first $-\infty$, 0, $+\infty$ in Sturm's functions, so as to obtain the whole number of negative and of positive roots. To separate the negative roots, the integers -1, -2, -3, &c., are to be substituted in succession till we reach the same series of signs as results from the substitution of $-\infty$; and to separate the positive roots we substitute 1, 2, 3, &c., till the signs furnished by $+\infty$ are reached.

EXAMPLES.

1. Find the number and situation of the real roots of the equation

$$f(x) = x^3 - 2x - 5 = 0.$$

We find $\quad f_1(x) = 3x^2 - 2, \quad f_2(x) = 4x + 15, \quad f_3(x) = -643.$

Corresponding to the values $-\infty$, 0, $+\infty$ of x, we have

$$
\begin{array}{llll}
(-\infty) & \quad - \ + \ - \ -, \\
(0) & \quad - \ - \ + \ -, \\
(+\infty) & \quad + \ + \ + \ -.
\end{array}
$$

Hence there is only one real root, and it is positive.

---

[*] The student often finds a difficulty in perceiving in what way a number of changes of sign can be lost in Sturm's series, since the only loss of sign takes place between the first two functions, $f(x)$ and $f_1(x)$. To remove this difficulty we observe, that as x increases from one root a of $f(x) = 0$ to a second b, although no alteration takes place in the *number* of changes of sign, the *distribution* of the signs among $f_1(x)$ and the following functions alters in such a way that the signs of $f(x)$ and $f_1(x)$, which were the same immediately after the passage of x through a, become again different immediately before the passage through b.

N

Again, corresponding to values 1, 2, 3 of x, we have

$$
\begin{array}{llcccc}
(1) & & - & + & + & -, \\
(2) & & - & + & + & -, \\
(3) & & + & + & + & -.
\end{array}
$$

The real root, therefore, lies between 2 and 3.

2. Find the number and situation of the real roots of the equation

$$x^3 - 7x + 7 = 0.$$

We easily find

$$
\begin{aligned}
f_1(x) &= 3x^2 - 7, \\
f_2(x) &= 2x - 3, \\
f_3(x) &= 1;
\end{aligned}
$$

whence

$$
\begin{array}{lcccc}
(-\infty) & - & + & - & +, \\
(0) & + & - & - & +, \\
(+\infty) & + & + & + & +.
\end{array}
$$

Hence all the roots are real: one negative, and two positive.

We have, further, for the following values of x:—

$$
\begin{array}{lcccc}
(-4) & - & + & - & +, \\
(-3) & + & + & - & +, \\
(-2) & + & + & - & +, \\
(-1) & + & - & - & +, \\
(1) & + & - & - & +, \\
(2) & + & + & + & +.
\end{array}
$$

Here -4 and $+2$ give the same series of signs as $-\infty$ and $+\infty$; hence we stop at these. The negative root lies between -4 and -3; and the two positive roots between 1 and 2.

This example illustrates the superiority of Sturm's method over that of Fourier.

The substitution of 1 and 2 in Fourier's functions gives, as can be immediately verified, the following series of signs:—

$$
\begin{array}{lcccc}
(1) & + & - & + & +, \\
(2) & + & + & + & +.
\end{array}
$$

From Fourier's theorem we are authorised to conclude only that there *cannot be more than* two roots between 1 and 2. From Sturm's we conclude that there *are* two roots between 1 and 2. If we have occasion to separate these two roots, we must, of course, make further substitutions in $f(x)$.

3. Find the number and situation of the real roots of the equation

$$x^4 - 2x^3 - 3x^2 + 10x - 4 = 0.$$

We obtain, removing the factor 2 from the derived,

$$f_1(x) = 2x^3 - 3x^2 - 3x + 5,$$
$$f_2(x) = 9x^2 - 27x + 11,$$
$$f_3(x) = -8x - 3,$$
$$f_4(x) = -1433.$$

[N. B.—In forming Sturm's functions we may, as is evident from the equations (1), Art. 89, introduce or suppress numerical factors just as in the process of finding the G. C. M.; taking care, however, that these are *positive*, so that the signs of the remainders are not thereby altered.]

We obtain the following series of signs :—

$$(-\infty) \qquad + \ - \ + \ + \ -,$$
$$(0) \qquad - \ + \ + \ - \ -,$$
$$(+\infty) \qquad + \ + \ + \ - \ -.$$

Hence there are two real roots, one positive, and one negative; and two imaginary roots. To find the position of the real roots, it is sufficient to substitute positive and negative integers successively in $f(x)$ alone, since there is only *one* positive, and *one* negative root; and we easily find that the negative root lies between -2 and -3, and the positive root between 0 and 1.

90. Sturm's Theorem.   Equal Roots.—Let the operation for finding the greatest common measure of $f(x)$ and $f'(x)$ be performed, the signs of the successive remainders being changed as before. The last of Sturm's functions will not now be numerical, for since $f(x)$ and $f'(x)$ are here supposed to contain a common measure involving x, this will now be the last function arrived at by the process. Let the series of functions be :—

$$f(x), \ f_1(x), \ f_2(x), \ \ldots \ldots \ldots, \ f_r(x).$$

During the passage of x through any value except a multiple root of $f(x) = 0$, the conclusions of the last Article are still true with respect to the present series, since no value except such a root can cause any consecutive pair of the series to vanish. When x passes through a multiple root of $f(x) = 0$, there is, by the Cor., Art. 75, one change of sign lost between f and f_1; and we proceed to prove that no change of sign is lost or gained in the rest of the series, i. e. $f_1, f_2, \ldots . f_r$. Suppose there exists an m-multiple root a of $f(x)$. It is evident from the equations (1) of Art. 89, that $(x-a)^{m-1}$ is a factor in each of the functions $f_1, f_2, \ldots f_r$. Let the remaining factors in these functions be, respectively, $\phi_1, \phi_2, \ldots \phi_r$. By dividing each of the equations (1) by $(x-a)^{m-1}$, we get a series of equations which establish by the reasoning of the last Article that, owing to a passage through a, no change of signs is lost or gained in the series $\phi_1, \phi_2, \ldots \phi_r$. Neither, therefore, is any change lost or gained in the series $f_1, f_2, \ldots f_r$; for the effect of the factor $(x-a)^{m-1}$ in the passage of x from a value $a-h$ to a value $a+h$ is either to change the signs of all (when $m-1$ is odd) or of none (when $m-1$ is even) of the functions $\phi_1, \phi_2, \ldots \phi_r$; and changing the signs of all these functions cannot increase or diminish the number of variations.

We have, therefore, proved that when x passes through a multiple root of $f(x) = 0$ one change of sign is lost between f and f_1, and none either lost or gained in any other part of the series. It remains true, of course, that when x passes through a single root of $f(x) = 0$ a change of sign is lost as before. We may thus state the theorem as follows for the case of equal roots :—

The difference between the number of changes of sign when a and b are substituted in the series

$$f, \; f_1, \; f_2, \; \ldots \ldots f_r,$$

the last of these being the greatest common measure of f and f_1, is equal to the number of real roots between a and b, each multiple root counting only once.

EXAMPLES.

1. Find the nature of the roots of the equation

$$x^4 - 5x^3 + 9x^2 - 7x + 2 = 0.$$

We easily obtain

$$f_1(x) = 4x^3 - 15x^2 + 18x - 7,$$

$$f_2(x) = x^2 - 2x + 1 ;$$

$f_2(x)$ divides $f_1(x)$ without remainder; hence in this case Sturm's series stops at $f_2(x)$, thus establishing the existence of equal roots.

To find the number of real roots of the equation, we substitute $-\infty$ and $+\infty$ for x in the series of functions f, f_1, f_2. The result is

$$(-\infty) \quad + \quad - \quad +,$$
$$(+\infty) \quad + \quad + \quad +.$$

Hence the equation has only two real distinct roots; but one of these is a triple root, as is evident from the form of $f_2(x)$, which is equal to $(x-1)^2$.

2. Find the nature of the roots of the equation

$$x^4 - 6x^3 + 13x^2 - 12x + 4 = 0,$$
$$f_1(x) = 4x^3 - 18x^2 + 26x - 12,$$
$$f_2(x) = x^2 - 3x + 2 ;$$

$f_2(x)$ is the last Sturmian function; so the equation has equal roots.

$$(-\infty) \quad + \quad - \quad +,$$
$$(+\infty) \quad + \quad + \quad +.$$

There are only two real distinct roots. In fact, since $f_2(x) \equiv (x-1)(x-2)$, each of the roots 1, 2 is a double root.

3. Find the nature of the roots of the equation

$$x^5 + 2x^4 + x^3 - x^2 - 2x - 1 = 0.$$

Here

$$f_1 = 5x^4 + 8x^3 + 3x^2 - 2x - 2,$$
$$f_2 = 2x^3 + 7x^2 + 12x + 7,$$
$$f_3 = - x^2 - 6x - 5,$$
$$f_4 = - x - 1,$$
$$f_5 = 0.$$

Since $f_5 = 0$, $x + 1$ is a common measure of f and f_1, and $f(x)$ has a double root -1. We have also

$$(-\infty) \quad - \quad + \quad - \quad - \quad +,$$
$$(+\infty) \quad + \quad + \quad + \quad - \quad -.$$

Hence there are two real distinct roots. The equation has, therefore, beside the double root, one other real root, and two imaginary roots.

4. Find the nature of the roots of the equation

$$x^6 - 7x^5 + 15x^4 - 40x^2 + 48x - 16 = 0.$$

Here

$$f_1(x) = 6x^5 - 35x^4 + 60x^3 - 80x + 48,$$
$$f_2(x) = 13x^4 - 84x^3 + 192x^2 - 176x + 48,$$
$$f_3(x) = x^3 - 6x^2 + 12x - 8 \equiv (x-2)^3.$$

Ans. There are three real distinct roots, one of them being quadruple.

91. **Application of Sturm's Theorem.**—In the case of equations of high degrees the calculation of Sturm's auxiliary functions becomes often very laborious. It is important, therefore, to pay attention to certain observations which tend somewhat to diminish this labour.

(1). In calculating the final remainder when it is numerical, since its sign is all we are concerned with, the labour of the last operation of division can be evaded by the consideration that the value of x which causes f_{n-1} to vanish must give opposite signs to f_{n-2} and f_n. It is in general possible to tell without any calculation what would be the sign of the result if the root of $f_{n-1}(x) = 0$ were substituted in $f_{n-2}(x)$. Thus in Ex. 3, Art. 89, if the value $-\dfrac{3}{8}$, which is the root of $f_3(x) = 0$, be substituted for x in $9x^2 - 27x + 11$, the result is evidently positive; hence the sign of $f_n(x)$ is $-$, and there is no occasion to calculate the value -1433 given for $f_n(x)$ in the example in question.

(2). When it is possible in any way to recognize that all the roots of any one of Sturm's functions are imaginary, we need not proceed to the calculation of any function beyond that one; for since that function retains constantly the same sign for all values of the variable (Cor. Art. 12), no alteration in the number of changes of sign presented by it and the following functions can ever take place, so that the difference in the number of changes when two quantities a and b are substituted is independent of whatever variations of sign may exist in that part of the series which consists of the function in question and those following it.

With a view to the application of this observation it is always well, when we arrive at the quadratic function ($ax^2 + bx + c$, suppose), to examine, in case the term containing x^2 and the absolute term have the same sign (otherwise the roots could not be imaginary), whether the condition $4ac > b^2$ is fulfilled; if so, we know that the roots are imaginary, and the calculation need not proceed farther.

EXAMPLES.

1. Analyse the equation

$$x^4 + 3x^3 + 7x^2 + 10x + 1 = 0.$$

We find

$$f_2(x) = -29x^2 - 78x + 14,$$

$$f_3(x) = -1086x - 481,$$

$$f_4(x) = \qquad -\ .$$

Here we see immediately that the value of x given by the equation $f_3(x) = 0$, which differs little from $-\frac{1}{2}$, makes $f_2(x)$ positive; hence $f_4(x)$ is negative. There are two real, and two imaginary roots. The real roots lie in the intervals $\{-2, -1\}$, $\{-1, 0\}$.

2. Analyse the equation

$$x^4 - 4x^3 - 3x + 23 = 0.$$

We find

$$f_2(x) = 12x^2 + 9x - 89,$$

$$f_3(x) = -491x + 1371,$$

$$f_4(x) = \qquad -\ .$$

Here $f_3(x) = 0$ gives $x = \dfrac{1371}{491} > \dfrac{1371}{500} > 2 \cdot 74 > \dfrac{5}{2}$, and $x = \dfrac{5}{2}$ makes $f_2(x)$ positive; hence the root of $f_3(x)$ makes it positive also.

There are two real and two imaginary roots.

The real roots lie in the intervals $\{2, 3\}$, $\{3, 4\}$.

3. Analyse the equation

$$2x^4 - 13x^2 + 10x - 19 = 0.$$

Here

$$f_1(x) = 4x^3 - 13x + 5,$$

$$f_2(x) = 13x^2 - 15x + 38.$$

Since $4 \times 13 \times 38 > 15^2$, the roots of $f_2(x)$ are imaginary, and we proceed no farther with the calculation of Sturm's remainders.

Substituting $-\infty$, 0, $+\infty$, we obtain

$$
\begin{array}{cccc}
(-\infty) & + & - & +, \\
(0) & - & + & +, \\
(+\infty) & + & + & +.
\end{array}
$$

There are two real roots, one positive, the other negative.

4. Analyse the equation

$$f(x) = x^5 + 2x^4 + x^3 - 4x^2 - 3x - 5 = 0.$$

Here

$$f_1(x) = 5x^4 + 8x^3 + 3x^2 - 8x - 3,$$

$$f_2(x) = 6x^3 + 66x^2 + 44x + 119,$$

$$f_3(x) = -116x^2 - 57x - 223,$$

Since $4 \times 116 \times 223 > 57^2$,   we may stop the calculation here.   We find, on substituting $-\infty$,   0, $+\infty$,

$$(-\infty) \qquad - \quad + \quad - \quad -,$$

$$(0) \qquad - \quad - \quad + \quad -,$$

$$(+\infty) \qquad + \quad + \quad + \quad -.$$

There are four imaginary roots, and one real positive root.

5. Find the number and situation of the real roots of the equation

$$x^4 - 2x^3 - 7x^2 + 10x + 10 = 0.$$

> *Ans.* The roots are all real, and are situated in the intervals
> $\{-3, -2\}$, $\{-1, 0\}$, and two between $\{2, 3\}$.

6. Analyse the equation

$$x^5 + 3x^4 + 2x^3 - 3x^2 - 2x - 2 = 0.$$

It will be found that the calculation may cease with the quadratic remainder.

> *Ans.* There is only one real root; in the interval $\{1, 2\}$.

7. Analyse the equation

$$x^3 + 11x^2 - 102x + 181 = 0.$$

We find

$$f_2(x) = 854x - 2751,$$

$$f_3(x) = 441.$$

In some examples, of which the present is an instance, it is not easy to tell immediately what sign the root of the penultimate function gives to the preceding function. We have here calculated $f_3(x)$, and it turns out to be a much smaller figure than might have been expected from the magnitude of the coefficients in $f_2(x)$. In fact when the root of $f_2(x)$ is substituted in $f_1(x)$ the positive part is nearly equal to the negative part. This is always an indication that *two roots of the proposed equation are nearly equal*. There are in the present instance two positive roots between 3 and 4. Subdividing the intervals, we find the two roots still to lie between $3 \cdot 2$ and $3 \cdot 3$; so that they are very close together. We see here another illustration of the continuity which exists between real and imaginary roots. If $f_3(x)$ turned out to be zero, the roots would be actually equal. If it turned out to be small negative number, the two nearly equal roots would be imaginary.

8. Analyse the equation

$$x^5 + x^4 + x^3 - 2x^2 + 2x - 1 = 0.$$

The quadratic function is found to have imaginary roots.

> *Ans.* One real root between $\{0, 1\}$; four imaginary roots.

9. Analyse the equation

$$x^6 - 6x^5 - 30x^2 + 12x - 9 = 0.$$

We find

$$f_2(x) = 5x^4 + 20x^2 + 7 ;$$

and as this has plainly all imaginary roots, the calculation may stop here.

Ans. Two real roots; in the intervals $\{-2, -1\}$, $\{6, 7\}$.

10. Analyse the equation

$$2x^6 - 18x^5 + 60x^4 - 120x^3 - 30x^2 + 18x - 5 = 0.$$

We find

$$f_2(x) = 5x^4 + 220x^2 + 1 ;$$

and the calculation may stop.

Ans. Two real roots; in the intervals $\{-1, 0\}$, $\{5, 6\}$.

11. Examine how the roots of the equation

$$2x^3 + 15x^2 - 84x - 190 = 0$$

are situated in the several intervals between the numbers $-\infty$, -7, 6, $+\infty$.

Here

$$f_1(x) = x^2 + 5x - 14,$$

$$f_2(x) = 27x + 40,$$

$$f_3(x) = +.$$

The substitution of the above quantities gives

$(-\infty)$	$-$	$+$	$-$	$+,$
(-7)	$+$	0	$-$	$+,$
(6)	$+$	$+$	$+$	$+,$
$(+\infty)$	$+$	$+$	$+$	$+.$

Whenever, as in this example, any quantity makes one of the auxiliary functions vanish (here -7 satisfies $f_1(x) = 0$), the zero may be disregarded in counting the number of changes of sign in the corresponding row; for, since the signs on each side of it are different, no alteration in the number of changes of sign in the row could take place, whatever sign be supposed attached to the vanishing quantity.

The roots are all real. There is one root between $-\infty$ and -7; and two between -7 and 6.

92. Conditions for the Reality of the Roots of an Equation.—The number of Sturm's functions, including $f(x), f'(x)$, and the $n-1$ remainders, will in general be $n+1$. In certain cases, owing to the absence of terms in the proposed function, some of the remainders will be wanting. This can occur only when the proposed equation has imaginary roots; for it is plain that, in order to ensure a loss of n changes of sign in the series of functions during the passage of x from $-\infty$ to $+\infty$ (namely, in order that the equation should have all its roots real), all the functions must be present. And, moreover, they must all take the same sign when $x = +\infty$; and alternating signs when $x = -\infty$. Since the leading term of an equation is always taken with a positive sign, we may state the condition for the reality of all the roots of any equation as follows :—*In order that all the roots of an equation of the nth degree should be real, the leading coefficients of all Sturm's remainders, in number $n-1$, must be positive.*

EXAMPLES.

1. Find the condition that the roots of the equation

$$ax^2 + 2bx + c = 0$$

should be real.

Here

$$f_1(x) = ax + b,$$

$$f_2(x) = b^2 - ac.$$

The condition is, therefore, $b^2 - ac > 0$.

2. Find the conditions that the roots of the cubic

$$z^3 + 3Hz + G = 0$$

should be all real.

When this cubic has its roots all real, it is evident that the general binomial cubic from which it is derived (Art. 37) has also its roots all real; so that, in investigating the conditions for the reality of the roots of a cubic in general, it is sufficient to discuss the form here written.

We find

$$f_1(z) = z^2 + H,$$

$$f_2(z) = -2Hz - G,$$

$$f_3(z) = -(G^2 + 4H^3).$$

[In calculating these, before dividing $f_1(z)$ by $f_2(z)$ multiply the former by the positive factor $2H^2$.]

Hence the two conditions are

H negative, $G^2 + 4H^3$ negative.

These can be expressed as one condition, i. e. $G^2 + 4H^3$ negative, since this implies the former (cf. Art. 42).

3. Calculate Sturm's remainders for the biquadratic

$$z^4 + 6Hz^2 + 4Gz + a^2I - 3H^2 = 0.$$

We find

$$f_2(z) = -3Hz^2 - 3Gz - (a^2I - 3H^2),$$

$$f_3(z) = -(2HI - 3aJ)z - GI,$$

$$f_4(z) = I^3 - 27J^2.$$

These are obtained without much difficulty by aid of the relation

$$G^2 + 4H^3 = a^2(HI - aJ).$$

Before dividing f_1 by f_2, multiply by the positive factor $3H^2$; and when the remainder is found, remove the positive factor a^2. Before dividing f_2 by f_3, multiply by the positive factor $(2HI - 3aJ)^2$; and when the remainder is found, remove the positive factor a^2H^2.

93. Criterion of the Nature of the Roots of the Biquadratic.—We are now in a position to resume the discussion in Art. 68, and to give complete criteria of the nature of the roots of the general algebraic equation of the fourth degree. For this purpose it is sufficient to consider the biquadratic of Ex. 3 in the last Article. The leading coefficients of Sturm's remainders are

$$-H, \quad -(2HI - 3aJ), \quad I^3 - 27J^2.$$

When the roots are all real we must have, in addition to the condition Δ positive, the functions H and $2HI - 3aJ$ *both negative*. If either, or both, of these be positive when Δ is positive, the roots will be all imaginary. The student will have no difficulty in verifying, by means of Sturm's remainders, the conclusions of cases (1), (3), and (4) of Art. 68.

EXAMPLES.

1. Prove that when the biquadratic

$$ax^4 + 4bx^3 + 6cx^2 + 4dx + e$$

has a triple root, it consists of the factors

$$\{ ax + b + \sqrt{-H} \}^3 \{ ax + b - 3\sqrt{-H} \}.$$

2. Prove that when $\Delta = 0$, $G = 0$, and $2HI - 3aJ = 0$, the biquadratic has two distinct pairs of equal roots, and that it is then the square of the quadratic

$$(ax + b)^2 + 3H;$$

and verify the conditions of Ex. 12, p. 139.

3. Show that the conditions that all the roots of a biquadratic should be equal are $H = 0$, $I = 0$, $J = 0$; or, otherwise expressed,

$$\frac{a}{b} = \frac{b}{c} = \frac{c}{d} = \frac{d}{e}.$$

Miscellaneous Examples.

1. Determine the number and position of the real roots of the equation

$$x^4 - 12x^3 + 13x^2 + 24x - 30 = 0.$$

2. Determine the number and signs of the real roots of the equation

$$x^4 - 5x^3 + 10x^2 - 5x - 21 = 0.$$

3. Apply Sturm's theorem to the analysis of the equation

$$x^4 - 4x^3 + 7x^2 - 6x - 4 = 0.$$

4. Find the number and position of the real roots of

$$x^5 - 10x^3 + 6x + 1 = 0.$$

5. Prove that the roots of the equation

$$x^3 - (a^2 + b^2 + c^2)\, x - 2abc = 0$$

are all real, and solve it when two of the quantities a, b, c become equal.

6. If, in the following, the sequences of signs are those of the leading coefficients of Sturm's remainders for a biquadratic, prove

$$+ \ + \ +\quad \text{four real roots;}\qquad \left.\begin{array}{ccc} + & + & - \\ + & - & - \\ - & - & - \end{array}\right\}\ \text{two real roots;}$$

$$\left.\begin{array}{ccc} - & + & + \\ + & - & + \\ - & - & + \end{array}\right\}\ \text{no real root;}\qquad - \ + \ -\quad \text{cannot occur.}$$

7. If the signs of the leading coefficients of the first two of Sturm's remainders for a quintic be $- +$, prove that the number of real roots is determined.

Ans. One real root only.

8. If H and J are both positive, prove that all the roots of the binomial biquadratic are imaginary; and that under the same conditions the binomial quintic has only one real root. Mr. M. Roberts, *Dublin Exam. Papers*, 1862.
Compare Ex. 11, p. 139.

9. Prove that, if c has any value except unity, the equation

$$c^2 x^4 - 2c^2 x^3 + 2x - 1 = 0$$

has a pair of imaginary roots.

10. Apply Budan's method to separate the roots of the equation

$$x^4 - 16x^3 + 69x^2 - 70x - 42 = 0.$$

Ans. Roots in intervals $\{-1,\ 0\}$, $\{2,\ 3\}$, $\{4,\ 5\}$, $\{9,\ 10\}$.

CHAPTER X.

94. Algebraical and Numerical Equations.—There is an essential distinction between the solutions of algebraical and numerical equations. In the former the result is a general formula of a purely symbolical character, which, being the general expression for a root, must represent all the roots indifferently. It must be such that, when for the functions of the coefficients involved in it the corresponding symmetric functions of the roots are substituted, the operations represented by the radical signs $\sqrt{\ }$, $\sqrt[3]{\ }$ become practicable; and when the square and cube roots of these symmetric functions are extracted, the whole expression in terms of the roots will reduce down to one root: the different roots resulting from the different combinations $\pm \sqrt{\ }$ of square roots, and $\sqrt[3]{\ }$, $\omega \sqrt[3]{\ }$, $\omega^2 \sqrt[3]{\ }$ of cube roots. For a simple illustration of this we refer to the case of the quadratic in Art. 55. In Articles 59 and 66 we have similar illustrations for the cubic and biquadratic. It is to be observed, also, that the formula which represents the root of an algebraic equation holds good even when the coefficients are imaginary quantities.

In the case of numerical equations the roots are determined separately by the methods we are about to explain; and before attempting the approximation to any individual root, it is in general essential to know that it is situated in an interval which contains no other real root.

The real roots of numerical equations may be either commensurable or incommensurable; the former class including integers, fractions, and terminating or repeating decimals which are reducible to fractions; the latter consisting of interminable decimals. The roots of the former class can be found exactly; and those of the latter approximated to with any degree of accuracy we choose, by the methods we are about to explain.

We shall commence by establishing a theorem which enables us to reduce the determination of the former class of roots to that of *integral roots* alone.

95. **Theorem.**—*An equation in which the coefficient of the first term is unity, and the coefficients of the other terms whole numbers, cannot have a commensurable root which is not a whole number.*

For, if possible, let $\dfrac{a}{b}$, a fraction in its lowest terms, be a root of the equation

$$x^n + p_1 x^{n-1} + p_2 x^{n-2} + \ldots + p_{n-1} x + p_n = 0 \, ;$$

we have then

$$\left(\frac{a}{b}\right)^n + p_1 \left(\frac{a}{b}\right)^{n-1} + \ldots + p_{n-1} \left(\frac{a}{b}\right) + p_n = 0 \, ;$$

from which, multiplying by b^{n-1}, we obtain

$$-\frac{a^n}{b} = p_1 a^{n-1} + p_2 a^{n-2} b + \ldots + p_{n-1} a b^{n-2} + p_n b^{n-1} :$$

now a^n is not divisible by b, and each term on the right-hand side of the equation is an integer. We have, therefore, a fraction in its lowest terms equal to an integer, which is impossible. Hence $\dfrac{a}{b}$ cannot be a root of the equation. The real roots of the equation, therefore, are either integers or incommensurable quantities.

Every equation whose coefficients are finite numbers, fractional or not, can be reduced to the form in which the coefficient of the first term is unity, and those of the other terms whole numbers (Art. 31); so that in this way, by the aid of a simple transformation, the determination of the commensurable roots in general can be reduced to that of integral roots.

We proceed to explain Newton's process, called the Method of Divisors, of obtaining the integral roots of an equation whose coefficients are all integers.

96. Newton's Method of Divisors.—Suppose h to be an integral root of the equation

$$a_0 x^n + a_1 x^{n-1} + \ldots + a_{n-1} x + a_n = 0. \tag{1}$$

Let the quotient, when the polynomial is divided by $x - h$, be

$$b_0 x^{n-1} + b_1 x^{n-2} + \ldots + b_{n-2} x + b_{n-1},$$

in which b_0, b_1, &c., are plainly all integers.

Proceeding as in Art. 8, we obtain the following equations:—

$$a_0 = b_0, \quad a_1 = b_1 - h b_0, \quad a_2 = b_2 - h b_1, \ldots$$

$$a_{n-2} = b_{n-2} - h b_{n-3}, \quad a_{n-1} = b_{n-1} - h b_{n-2}, \quad a_n = - h b_{n-1}.$$

The last of these equations proves that a_n is divisible by h, the quotient being $- b_{n-1}$. The second last, which is the same as

$$a_{n-1} + \frac{a_n}{h} = - h b_{n-2},$$

proves that the sum of the quotient thus obtained and the second last coefficient is again divisible by h, the quotient being $- b_{n-2}$; and so on.

Continuing the process, the last quotient obtained in this way will be $- b_0$, which is equal to $- a_0$.

If we perform the process here indicated with all the divisors of a_n which lie within the limits of the roots, those which satisfy the above conditions, giving integral quotients at each step, and a final quotient equal to $- a_0$, are roots of the proposed equation. Those which at any stage of the process give a fractional quotient are to be rejected.

When the coefficient $a_0 = 1$, we know by the theorem of the last Article that the integral roots determined in this way are all the commensurable roots of the proposed equation. If a_0 be not $= 1$, the process will still give the integral roots of the equation as it stands; but to be sure of determining in this way all the commensurable roots, the equation must be first transformed to one which shall have the coefficient of the highest term equal to unity.

97. **Application of the Method of Divisors.**—With a view to the most convenient mode of applying the Method of Divisors, we write the series of operations as follows, in a manner analogous to Art. 8 :—

$$a_n \qquad a_{n-1} \qquad a_{n-2} \ldots \ldots a_2 \qquad a_1 \qquad a_0$$

$$-b_{n-1} \quad -b_{n-2} \qquad -b_2 \quad -b_1 \quad -b_0$$

$$-h\,b_{n-2} \ -h\,b_{n-3} \qquad -h\,b_1 \ -h\,b_0 \ldots 0$$

The first figure in the second line $(-b_{n-1})$ is obtained by dividing a_n by h. This is to be added to a_{n-1} to obtain the first figure in the third line $(-h\,b_{n-2})$. This is to be divided by h to obtain the second figure in the second line $(-b_{n-2})$; this to be added to a_{n-2}; and so on. If h be a root, the last figure in the second line thus obtained will be $-a_0$.

When we succeed in proving in this manner that any integer h is a root, our next operation with any divisor may be performed, not on the original coefficients a_n, a_{n-1}, ..., but on those of the second line with their signs changed, for these are the coefficients of the quotient when the original polynomial is divided by $x-h$. When any divisor gives at any stage a fractional result it is to be rejected at once, and the operation so far as it is concerned stopped.

The numbers 1 and -1, which are always of course integral divisors of a_n, need not be included in the number of trial divisors. It is more convenient to determine by direct substitution whether either of them is a root.

EXAMPLES.

1. Find the integral roots of the equation

$$x^4 - 2x^3 - 13x^2 + 38x - 24 = 0.$$

By grouping the terms (see Art. 79) we observe without difficulty that all the roots lie between -5 and $+5$. The following divisors are possible roots :—

$$-4, \quad -3, \quad -2, \quad 2, \quad 3, \quad 4.$$

We commence with 4 :

$$
\begin{array}{rrrrr}
-24 & 38 & -13 & -2 & 1 \\
 & -6 & 8 & & \\
\hline
 & 32 & -5 & &
\end{array}
$$

The operation stops here, for since -5 is not divisible by 4, 4 cannot be a root. We proceed then with the number 3.

$$
\begin{array}{rrrrr}
-24 & 38 & -13 & -2 & 1 \\
 & -8 & 10 & -1 & -1 \\
\hline
 & 30 & -3 & -3 & 0 \, ;
\end{array}
$$

hence 3 is a root; and in proceeding with the next integer, 2, we make use, as above explained, of the coefficients of the second line with signs changed :

$$
\begin{array}{rrrr}
8 & -10 & 1 & 1 \\
 & 4 & -3 & -1 \\
\hline
 & -6 & -2 & 0 \, ;
\end{array}
$$

hence 2 also is a root; and we proceed with -2 ;

$$
\begin{array}{rrr}
-4 & 3 & 1 \\
 & 2 & \\
\hline
 & 5 & ;
\end{array}
$$

hence -2 is not a root, for it does not divide 5. -3 is plainly not a root, for it does not divide -4.

[We might at once have struck out -3 as not being a divisor of the absolute term 8 of the reduced polynomial. This remark will often be of use in diminishing the number of divisors.]

o

We proceed, then, with the last divisor, -4.

$$
\begin{array}{ccc}
-4 & 3 & 1 \\
& 1 & -1 \\
\hline
& 4 & 0
\end{array}
$$

Thus -4 is a root.

The equation has, therefore, the integral roots $3, 2, -4$; and the last stage of the operation shows that when the original polynomial is divided by the binomials $x-3$, $x-2$, $x+4$, the result is $x-1$; so that 1 is also a root. Hence the original polynomial is equivalent to

$$(x-1)(x-2)(x-3)(x+4).$$

2. Find the integral roots of

$$3x^4 - 23x^3 + 35x^2 + 31x - 30 = 0.$$

The roots lie between -2 and 8; hence we have only to test the divisors 2, 3, 5, 6.

We find immediately that 6 is not a root.

For 5 we have

$$
\begin{array}{ccccc}
-30 & 31 & 35 & -23 & 3 \\
& -6 & 5 & 8 & -3 \\
\hline
& 25 & 40 & -15 & 0 \; ;
\end{array}
$$

hence 5 is a root. For 3 we have

$$
\begin{array}{cccc}
6 & -5 & -8 & 3 \\
& 2 & -1 & -3 \\
\hline
& -3 & -9 & 0 \; ;
\end{array}
$$

hence 3 is a root; and we easily find that 2 is not a root.

The quotient, when the original polynomial is divided by $(x-5)(x-3)$, is, from the last operation,

$$3x^2 + x - 2:$$

of this, 1 is not a root, and -1 is a root. Hence all the integral roots of the proposed equation are $-1, 3, 5$.

The other root of the equation is $\frac{2}{3}$. It is a commensurable root; but, not being integral, is not given in the above operation.

3. Find all the roots of the equation

$$x^4 + x^3 - 2x^2 + 4x - 24 = 0.$$

Limits of the roots are $-4, 3$.

$$\textit{Ans.}\ \text{Roots} - 3,\ 2,\ \pm 2\sqrt{-1}.$$

4. Find all the roots of the equation

$$x^4 - 2x^3 - 19x^2 + 68x - 60 = 0.$$

The roots lie between -6 and 6.

We find that $2, 3, -5$ are roots, and that the factor left after the final division is $x - 2$; hence 2 is a double root. The polynomial is therefore equivalent to

$$(x - 2)^2 (x - 3) (x + 5).$$

In Art. 99 the case of multiple roots will be further considered.

98. Method of Limiting the Number of Trial Divisors.—It is possible of course to determine by direct substitution whether any of the divisors of a_n are roots of the proposed equation; but Newton's Method has the advantage, as the above examples show, that some of the divisors are rejected after very little labour. It has a further advantage which will now be explained. When the number of divisors of a_n within the limits of the roots is large, it is important to be able to diminish the number of these divisors which need be tested. This can be done as follows :—

If h is an integral root of $f(x) = 0$, then $f(x)$ is divisible by $x - h$, and the coefficients of the quotient are integers, as was above explained. If then we assign to x any integral value, the quotient of the corresponding value of $f(x)$ by the corresponding value of $x - h$ must be an integer. We take, for convenience, the simplest integers 1 and -1 ; and, before testing any divisor h, we subject it to the condition that $f(1)$ must be divisible by $1 - h$ (or, changing the sign, by $h - 1$) ; and that $f(-1)$ must be divisible by $-1 - h$ (or, changing the sign, by $1 + h$).

In applying this observation we first calculate $f(1)$ and $f(-1)$; if either of these vanishes, the corresponding integer is a root, and we proceed with the operation on the reduced polynomial whose coefficients have been ascertained in the process of finding the result of substituting the integer in question.

EXAMPLES.

1. $x^5 - 23x^4 + 160x^3 - 281x^2 - 257x - 440 = 0.$

The roots lie between -1 and 24.
We have the following divisors :—

$$2, \quad 4, \quad 5, \quad 8, \quad 10, \quad 11, \quad 20, \quad 22.$$

We easily find

$$f(1) = -840, \quad \text{and} \quad f(-1) = -648.$$

We, therefore, exclude all the above divisors which, when diminished by 1, do not divide 840 ; and which, when increased by 1, do not divide 648. The first condition excludes 10 and 20, and the second 4 and 22. Trying the remaining integers 2, 5, 8, 11 by the method of last Article, we find that 5, 8, and 11 are roots, and that the resulting quotient is $x^2 + x + 1$. Hence the given polynomial is equivalent to

$$(x - 5)(x - 8)(x - 11)(x^2 + x + 1).$$

2. $x^5 - 29x^4 - 31x^3 + 31x^2 - 32x + 60 = 0.$

The roots lie between -3 and 32.

Divisors : $-2, \ 2, \ 3, \ 4, \ 5, \ 6, \ 10, \ 12, \ 15, \ 20, \ 30,$

$f(1) = 0$; so 1 is a root.

$f(-1) = 124$; and the above condition excludes all the divisors except $-2, 3, 30$.

We easily find that -2 and 30 are roots, and that the final quotient is $x^2 + 1$. The given polynomial is equivalent to $(x-1)(x-30)(x+2)(x^2+1).$

99. Determination of Multiple Roots.—The Method of Divisors determines multiple roots when they are commensurable. In applying the method, when any divisor of a_n which is found to be a root is a divisor of the absolute term of the reduced polynomial, we must proceed to try whether it is also a root of the latter, in which case it will be a double root of the proposed equation. If it be found to be a root of the next reduced polynomial, it will be a triple root of the proposed ; and so on. Whenever in an equation of any degree there exists only *one* multiple root, r times repeated, it can be found in this way ; for the common measure of $f(x)$ and $f'(x)$ will then be of the form $(x - a)^{r-1}$, and the coefficients of this could not be commensurable if a were incommensurable.

Multiple roots of equations of the third, fourth, and fifth degrees can be completely determined without the use of the process of finding the greatest common measure, as will appear from the following observations :—

(1). *The Cubic.*—In this case multiple roots must be commensurable, since the degree is not high enough to allow of two distinct roots being repeated.

(2). *The Biquadratic.*—In this case, either the multiple roots are commensurable or the function is a perfect square. For the only way in which two distinct roots can be repeated is when the biquadratic is of the form $(x - a)^2 (x - \beta)^2$, *i.e.* the square of a quadratic. The roots of the quadratic may be incommensurable. If we find, therefore, that a biquadratic has no commensurable roots, we must try whether it is a perfect square in order to determine further whether it has equal incommensurable roots.

(3). *The Quintic.*—In this case, either the multiple roots are commensurable, or the function consists of a linear commensurable factor multiplied by the square of a quadratic factor. For, the only way in which two distinct roots can be repeated is when the function is of either of the forms

$$(x - a)^2 (x - \beta)^2 (x - \gamma), \quad (x - a)^2 (x - \beta)^3 :$$

in the latter case the roots cannot be incommensurable; but the former may correspond to the case of a commensurable factor multiplied by the square of a quadratic whose roots are incommensurable. If then a quintic be found to have no commensurable roots it can have no multiple roots. If it be found to have one commensurable root only, we must examine whether the remaining factor is a perfect square. If it have more than one commensurable root, the multiple roots will be found among the commensurable roots.

EXAMPLES.

1. Find all the commensurable roots of

$$2x^3 - 31x^2 + 112x + 64 = 0.$$

The roots lie between the limits -1, 16.   The divisors are 2, 4, 8.

64	112	-31	2
	8	15	-2
	——	——	—
	120	-16	0 ;

8 is therefore a root.   Proceed now with the reduced equation :

-8	-15	2
	-1	-2
	——	—
	-16	0 ;

8 is a root again, and the remaining factor is $2x + 1$.

$$\textit{Ans. } f(x) \equiv (2x + 1)(x - 8)^2.$$

2. Find the commensurable and multiple roots of

$$x^4 - x^3 - 30x^2 - 76x - 56 = 0.$$

The roots lie between the limits -6, 12.   (Apply method of Ex. 10, Art. 80).

$$\textit{Ans. } f(x) \equiv (x + 2)^3 (x - 7).$$

3. Find the commensurable and multiple roots of

$$9x^4 - 12x^3 - 71x^2 - 40x + 16 = 0.$$

The roots lie between the limits -2, 5.

The equation as it stands is found to have no integral root; but it may still have a commensurable root.   To test this we multiply the roots by 3 in order to get rid of the coefficient of x^4.   We find then

$$x^4 - 4x^3 - 71x^2 - 120x + 144 = 0.$$

Limits : -6, 15.

We find -4 to be a double root of this, and the function to be equivalent to $(x^2 - 12x + 9)(x + 4)^2$.   The original equation is therefore identical with the following :—

$$(x^2 - 4x + 1)(3x + 4)^2 = 0.$$

4. Find the commensurable and multiple roots of

$$x^4 + 12x^3 + 32x^2 - 24x + 4 = 0.$$

The roots lie between -12 and 1. The only divisors to be tested are, therefore, -4, -2, -1. We find that the equation has no commensurable root. We proceed to try whether the given function is a perfect square. This can be done by extracting the square root, or by applying Ex. 2, Art. 93. We find that it is the square of $x^2 + 6x - 2$ (cf. **Ex.** 1, p. 154). Hence the given equation has two pairs of equal roots, both incommensurable.

5. Find the commensurable and multiple roots of

$$f(x) \equiv x^5 - x^4 - 12x^3 + 8x^2 + 28x + 12 = 0.$$

The limits of the roots are -4, 4.

We find that -3 is a root, and that the reduced equation is

$$x^4 - 4x^3 + 8x + 4 = 0,$$

and that there is no other commensurable root.

The only case of possible occurrence of multiple roots is, therefore, when this latter function is a perfect square. It is found to be a perfect square, and we have

$$f(x) \equiv (x^2 - 2x - 2)^2 (x + 3).$$

6. Find the commensurable and multiple roots of

$$f(x) \equiv x^5 - 8x^4 + 22x^3 - 26x^2 + 21x - 18 = 0.$$

$$\textit{Ans. } f(x) \equiv (x^2 + 1)(x - 2)(x - 3)^2.$$

7. The following equation has only two different roots, find them :—

$$x^5 - 13x^4 + 67x^3 - 171x^2 + 216x - 108 = 0.$$

In general it is obvious that if an integral root h occurs twice, the last coefficient must contain h^2 as a factor, and the second last h; if the root occurs three times, h^3 must be a factor of the last, h^2 of the second last, and h of the third last coefficient. The last coefficient here $= 2^2 \cdot 3^3$. Hence, if neither -1 nor 1 is a root, the required roots must be 2 and 3. That these are the roots is easily verified.

8. The equation
$$800x^4 - 102x^2 - x + 3 = 0$$

has equal roots : find all the roots.

In this example it is convenient to change the roots into their reciprocals before applying the Method of Divisors.

$$\textit{Ans. } f(x) \equiv (10x - 3)(5x - 1)(4x + 1)^2.$$

100. Newton's Method of Approximation.—We proceed now to explain certain methods of approximation to the incommensurable roots of equations. The method of the present Article is commonly ascribed to Newton, although one very similar had been employed by Vieta.* The principle on which this method is founded is important in approximations gene-

---

rally, and is not confined in its application to algebraical equations. It will appear, also, that the most convenient practical method of approximating to the real roots of numerical equations (*i. e.* Horner's, Art. 101) is in some degree founded on Newton's principle.

In all methods of approximation the root we are seeking is supposed to be separated from the other roots, and to be situated in a known interval between close limits.

Let $f(x) = 0$ be a given equation, and suppose a value a to be known, differing by a small quantity h from a root of the equation. We have, then, since $a + h$ is a root of the equation, $f(a + h) = 0$; or

$$f(a) + f'(a)\, h + \frac{f''(a)}{1\,.\,2}\, h^2 + \ldots = 0.$$

Neglecting now, since h is small, all powers of h higher than the first, we have

$$f(a) + f'(a)\, h = 0,$$

giving, as a first approximation to the root, the value

$$a - \frac{f(a)}{f'(a)}.$$

Representing this value by b, and applying the same process a second time, we find as a closer approximation

$$b - \frac{f(b)}{f'(b)}.$$

By repeating this process the approximation can be carried to any degree of accuracy required.

EXAMPLE.

Find an approximate value of the positive root of the equation

$$x^3 - 2x - 5 = 0.$$

The root lies between 2 and 3 (Ex. 1, Art. 89). Narrowing the limits, the root is found to lie between 2 and 2·2. We take 2·1 as the quantity represented by a. It cannot differ from the true value $a + h$ of the root by more than 0·1. We find easily

$$\frac{f(a)}{f'(a)} = \frac{f(2{\cdot}1)}{f'(2{\cdot}1)} = \frac{{\cdot}061}{11{\cdot}23} = 0{\cdot}00543.$$

A first approximation is, therefore,

$$2 \cdot 1 - 0 \cdot 00543 = 2 \cdot 0946.$$

Taking this as b, and calculating the fraction $\dfrac{f(b)}{f'(b)}$, we obtain

$$b - \frac{f(b)}{f'(b)} = 2 \cdot 09455148$$

for a second approximation; and so on.

The approximation in Newton's method is, in general, rapid. When, however, the root we are seeking is accompanied by another nearly equal to it, the fraction $\dfrac{f(a)}{f'(a)}$ is not necessarily small, since the value of either of the nearly equal roots reduces $f'(x)$ to a small quantity. A case of this kind requires special precautions. We do not enter into any further discussion of the method, since for practical purposes it may be regarded as entirely superseded by Horner's method, which will now be explained.

101. Horner's Method of Solving Numerical Equations.—By this method both the commensurable and incommensurable roots can be obtained. The root is evolved figure by figure: first the integral part (if any), and then the decimal part, till the root terminates if it be commensurable, or to any number of places required if it be incommensurable. The process is similar to the known processes of extraction of the square and cube root, which are, indeed, only particular cases of the general solution by the present method of quadratic and cubic equations.

The main principle involved in Horner's method is the successive diminution of the roots of the given equation by known quantities, in the manner explained in Art. 34. The great advantage of the method is, that the successive transformations are exhibited in a compact arithmetical form, and the root obtained by one continuous process correct to any number of places of decimals required.

This principle of the diminution of the roots will be illustrated in the present Article by some simple examples. In the following Articles we shall proceed to certain considerations which tend to facilitate the practical application of the method.

EXAMPLES.

1. Find the positive root of the equation

$$2x^3 - 85x^2 - 85x - 87 = 0.$$

The first step, when any numerical equation is proposed for solution, is to find the *first figure* of the root. This can usually be done by a few trials; although in certain cases the methods of separation of the roots explained in Chap. IX. may have to be employed. In the present example there can be only one positive root; and it is found by trial to lie between 40 and 50. Thus the first figure of the root is 4. We now diminish the roots by 40. The transformed equation will have one root between 0 and 10. It is found by trial to lie between 3 and 4. We now diminish the roots of the transformed equation by 3; so that the roots of the proposed equation will be diminished by 43. The second transformed equation will have one root between 0 and 1. On diminishing the roots of this latter equation by ·5, we find that its absolute term is reduced to zero, *i.e.* the diminution of the roots of the proposed equation by 43·5 reduces *its* absolute term to zero. We conclude that 43·5 is a root of the given equation. The series of arithmetical operations is represented as follows :—

```
   2        − 85          − 85           − 87        (43·5
              80          − 200        − 11400
            ──────        ──────       ────────
            − 5           − 285        − 11487
              80          3000           9594
            ──────        ──────       ────────
             75           2715         − 1893
              80           483           1893
            ──────        ──────       ────────
            155           3198             0
              6            501
            ──────        ──────
            161           3699
              6            87
            ──────        ──────
            167           3786
              6
            ──────
            173
              1
            ──────
            174
```

The broken lines mark the conclusion of each transformation, and the figures in dark type are the coefficients of the successive transformed equations (see Art. 34). Thus

$$2x^3 + 155x^2 + 2715x - 11487 = 0$$

is the equation whose roots are each less by 40 than the roots of the given equation, and whose positive root is found to lie between 3 and 4. If the second transformed equation had not an exact root $\cdot5$; but one, we shall suppose, between $\cdot5$ and $\cdot6$, the first three figures of the root of the proposed equation would be $43\cdot5$; and to find the next figure we should proceed to a further transformation, diminishing the roots by $\cdot5$; and so on.

2. Find the positive root of the equation

$$4x^3 - 13x^2 - 31x - 275 = 0.$$

We first write down the arithmetical work, and proceed to make certain observations on it :—

4	-13	-31	-275	$(6\cdot25$
	24	66	210	
	11	35	-65	
	24	210	$51\cdot392$	
	35	245	$-13\cdot608$	
	24	$11\cdot96$	$13\cdot608$	
	59	$256\cdot96$	0	
	$\cdot8$	$12\cdot12$		
	$59\cdot8$	$269\cdot08$		
	$\cdot8$	$3\cdot08$		
	$60\cdot6$	$272\cdot16$		
	$\cdot8$			
	$61\cdot4$			
	$\cdot2$			
	$61\cdot6$			

We find by trial that the proposed equation has its positive root between 6 and 7. The first figure of the root is, therefore, 6. Diminish the roots by 6. The equation

$$x^3 + 59x^2 + 245x - 65 = 0$$

has, then, a root between 0 and 1. It is found by trial to lie between $\cdot2$ and $\cdot3$. The first two figures of the root of the proposed are therefore $6\cdot2$. Diminish the roots again by $\cdot2$. The transformed equation is found to have the root $\cdot05$. Hence $6\cdot25$ is a root of the proposed equation.

It is convenient in practice to avoid the use of the decimal points. This can easily be effected as follows :—When the decimal part of the root (suppose $\cdot abc \ldots$) is about to appear, multiply the roots of the corresponding transformed equation by 10, *i. e.* annex one zero to the right of the figure in the first column, two to the right of the figure in the second column, three to the right of that in the third; and so on, if there be more columns (as there will of course be in equations of a degree higher than the third). The root of the transformed equation is, then, not $\cdot abc \ldots$, but $a\cdot bc \ldots$ Diminish the roots by a. The transformed has then the root $\cdot bc \ldots$ Multiply the roots of this equation again by 10. The root becomes $b\cdot c \ldots$, and

the process is continued as before. To illustrate this, we repeat the above operation, omitting the decimal points. In all subsequent examples this simplification
will be adopted :—

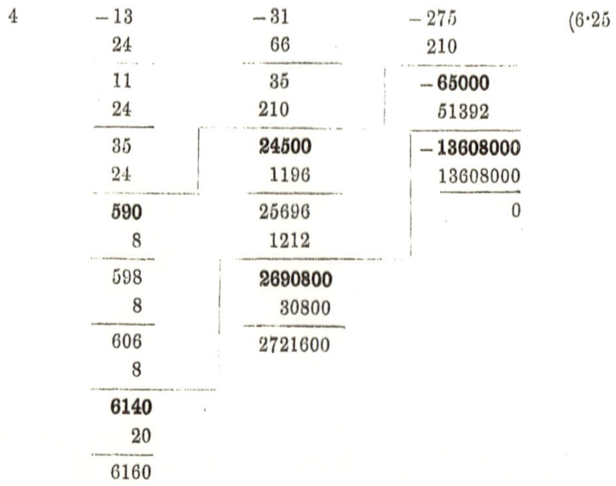

3. Find the positive root of the equation

$$20x^3 - 121x^2 - 121x - 141 = 0.$$

The root is easily found to lie between 7 and 8. It is, therefore, of the form
7 . ab ... After diminishing the roots by 7, and multiplying by 10, the resulting
equation is

$$20x^3 + 2990x^2 + 112500x - 57000 = 0.$$

The positive root of this is $a . b$ ...; and as the root plainly lies between 0 and 1,
we have $a = 0$. We therefore place zero as the first figure in the decimal part of
the root, and multiply the roots again by 10, before proceeding to the second transformation. 5 is easily seen to be a root of the equation thus transformed.

Ans. 7.05.

In the examples here considered the root terminates at an
early stage. When the calculation is of greater length, if it
were necessary to find the successive figures by substitution,
the labour of the process would be very great. This, however,
is not necessary, as will appear in the next Article; and one of
the most valuable practical advantages of Horner's method is,
that after the second, or third (sometimes even after the first)
figure of the root is found, the *transformed equation itself suggests
by mere inspection the next figure of the root.* The principle of
this simplification will now be explained.

102. Principle of the Trial-divisor.—We have seen in Art. 100 that when an equation is transformed by the substitution of $a + h$ for x, a being a number differing from the true root by a quantity h small in proportion to a, an approximate numerical value of h is obtained by dividing $f(a)$ by $f'(a)$. Now the successive transformed equations in Horner's process are the results of transformations of this kind, the last coefficient being $f(a)$, and the second last $f'(a)$ (see Art. 34). Hence, after two or three steps have been completed, so that the part of the root remaining bears a small ratio to the part already evolved, we may expect to be furnished with two or three more figures of the root correctly by mere division of the last by the second last coefficient of the final transformed equation. We might thus, if we pleased, at any stage of Horner's operations, apply Newton's method to get a further approximation to the root. In Horner's method this principle is employed to suggest the next following figure of the root after the figures already obtained. The second last coefficient of each transformed equation is called the *trial-divisor*. Thus, in the second example of the last Article, the number 5 is correctly suggested by the trial-divisor 2690800. In this example, indeed, the second figure of the root is correctly suggested by the trial-divisor of the first transformed equation, although, in general, this is not the case. In practice the student will have to estimate the probable effect of the previous coefficients of the equation; he will find, however, that the influence of these terms becomes less and less as the evolution of the root proceeds.

EXAMPLES.

1. Find the positive root of the equation

$$x^3 + x^2 + x - 100 = 0$$

correct to four decimal places.

We easily see that the root lies between 4 and 5. We write down the work, and proceed to make observations on it :—

1	1	1	− 100	(4·2644
	4	20	84	
	5	21	− 16000	
	4	36	11928	
	9	5700	− 4072000	
	4	264	3788376	
	130	5964	− 283624000	
	2	268	256071744	
	132	623200	− 27552256	
	2	8196		
	134	631396		
	2	8232		
	1360	63962800		
	6	55136		
	1366	64017936		
	6	55152		
	1372	64073088		
	6			
	13780			
	4			
	13784			
	4			
	13788			
	4			
	13792			

First diminish the roots by 4. As the decimal part is now about to appear, attach ciphers to the coefficients of the transformed equation as explained in Ex. 2, Art. 101. Since the coefficient 130 is small in proportion to 5700, we may expect that the trial-divisor will give a good indication of the next figure. The figure to be adopted in every case as part of the root is *that highest number which in the process of transformation will not change the sign of the absolute term.* Here 2 is the proper figure. In diminishing by 2 the roots of the transformed equation

$$x^3 + 130x^2 + 5700x - 16000 = 0,$$

the absolute term retains its sign (− 4072). If we had adopted the figure 3, the absolute term would have become positive, the change of sign showing that we had gone beyond the root. We must take care that, after the first transformation (the reason of this restriction will appear in the next example), the absolute term preserves its sign throughout the operation. If we were to take by mistake a number too small, the error would show itself, just as in ordinary division or evolution, by the next suggested number being greater than 9. Such a mistake, however, will rarely be made. The error which is most common is to take the number too large, and this will show itself in the work by the change of sign in the absolute term. In the above work it is evident, without performing the fifth transformation, that the corresponding figure of the root is 4, so that the correct root to four decimal places is 4·2644.

2. The equation $\qquad x^4 + 4x^3 - 4x^2 - 11x + 4 = 0$

has one root between 1 and 2; find its value correct to 4 decimal places.

1 4	−4	−11	4	(1·6369
1	5	1	−10	
5	1	−10	−60000	
1	6	7	50976	
6	7	−3000	−90240000	
1	7	11496	72690561	
7	1400	8496	−175494390000	
1	516	14808	152131052016	
80	1916	23304000	−23363337984	
6	552	926187		
86	2468	24230187		
6	588	935601		
92	305600	25165788000		
6	3129	189387336		
98	308729	25355175336		
6	3138	189766488		
1040	311867	25544941824		
3	3147			
1043	31501400			
3	63156			
1046	31564556			
3	63192			
1049	31627748			
3	63228			
10520	31690976			
6				
10526				
6				
10532				
6				
10538				
6				
10544				

We see without completing the fifth transformation that 9 is the next figure of the root.   The root is, therefore, 1·6369 correct to 4 decimal places.

The trial-divisor becomes effective after the second transformation, suggesting correctly the number 3, and all subsequent numbers.   The first transformed equation has its last two terms negative.   We may expect, therefore, that the influence of the preceding coefficients is greater than that of the trial-divisor, as in fact is here the case.   The number 6, the second figure of the root, must be found by substitution.   We have to determine what is the situation between 0 and 10 of the root of the equation

$$x^4 + 80x^3 + 1400x^2 - 3000x - 60000 = 0.$$

A few trials show that 6 gives a negative, and 7 a positive result.   Hence the root lies between 6 and 7 ; and 6 is the number of which we are in search.   In the subsequent trials we take those greatest numbers 3, 6, 9, in succession, which allow the absolute term to retain its negative sign.   In the first transformation, diminishing the roots by 1, there is a change of sign in the absolute term.   The meaning of this is, that we have passed over a root lying between 0 and 1, for 0 gives a positive result, 4 ; and 1 gives a negative result, -6.   In all subsequent transformations, so long as we keep below the root, the sign of the absolute term must be the same as the sign resulting from the substitution of 1.   This supposes of course that no root lies between 1 and that of which we are in search.   This supposition we have already made in the statement of the question.   In fact the proposed can have only two positive roots, and one is between 0 and 1, and therefore only one between 1 and 2.

When two roots exist between the limits employed in Horner's method, *i.e.* when the equation has a pair of roots nearly equal, certain precautions must be observed which will form the subject of a subsequent Article.

3. Find the root of the preceding equation between 0 and 1 to 4 decimal places.   Commence by multiplying by 10.   The coefficients are then

$$1, \ 40, \ -400, \ -11000, \ 40000 \, ;$$

the trial-divisor becomes effective at once in consequence of the comparative smallness of the leading coefficients.   The positive sign of the absolute term must be preserved throughout.   *Ans.* ·3373.

4. Find to three places of decimals the root situated between 9 and 10 of the equation

$$x^4 - 3x^2 + 75x - 10000 = 0.$$

Ans. 9·886.

[Supply the zero coefficient of x^3.]

In the examples hitherto considered the root has been found to a few decimal places only.   We proceed now to explain a method by which, after 3 or 4 places of decimals have been evolved as above, several more may be correctly obtained with great facility by a contracted process.

103. Contracted Method of applying Horner's Process.—In the ordinary process of contracted Division, when the given figures are exhausted, in place of appending ciphers to the successive dividends, we cut off figures successively from the right of the divisor, so that the divisor itself becomes exhausted after a number of steps depending on the number of figures it contains. The resulting quotient will differ from the true quotient in the last figure only, or at most in the last two figures. In Horner's contracted method the principle is the same. We retain those figures only which are effective in contributing to the result to the degree of approximation desired. When the contracted process commences, in place of appending ciphers to the successive coefficients of the transformed equation in the way before explained, we cut off one figure from the right of the last coefficient but one, two from the right of the last coefficient but two, three from the right of the last coefficient but three; and so on. The effect of this is to retain in their proper places the important figures in the work, and to banish altogether those which are of little importance.

The student will do well to compare the first transformation by the contracted process in the first of the following examples with the corresponding step in the second example of the last Article, where the transformation is exhibited in full. He will then observe how the leading figures (those which are most important in contributing to the result) coincide in both cases, and retain their relative places; while the figures of little importance are entirely dispensed with.

In addition to the contraction now explained other abbreviations of Horner's process are sometimes recommended; but as the advantage to be derived from them is small, and as they increase the chances of error, we do not think it necessary to give any account of them. The contraction here explained is so important that it must not be overlooked.

EXAMPLES.

1. Find the root between 1 and 2 of the equation in Ex. 2 of the last Article correct to 8 or 9 decimal places.

Assuming the result of the Example in question, we shall commence the contracted process after the third transformation has been completed. The subsequent work stands as follows :—

```
1052   315014        2516578̸8        − 17549439   (1·636913575
          6            18936            15213090
       ────────        ───────        ──────────
       3156           2535515          − 2336349
          6            18972            2301597
       ────────        ───────        ──────────
       3162           2554487̸          − 34752
          6             285             25601
       ────────        ───────        ──────────
       3168           255733           − 9151
                        285             7680
                      ────────        ──────────
       81             256018           − 1471
                                        1280
                                       ──────────
                                        − 191
                                         179
                                       ──────────
                                         12
```

Here the effect of the first cutting off of figures, namely, 8 from the second last coefficient, 14 from the third last, and 052 from the fourth last, is to banish altogether the first coefficient of the biquadratic. We proceed to diminish the roots by 6 as if the coefficients 1, 3150, 2516578, − 17549439 which are left were those of a cubic equation. In multiplying by the corresponding figure of the root the figures cut off should be multiplied mentally, and account taken of the number to be carried, just as in contracted division.

After the diminution by 6 has been completed, we cut off again in the transformed cubic 7 from the last coefficient but one, 68 from the last but two, and the first coefficient disappears altogether. The work then proceeds as if we were dealing with the coefficients 31, 255448, − 2336349 of a quadratic. The effect of the next process of cutting off is to banish altogether the leading coefficient 31. The subsequent work coincides with that of contracted division. When the operation terminates, the number of decimals in the quotient may be depended on up to the last, or last couple of figures. The extent to which the evolution of the root must be carried before the contracted process is commenced depends on the number of decimal places required; for after the contraction commences we shall be furnished, in addition to the figures already evolved, with as many more as there are figures in the trial-divisor, less one.

2. Find to 8 or 9 decimal places the root of the equation

$$x^4 - 12x + 7 = 0$$

which lies between 2 and 3.

This equation can have only two positive roots : one lies between 0 and 1, and the other between 2 and 3.   For the evolution of the latter we have the following :—

1	0	0	− 12	7 (2·047275671
	2	4	8	− 8
	2	4	−4	− 1,00000000
	2	8	24	83891456
	4	12	20,000000	− 16108544
	2	12	.972864	15493401
	6	24,0000	20972864	− 615143
	2	3216	985792	446262
	8,00	243216	21958656	− 168881
	4	3232	17478	156226
	804	246448	2213343	− 12655
	4	3248	17478	11159
	808	249696	2230821	− 1496
	4	.2496	49	1338
	812		223131	− 158
	4		49	156
	816	24	223180	2

On this we remark, that after diminishing the roots by 2, and multiplying the roots of the transformed equation by 10, we find that the trial-divisor 20000 will not "go into" the absolute term 10000 ; we put, therefore, zero in the quotient, and multiply again by 10, and then proceed as before.

3. Find to 8 or 9 decimal places the root of the same equation between 0 and 1.

Ans. ·593685829.

4. Find the positive root of the equation

$$x^3 + 24 \cdot 84 x^2 - 67 \cdot 613 x - 3761 \cdot 2758 = 0.$$

[When the coefficients of the proposed equation contain decimal points, it will be found that they soon disappear in the work in consequence of the multiplications by 10 after the decimal part of the root begins to appear.]

Ans. 11·1973222.

5. Find the negative root of the equation

$$x^4 - 12x^2 + 12x - 3 = 0$$

to 7 places of decimals.

When a negative root has to be found, it is convenient to change the sign of x and find the corresponding positive root of the transformed equation.

Ans. − 3·9073785.

104. Application of Horner's Method to Cases where Roots are nearly Equal.— We have seen in Art. 100 that the method of approximation there explained fails when the proposed equation has two roots nearly equal. Examples of this nature are those which present most difficulties, both in their analysis (see Ex. 7, Art. 91) and in their solution. Horner's method enables us, with very little more labour than is necessary in other cases, to effect the solution of such equations. So long as the leading figures of the two roots are the same, certain precautions must be observed, which will be illustrated by the following examples. After the two roots have been separated, the subsequent operations proceed for each root separately, just as in the examples of the previous Articles. It is evident, from the explanation of the trial-divisor given in Art. 102, that for the same reason as that which explains the failure of Newton's method (see Art. 100), it will not become effective till the first or second stage after the roots have been separated.

EXAMPLES.

1. The equation

$$x^3 - 7x + 7 = 0$$

has two roots between 1 and 2 (see Ex. 2, Art. 89) ; find each of them to 8 decimal places.

Diminishing the roots by 1, we find that the transformed equation (after its roots are multiplied by 10) *i. e.*

$$x^3 + 30x^2 - 400x + 1000 = 0,$$

must have two roots between 0 and 10. We find that these roots lie, one between 3 and 4, and the other between 6 and 7. The roots are now separated, and we proceed with each separately in the manner already explained. If the roots were not separated at this stage, we should find the leading figure common to the two, and, having diminished the roots by it, find in what intervals the roots of the resulting equation were situated ; and so on.

Ans.   1·35689584,   1·69202147.

2. Find the two roots of the equation

$$x^3 - 49x^2 + 658x - 1379 = 0$$

which lie between 20 and 30.

We shall exhibit the complete work of approximation to the smaller of the two roots to 7 places ; and then make certain observations which will be a guide to the student in all cases of the kind.

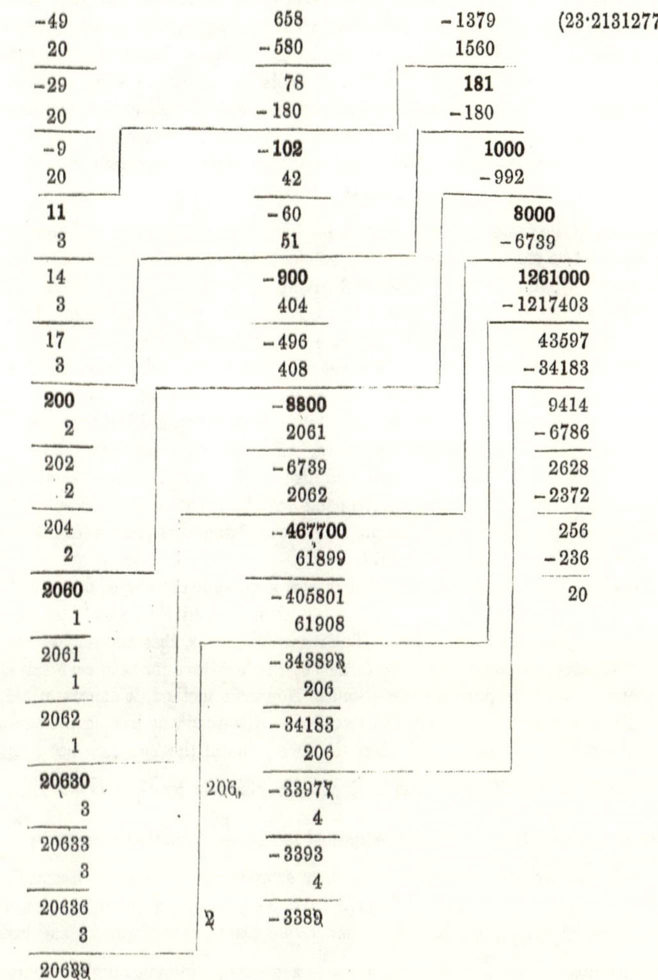

The diminution of the roots by 20 changes the sign of the absolute term. This is an indication that a root exists between 0 and 20, with which we are not at present concerned. The roots of the first transformed equation

$$x^3 + 11x^2 - 102x + 181 = 0$$

are not yet separated, lying both between 3 and 4. The substitution of each of these numbers gives a positive result, so that we have not here the same criterion to guide us in our search for the proper figure as in former cases, viz., a change of sign in the absolute term. We have, however, a different criterion which enables us to find by mere substitution the interval within which the two roots lie. If we diminish the roots of $x^3 + 11x^2 - 102x + 181 = 0$ by 4, the resulting equation is $x^3 + 23x^2 + 34x + 13 = 0$, which has no change of sign. Hence the two roots must lie between 0 and 4. If we diminish its roots by 3, the resulting equation (as in the above work) has the same number of changes of sign as the equation itself. Hence the two roots lie between 3 and 4. They are, therefore, not yet separated; and we proceed to diminish by 3. The next transformed equation

$$x^3 + 200x^2 - 900x + 1000 = 0$$

is found in the same way to have both its roots between 2 and 3: the diminution by 2 leaving two changes of sign in the coefficients of the transformed equation (as in the above work), and the diminution by 3 giving all positive signs. So far, then, the two roots agree in their first three figures, *i. e.* 23·2. We diminish again by 2. The resulting equation $x^3 + 2060x^2 - 8800x + 1261000 = 0$ has one root only between 1 and 2; 1 giving a positive, and 2 a negative result: its other root lies between 2 and 3; 3 giving a positive result. The roots are now separated. We proceed, as in the above work, to approximate to the lesser root, by diminishing the roots of this equation by 1; the trial-divisor becoming effective at the next step. To approximate to the greater root, we must diminish by 2 the roots of the same equation, taking care that in the subsequent operations the negative sign, to which the previously positive sign of the absolute term now changes, is preserved. The second root will be found to be 23·2295212.

So long as the two roots remain together, a guide to the proper figure of the root may be obtained by dividing twice the last coefficient by the second last, or the second last by twice the third last. The reason of this is, that the proposed equation approximates now to the quadratic formed by the last three terms in each transformed equation, just as in previous cases, and in Newton's method, it approximated to the simple equation formed by the last two terms, this quadratic having the two nearly equal roots for its roots; and when the two roots of the equation $ax^2 + bx + c = 0$ are nearly equal, either of them is given approximately by $\dfrac{-2c}{b}$ or $\dfrac{-b}{2a}$. Thus, in the above example, the number 3 is suggested by $\dfrac{2 \times 181}{102}$, and the number 2 by $\dfrac{2 \times 1000}{900}$. In this way we can generally, at the first attempt, find the two integers between which the pair of roots lies. We shall have, also, an indication of the separation of the roots by observing when the numbers suggested in this way by the last three coefficients become different, *i. e.* when $\dfrac{2c}{b}$ suggests a different number from $\dfrac{b}{2a}$.

3. Calculate to three decimal places each of the roots lying between 4 and 5 of the equation

$$x^4 + 8x^3 - 70x^2 - 144x + 936 = 0.$$

Ans. 4·242; 4·246.

4. Find the two roots between 2 and 3 of the equation

$$64x^3 - 592x^2 + 1649x - 1445 = 0.$$

Ans. The roots are both = 2·125.

Here we find that the two roots are not separated at the third decimal place. When we diminish by 5 the absolute term vanishes, showing that 2·125 is a root; and proceeding with this diminution the second last coefficient also vanishes. Hence 2·125 is a double root.

When an equation contains more than two nearly equal roots, they can be all found by Horner's process in a manner similar to that now explained. Such cases are, however, of rare occurrence in practice. The principles already laid down will be a sufficient guide to the student in all cases of the kind.

105. **Lagrange's Method of Approximation.**—Lagrange has given a method of expressing the root of a numerical equation in the form of a continued fraction. As this method is, for practical purposes, much inferior to that of Horner, we shall content ourselves with a brief account of it.

Let the equation $f(x) = 0$ have one root, and only one root, between the two consecutive integers a and $a + 1$. Substitute $a + \dfrac{1}{y}$ for x in the proposed equation. The transformed equation in y has one positive root. Let this be determined by trial to lie between the integers b and $b + 1$. Transform the equation in y by the substitution $y = b + \dfrac{1}{z}$. The positive root of the equation in z is found by trial to lie between c and $c + 1$. Continuing this process, an approximation to the root is obtained in the form of a continued fraction, as follows :—

$$a + \cfrac{1}{b + \cfrac{1}{c + 1 \ldots}}$$

EXAMPLES.

1. Find in the form of a continued fraction the positive root of the equation

$$x^3 - 2x - 5 = 0.$$

The root lies between 2 and 3.

To make the transformation $x = 2 + \dfrac{1}{y}$, we first employ the process of Art. 34, diminishing the roots by 2. We then find the equation whose roots are the reciprocals of those of the transformed.

The equation in y is in this way found to be

$$y^3 - 10y^2 - 6y - 1 = 0.$$

This has a root between 10 and 11.

Make now the substitution $\quad y = 10 + \dfrac{1}{z}$.
The equation in z is

$$61z^3 - 94z^2 - 20z - 1 = 0.$$

This has a root between 1 and 2.   Take $z = 1 + \dfrac{1}{u}$.
The equation in u is

$$54u^3 + 25u^2 - 89u - 61 = 0,$$

which has a root between 1 and 2 ; and so on.

We have, therefore, the following expression for the root

$$x = 2 + \cfrac{1}{10 + \cfrac{1}{1 + \cfrac{1}{1 + \ldots}}}$$

2. Find in the form of a continued fraction the positive root of

$$x^3 - 6x - 13 = 0.$$

$$Ans. \quad 3 + \cfrac{1}{5 + \cfrac{1}{1 + \cfrac{1}{1 + \ldots}}}$$

106. Numerical Solution of the Biquadratic.

—It is proper, before closing the subject of the solution of numerical equations, to illustrate the practical uses which may be made of the methods of solution of Chap. VI. Although, as before observed, the numerical solution of equations is in general best effected by the methods of the present Chapter, there are certain cases in which it is convenient to employ the methods of Chap. VI. for the resolution of the biquadratic. When a biquadratic equation leads to a reducing cubic which has a commensurable root, this root can be readily found, and the solution of the biquadratic completed. We proceed to solve a few examples of this kind, using Descartes' method (Art. 64), which will usually be found the most convenient in practice.

EXAMPLES.

1. Resolve the quartic

$$x^4 - 6x^3 + 3x^2 + 22x - 6$$

into quadratic factors.

Making the assumption of Art. 64, we easily obtain

$$p + p' = -3, \quad q + q' + 4pp' = 3, \quad pq' + p'q = 11, \quad qq' = -6.$$

Also

$$\phi = \frac{1}{2} - pp' = \frac{1}{4}(q + q' - 1),$$

and, calculating I and J, the equation for ϕ is

$$4\phi^3 - \frac{111}{4}\phi - \frac{225}{8} = 0.$$

Multiplying the roots by 4, we have, if $4\phi = t$,

$$t^3 - 111t - 450 = 0.$$

By the Method of Divisors this is easily found to have a root -6; hence $\phi = -\frac{3}{2}$, giving

$$pp' = 2, \quad q + q' = -5.$$

From these, combined with the preceding equations, we get

$$p = -2, \quad p' = -1, \quad q = 1, \quad q' = -6.$$

When the values of q and q' are found, the equation giving the value of $pq' + p'q$ determines which value of q goes with p, and which with p', in the quadratic factors. The quartic is resolved, therefore, into the factors

$$(x^2 - 4x + 1)(x^2 - 2x - 6).$$

By means of the other two values of ϕ we can resolve the quartic into quadratic factors in two other ways; or we can do the same thing by solving the two quadratics already obtained.

2. Resolve into factors the quartic

$$x^4 - 8x^3 - 12x^2 + 60x + 63.$$

The equation for ϕ is found to be

$$4\phi^3 - 195\phi - 475 = 0.$$

This has a root $= -5$, and we easily find

$$pp' = 3, \quad q + q' = -24,$$

giving the following values:—

$$p = -1, \quad p' = -3, \quad q = -3, \quad q' = -21.$$

The quartic is, therefore, equivalent to

$$(x^2 - 2x - 3)(x^2 - 6x - 21).$$

3. Resolve into factors

$$f(x) \equiv x^4 - 17x^2 - 20x - 6.$$

The reducing cubic is found to be

$$4\phi^3 - \frac{217}{12}\phi + \frac{3185}{216} = 0;$$

or, multiplying the roots by 6,

$$4t^3 - 651t + 3185 = 0.$$

This has a root $= 7$; hence $\phi = \dfrac{7}{6}$, and we obtain, finally,

$$f(x) \equiv (x^2 + 4x + 2)(x^2 - 4x - 3).$$

4. Resolve into factors

$$f(x) \equiv x^4 - 6x^3 - 9x^2 + 66x - 22.$$

The reducing cubic is

$$4\phi^3 - \frac{335}{4}\phi - \frac{897}{8} = 0;$$

hence

$$\phi = -\frac{3}{2}.$$

$$\textit{Ans. } f(x) \equiv (x^2 - 11)(x^2 - 6x + 2).$$

5. Resolve into quadratic factors

$$f(x) \equiv x^4 - 8x^3 + 21x^2 - 26x + 14.$$

$$\textit{Ans. } f(x) \equiv (x^2 - 2x + 2)(x^2 - 6x + 7).$$

Miscellaneous Examples.

1. Find the positive root of
$$x^3 - 6x - 13 = 0.$$
Ans. 3·176814393.

2. Find the positive root of
$$x^3 - 2x - 5 = 0$$
correct to 9 or 10 places.
Ans. 2·0945514845.

3. The equation
$$2x^3 - 650·8x^2 + 5x - 1627 = 0$$
has a root between 300 and 400.   Find it.
Ans. Commensurable root, 325·4.

4. Find the root between 20 and 30 of the equation
$$x^3 - 180x^2 + 1896x - 457 = 0.$$
Ans. 28·521277389.

5. Find the root between 2 and 3 of
$$x^3 - 49x^2 + 658x - 1379 = 0.$$
Ans. 2·557351.

6. Find the root between 2 and 3 of the equation
$$x^4 - 12x^2 + 12x - 3 = 0$$
to six places of decimals.
Ans. 2·858033.

7. Find the positive root of the equation
$$x^3 + 2x^2 - 23x - 70 = 0$$
correct to about 12 decimal places.
Ans. 5·134578725282.

8. Find the cube root of 673373097125.
Ans. 8765.

9. Find the fifth root of 537824.
Ans. 14.

10. Find all the roots of the cubic equation
$$x^3 - 3x + 1 = 0.$$

The equation $x^6 + x^3 + 1 = 0$, of Ex. 7, p. 98, reduces to this.
Ans. − 1·87938, ·34729, 1·53209.

Note.—The smaller positive root furnishes the solution of the problem—To divide a hemisphere whose radius is unity into two equal parts by a plane parallel to the base.

11. Find all the roots of the cubic
$$x^3 + x^2 - 2x - 1 = 0.$$

(See Ex. 1, p. 98.)
Ans. − 1·80194, − ·44504, 1·24698.

12. Find to five decimal places the negative root between -1 and 0 (see **Ex.** 3, p. 98) of the equation

$$x^5 + x^4 - 4x^3 - 3x^2 + 3x + 1 = 0.$$

Ans. $-\cdot28463.$

13. Solve the equation

$$x^3 - 315x^2 - 19684x + 2977260 = 0.$$

We here find that a root exists between 70 and 80.   By Horner's process it is found to be 78.   The depressed equation furnishes two roots, which, increased by 78, are the other roots of the cubic.

Ans. 78,   347, $-110.$

14. Find the two real roots of the equation

$$x^4 - 11727x + 40385 = 0.$$

Ans. $3\cdot45592$, $21\cdot43067.$

This equation is given by Mr. G. H. Darwin in a paper *On the Precession of a Viscous Spheroid, and on the Remote History of the Earth.   Phil. Trans.*, Part ii., 1879, p. 508.   The roots are "the two values of the cube root of the earth's rotation for which the earth and moon move round as a rigid body."

15. Find all the roots of the cubic equation

$$20x^3 - 24x^2 + 3 = 0.$$

Ans. $-0\cdot31469$, $0\cdot44603$, $1\cdot06865.$

This equation occurs in the solution by Professor Ball of a problem of Professor Townsend's in the *Educational Times* of Dec. 1878, to determine the deflection of a beam uniformly loaded and supported at its two ends and points of trisection.

16. Find the positive root of the equation

$$14x^3 + 12x^2 - 9x - 10 = 0.$$

Ans. $0\cdot85906.$

Note.—The equations of this and the next example occur also in the investigation of questions relative to beams supported by props.

17. Find the positive root of the equation

$$7x^4 + 20x^3 + 3x^2 - 16x - 8 = 0.$$

Ans. $0\cdot91336.$

CHAPTER XI.

107. Elementary Notions and Definitions.—This chapter will be occupied with a discussion of an important class of functions which constantly present themselves in analysis. These functions possess remarkable properties, by a knowledge of which much simplification may be introduced into many mathematical operations.

The function $\quad a_1 b_2 + a_2 b_1, \quad$ of the four quantities

$$a_1, \quad b_1,$$
$$a_2, \quad b_2,$$

is obtained by assigning to a and b, written in aphabetical order, the suffixes 1, 2, and 2, 1, corresponding to the two permutations of the numbers 1, 2, and adding the two products so formed.

Similarly the function

$$a_1 b_2 c_3 + a_1 b_3 c_2 + a_2 b_3 c_1 + a_2 b_1 c_3 + a_3 b_1 c_2 + a_3 b_2 c_1, \tag{1}$$

of the nine quantities

$$a_1, \quad b_1, \quad c_1,$$
$$a_2, \quad b_2, \quad c_2,$$
$$a_3, \quad b_3, \quad c_3,$$

is obtained by adding all the products abc which can be formed by assigning to the letters (retained in their alphabetical order) suffixes corresponding to all the permutations of the numbers 1, 2, 3. The whole expression might be represented by (abc), or any other convenient notation, from which all the terms could be written down.

The notation $(abcd)$ might be employed to represent a similar function of the 16 quantities a_1, b_1, c_1, d_1, a_2, &c.; consisting of 24 terms, which can all be written down by the aid of the 24 permutations of the numbers 1, 2, 3, 4.

And, in general, taking n letters a, b, c, ... l, we can write down a similar function consisting of $n(n-1)(n-2)\ldots.3.2.1$ terms, this being the number of permutations of the first n numbers 1, 2, 3 ... n.

Now the functions above referred to, which are of such frequent occurrence in mathematical analysis, differ from those just described in one respect only; namely, of the $1.2.3\ldots n$ (which is an even number) terms, half are affected with positive, and half with negative signs, instead of being all positive, as in the examples just given.

We shall now give some instances of these latter functions. They occur most frequently as the result of elimination of the variables from linear equations. If, for example, x and y be eliminated from the equations

$$a_1 x + b_1 y = 0,$$

$$a_2 x + b_2 y = 0,$$

the result is $a_1 b_2 - a_2 b_1 = 0$.

Again, the result of eliminating x, y, z from the equations

$$a_1 x + b_1 y + c_1 z = 0,$$

$$a_2 x + b_2 y + c_2 z = 0,$$

$$a_3 x + b_3 y + c_3 z = 0,$$

is, as the student will readily perceive by solving from two of them and substituting in the third,

$$a_1 b_2 c_3 - a_1 b_3 c_2 + a_2 b_3 c_1 - a_2 b_1 c_3 + a_3 b_1 c_2 - a_3 b_2 c_1 = 0 ; \qquad (2)$$

and this function differs from (1) above written only in having three of its terms negative, instead of having the six terms positive.

Similarly the elimination of four variables from four linear equations gives rise to a function of the 16 quantities

$$a_1, \; b_1, \; c_1, \; d_1, \; a_2, \; b_2, \; \&c.,$$

which differs from the function above represented by ($abcd$) only in having 12 of its terms negative.

Expressions of the kind here described are called *Determinants.*[*] The notation by which they are usually represented was first employed by Cauchy, and possesses many advantages in the treatment of these expressions. The quantities of which the function consists are arranged in a square between two vertical lines. Thus

$$\begin{vmatrix} a_1 & b_1 \\ a_2 & b_2 \end{vmatrix}$$

represents the determinant $a_1 b_2 - a_2 b_1$.

Similarly, the expression on the left-hand side of equation (2) is represented by the notation

$$\begin{vmatrix} a_1 & b_1 & c_1 \\ a_2 & b_2 & c_2 \\ a_3 & b_3 & c_3 \end{vmatrix}$$

And, in general, the determinant of the n^2 quantities $a_1, \; b_1, \; c_1 \ldots l_1, \; a_2, \; b_2, \; \&c.,$ is represented by

$$\begin{vmatrix} a_1 & b_1 & c_1 & \cdot & \cdot & \cdot & l_1 \\ a_2 & b_2 & c_2 & \cdot & \cdot & \cdot & l_2 \\ a_3 & b_3 & c_3 & \cdot & \cdot & \cdot & l_3 \\ \cdot & \cdot & \cdot & \cdot & \cdot & \cdot & \cdot \\ a_n & b_n & c_n & \cdot & \cdot & \cdot & l_n \end{vmatrix}. \qquad (3)$$

By taking the n letters in alphabetical order, and assigning to them suffixes corresponding to the $n(n-1)(n-2) \ldots 3 . 2 . 1$ permutations of the numbers $1, 2, 3, \ldots n$, all the terms of the

---

determinant can be written down. Half of the terms must receive positive and half negative signs. In the next Article the rule will be explained by which the positive and negative terms are distinguished.

The individual letters a_1, b_1, c_1, ... a_2, ... &c., of which a determinant is composed are called its *constituents*.

The several terms $a_1 b_2 c_3 \ldots l_n$, &c., consisting each of the product of n constituents, are called *elements*.

Any series of constituents such as a_1, b_1, c_1, ... l_1, arranged horizontally, form a *row* of the determinant; and any series such as a_1, a_2, $a_3 \ldots a_n$, arranged vertically, form a *column*.

The term *line* will sometimes be used to express a row or column indifferently.

108. **Rule with regard to Signs.**—It is evident from the preceding Article that each element of the determinant will, since it contains all the letters, contain one constituent (and only one) from every column; and will also, since the suffixes in each term comprise all the numbers, contain one constituent (and only one) from every row. We may thus regard the square array (3) of Art. 107 as the symbolical representation of a function consisting in general of $n(n-1)(n-2) \ldots 3 . 2 . 1$ terms, comprising all possible products which can be formed by taking one constituent, and one only, from each row, and one constituent, and one only, from each column. All that is required to give perfect definiteness to the function is to fix the sign to be attached to any particular element. For this purpose the following two rules are to be observed :—

(1). *The element $a_1 b_2 c_3 \ldots l_n$, formed by the constituents situated in the diagonal drawn from the left-hand top corner to the right-hand bottom corner, is positive.*

This element is called the *leading* or *principal* element. In it the suffixes and letters both occur in their natural order; and from it the sign of any other element is determined by means of the following rule.

(2). *Any interchange of two suffixes, the letters retaining their order, alters the sign of an element.*

This rule may be otherwise expressed thus :—*Any interchange of two letters, the suffixes retaining their order, alters the sign of an element.* For if two letters be interchanged, and then the two corresponding constituents interchanged, the process is equivalent to an interchange of suffixes. If, for example, in $a_1 b_2 c_3 d_4 e_5$ the letters b and e be interchanged, we get $a_1 e_2 c_3 d_4 b_5$, which is equal to $a_1 b_5 c_3 d_4 e_2$, and this is derived from the given element by an interchange of the suffixes 2 and 5.

In applying this rule it is evident that an even number of interchanges will not alter the sign of an element, and that an odd number will.

EXAMPLES.

1. What is the sign of the element $a_3 b_4 c_2 d_5 e_1$ in the determinant of the 5th order ?

The question is, How many interchanges will change the order 12345 into 34251 ? Here, when 3 is interchanged with 2, and afterwards with 1, it comes into the leading place, the order becoming 31245. Again, the interchange in 31245 of 4 with 2, and afterwards with 1, presents the order 34125. The interchange of 2 with 1 gives the order 34215 ; and finally the interchange of 5 with 1 gives the required order 34251. Thus there are in all six interchanges ; and therefore the required sign is positive.

The general mode of proceeding may plainly be stated as follows :—Take the figure which stands first in the required order, and move it from its place in the natural order 1234 . . . into the leading place, counting one displacement for each figure passed over. Take then the figure which stands second in the required order, and move it from its place in the natural order into the second place ; and so on. If the number of displacements in this process be even, the sign is positive ; if it be odd, the sign is negative.

2. What sign is to be attached to the element $a_3 b_7 c_6 d_5 e_1 f_4 g_2$ in the determinant of the 7th order ?

Here two displacements bring 3 to the leading place ; five displacements then bring 7 to the second place ; four then bring 6 to the third place ; three then bring 5 to the fourth place ; the figure 1 is in its place ; and finally, one displacement brings 4 into the sixth place. Thus there are in all fifteen displacements ; and the sign of the element is therefore negative.

3. Write down all the terms of the determinant

$$\begin{vmatrix} a_1 & b_1 & c_1 & d_1 \\ a_2 & b_2 & c_2 & d_2 \\ a_3 & b_3 & c_3 & d_3 \\ a_4 & b_4 & c_4 & d_4 \end{vmatrix}$$

Q

The six permutations of suffixes in which the figure 1 occurs first are

$$1234, \quad 1243, \quad 1324, \quad 1342, \quad 1423, \quad 1432.$$

The six corresponding elements are, as the student will easily see by applying the Rule (2), as in the previous examples,

$$a_1 b_2 c_3 d_4 - a_1 b_2 c_4 d_3 + a_1 b_3 c_4 d_2 - a_1 b_3 c_2 d_4 + a_1 b_4 c_2 d_3 - a_1 b_4 c_3 d_2.$$

The other 18 terms, corresponding to the permutations in which the figures 2, 3, 4, respectively, stand first, are as follows :—

$$a_2 b_1 c_4 d_3 - a_2 b_1 c_3 d_4 + a_2 b_3 c_1 d_4 - a_2 b_3 c_4 d_1 + a_2 b_4 c_3 d_1 - a_2 b_4 c_1 d_3$$

$$+ a_3 b_1 c_2 d_4 - a_3 b_1 c_4 d_2 + a_3 b_2 c_4 d_1 - a_3 b_2 c_1 d_4 + a_3 b_4 c_1 d_2 - a_3 b_4 c_2 d_1$$

$$+ a_4 b_1 c_3 d_2 - a_4 b_1 c_2 d_3 + a_4 b_2 c_1 d_3 - a_4 b_2 c_3 d_1 + a_4 b_3 c_2 d_1 - a_4 b_3 c_1 d_2.$$

It will be observed here that the number of positive terms is equal to the number of negative terms. The same must be true in general; for, corresponding to any positive term there exists a negative term obtained by simply interchanging the last two suffixes.

4. Show that if any two adjacent figures are moved together over any number m of figures, the sign is unaltered.

For if they be moved separately, the whole process is equivalent to a movement over $2m$ figures.

5. Determine the sign to be attached to the second diagonal element, *i.e.* $a_n b_{n-1} c_{n-2} \ldots k_2 l_1$, in the determinant of the nth order.

Here the number of displacements required to change the natural order to the required order is plainly

$$(n - 1) + (n - 2) + (n - 3) + \ldots + 2 + 1 = \frac{n(n-1)}{2}.$$

Hence the required sign is $(-1)^{\frac{n(n-1)}{2}}$.

109. In this and the following Articles will be proved those properties of determinants which, by the aid of Cauchy's notation above described, render the employment of these functions of such practical advantage.

PROP. I.—*If any two rows, or any two columns, of a determinant be interchanged, the sign of the determinant is changed.*

This follows at once from the mode of formation (Rule (2), Art. 108), for an interchange of two rows is the same as an interchange of two suffixes, and an interchange of two columns is the same as an interchange of two letters ; so that in either case the sign of every element of the determinant is changed.

This proposition enables us to state the rule for determining the sign of any element in terms of displacements of the rows (or columns) ; and this will be found convenient for practical purposes. The student will readily perceive that the general mode of procedure explained in Ex. 1, Art. 108, is equivalent to the following :—*Bring, by movements of rows (or columns), the element whose sign is required into the position of the leading diagonal term. Its sign will be positive or negative according as the number of displacements is even or odd.*

EXAMPLE.

What sign is to be attached to the element $\lambda\beta nx$ in the determinant

$$\begin{vmatrix} a & b & c & \underline{x} \\ \alpha & \underline{\beta} & \gamma & y \\ l & m & \underline{n} & z \\ \underline{\lambda} & \mu & \nu & 0 \end{vmatrix} \ ?$$

Here a movement of the fourth row over three rows (*i.e.* three displacements) brings λ into the leading place. One displacement of the original second row upwards brings β into the required place in the diagonal term. And one further displacement of the original third row upwards effects the required arrangement, bringing $\lambda\beta nx$ into the diagonal place. Thus the number of displacements being odd, the required sign is negative.

110. Prop. II.—*When, in any determinant, two rows or two columns are identical, the determinant vanishes.*

For, by Prop. I., the interchange of these two lines ought to change the sign of the determinant Δ ; but the interchange of two identical rows or columns cannot alter the determinant in any way ; hence $\Delta = -\Delta$, or $\Delta = 0$.

111. Prop. III.—*The value of a determinant is not altered if the rows be written as columns, and the columns as rows.*

For all the elements, formed by taking one constituent from each row and one from each column, are plainly the same in value in both cases ; the principal element is identically the same ;

Q 2

and to determine the sign of any other element (by Prop. I.) the number of displacements of rows necessary to bring it into the leading diagonal in the first case is the same as the number of displacements of columns necessary in the second place.

EXAMPLE.

$$\begin{vmatrix} a_1 & b_1 & c_1 & d_1 \\ a_2 & b_2 & c_2 & d_2 \\ a_3 & b_3 & c_3 & d_3 \\ a_4 & b_4 & c_4 & d_4 \end{vmatrix} = \begin{vmatrix} a_1 & a_2 & a_3 & a_4 \\ b_1 & b_2 & b_3 & b_4 \\ c_1 & c_2 & c_3 & c_4 \\ d_1 & d_2 & d_3 & d_4 \end{vmatrix}.$$

Here the sign of any element, e. g. $a_2 b_4 c_1 d_3$, is the same in both determinants. For three displacements of rows are required to bring this element into the leading position in the first determinant; and the same number of displacements of columns is required to bring the same element into the leading position in the second determinant.

112. PROP. IV.—*If every constituent in any line be multiplied by the same factor, the determinant is multiplied by that factor.*

For every element of the determinant must contain one, and only one, constituent from any row or any column.

Cor. 1. If the constituents in any line differ only by the same factor from the constituents in another line, the determinant vanishes.

Cor. 2. If the signs of all the constituents in any line be changed, the sign of the determinant is changed. For this is equivalent to multiplying by the factor -1.

EXAMPLES.

1.
$$\begin{vmatrix} ka_1 & b_1 & c_1 \\ ka_2 & b_2 & c_2 \\ ka_3 & b_3 & c_3 \end{vmatrix} = k \begin{vmatrix} a_1 & b_1 & c_1 \\ a_2 & b_2 & c_2 \\ a_3 & b_3 & c_3 \end{vmatrix}.$$

2.
$$\begin{vmatrix} a_1 & ma_1 & a_2 \\ \beta_1 & m\beta_1 & \beta_2 \\ \gamma_1 & m\gamma_1 & \gamma_2 \end{vmatrix} = m \begin{vmatrix} a_1 & a_1 & a_2 \\ \beta_1 & \beta_1 & \beta_2 \\ \gamma_1 & \gamma_1 & \gamma_2 \end{vmatrix} = 0.$$

3. Show that the following determinant vanishes :—

$$\begin{vmatrix} 3 & 1 & 5 & 2 \\ 2 & 5 & 7 & 3 \\ 8 & 9 & 1 & 4 \\ 6 & 15 & 21 & 9 \end{vmatrix}$$

When the constituents of the last row are divided by 3, they become identical with those of the second row.

4. Prove the identity

$$\begin{vmatrix} bc & a & a^2 \\ ca & b & b^2 \\ ab & c & c^2 \end{vmatrix} \equiv \begin{vmatrix} 1 & a^2 & a^3 \\ 1 & b^2 & b^3 \\ 1 & c^2 & c^3 \end{vmatrix}.$$

Represent the first determinant by Δ, and multiply the rows by a, b, c, respectively. We have then

$$abc\,\Delta = \begin{vmatrix} abc & a^2 & a^3 \\ abc & b^2 & b^3 \\ abc & c^2 & c^3 \end{vmatrix};$$

and, dividing the first column by abc, the result follows.

5. Prove the identity

$$\begin{vmatrix} \beta\gamma\delta & \alpha & \alpha^2 & \alpha^3 \\ \gamma\delta\alpha & \beta & \beta^2 & \beta^3 \\ \delta\alpha\beta & \gamma & \gamma^2 & \gamma^3 \\ \alpha\beta\gamma & \delta & \delta^2 & \delta^3 \end{vmatrix} \equiv \begin{vmatrix} 1 & \alpha^2 & \alpha^3 & \alpha^4 \\ 1 & \beta^2 & \beta^3 & \beta^4 \\ 1 & \gamma^2 & \gamma^3 & \gamma^4 \\ 1 & \delta^2 & \delta^3 & \delta^4 \end{vmatrix}.$$

6. Prove the identity

$$\begin{vmatrix} 2 & 1 & -7 \\ -4 & -3 & 8 \\ 6 & 5 & -9 \end{vmatrix} = 2\begin{vmatrix} 1 & 1 & 7 \\ 2 & 3 & 8 \\ 3 & 5 & 9 \end{vmatrix}.$$

Change all the signs of the second row, and afterwards of the third column.

7. Prove the identity

$$\begin{vmatrix} \alpha & \beta & \gamma \\ \alpha' & \beta' & \gamma' \\ \alpha'' & \beta'' & \gamma'' \end{vmatrix} = \frac{1}{\alpha\beta\gamma}\begin{vmatrix} 1 & 1 & 1 \\ \alpha'\beta\gamma & \beta'\gamma\alpha & \gamma'\alpha\beta \\ \alpha''\beta\gamma & \beta''\gamma\alpha & \gamma''\alpha\beta \end{vmatrix}.$$

This is easily proved by multiplying the columns of the first determinant by $\beta\gamma$, $\gamma\alpha$, $\alpha\beta$, respectively ; and then dividing the first row by $\alpha\beta\gamma$.

It is evident that a similar process may be employed in general to reduce any determinant to one in which all the constituents of any selected row or column shall be units.

8. Reduce the following determinant to one in which the first row shall consist of units :

$$\Delta \equiv \begin{vmatrix} 4 & 2 & 5 & 10 \\ 1 & 1 & 6 & 3 \\ 7 & 3 & 0 & 5 \\ 0 & 2 & 5 & 8 \end{vmatrix}.$$

Since 20 is the least common multiple of 4, 2, 5, 10, it is sufficient to multiply the columns in order by 5, 10, 4, 2 : we thus obtain

$$\Delta = \frac{1}{5 \cdot 10 \cdot 4 \cdot 2} \begin{vmatrix} 20 & 20 & 20 & 20 \\ 5 & 10 & 24 & 6 \\ 35 & 30 & 0 & 10 \\ 0 & 20 & 20 & 16 \end{vmatrix}.$$

Taking out the multiplier 20 from the first row, 5 from the third row, and 4 from the fourth row, we get finally

$$\Delta = \begin{vmatrix} 1 & 1 & 1 & 1 \\ 5 & 10 & 24 & 6 \\ 7 & 6 & 0 & 2 \\ 0 & 5 & 5 & 4 \end{vmatrix}.$$

9. Prove the identity

$$\begin{vmatrix} 1 & 1 & 1 \\ \alpha & \beta & \gamma \\ \alpha^2 & \beta^2 & \gamma^2 \end{vmatrix} \equiv (\beta - \gamma)(\gamma - \alpha)(\alpha - \beta).$$

Since if β were equal to γ, two columns would become identical ; $\beta - \gamma$ must be a factor in the determinant. Similarly, $\gamma - \alpha$ and $\alpha - \beta$ must be factors in it. Hence the product of the three differences can differ by a numerical factor only from the value of the determinant, since both functions are of the third degree in α, β, γ ; and by comparing the term $\beta\gamma^2$ we observe that this factor is $+ 1$.

10. Prove similarly the identity

$$\begin{vmatrix} 1 & 1 & 1 & 1 \\ \alpha & \beta & \gamma & \delta \\ \alpha^2 & \beta^2 & \gamma^2 & \delta^2 \\ \alpha^3 & \beta^3 & \gamma^3 & \delta^3 \end{vmatrix} = -(\beta - \gamma)(\alpha - \delta)(\gamma - \alpha)(\beta - \delta)(\alpha - \beta)(\gamma - \delta).$$

It is evident that a similar proof shows in general that the value of the determinant of this form, constituted by the n quantities α, β, $\gamma \ldots \lambda$, is the product of the $\frac{1}{2}n(n-1)$ differences which can be formed with these n quantities.

113. Minor Determinants. Definitions.—When in a determinant any number of rows, and the same number of columns, are erased, the determinant formed by the remaining constituents (maintaining their relative positions) is called a *minor determinant*.

If one row and one column only be suppressed, the corresponding minor is called a *first minor*. If two rows and two columns be suppressed, the minor is called a *second minor ;* and so on. The suppressed rows and columns have common constituents forming a determinant; and the minor which remains is said to be *complementary* to this determinant.

It is usual to denote a determinant in general by Δ. We shall denote by Δ_α the minor obtained by suppressing in Δ the row and column which contain any constituent α ; by $\Delta_{\alpha,\beta}$ that obtained by suppressing the two rows and two columns which contain α and β ; and so on.

The determinant Δ, formed by the constituents a_1, b_1, c_1, &c., is often denoted for brevity by placing the leading term within brackets, as follows: $\Delta = (a_1 b_2 c_3 \ldots\ldots l_n)$. The notation $\Sigma \pm a_1 b_2 c_3 \ldots l_n$ is also used to represent Δ ; this expressing its constitution as consisting of the sum of a number of elements (with their proper signs attached) formed by taking all possible permutations of the n suffixes.

114. Development of Determinants.—Since every term of any determinant contains one, and only one, constituent from each row and from each column, it follows that Δ *is a linear and homogeneous function of the constituents of any one row or any one column.* Thus we may write

$$\Delta = a_1 A_1 + a_2 A_2 + a_3 A_3 + \text{&c.}$$

$$\Delta = b_1 B_1 + b_2 B_2 + b_3 B_3 + \text{&c.}$$

or, again,

$$\Delta = a_1 A_1 + b_1 B_1 + c_1 C_1 + \text{&c.}$$

The student, on referring to Ex. 3, Art. 108, will observe that the determinant of the fourth order there written at length is constituted in the way here described, namely

$$\Delta = a_1 \begin{vmatrix} b_2 & c_2 & d_2 \\ b_3 & c_3 & d_3 \\ b_4 & c_4 & d_4 \end{vmatrix} + a_2 \begin{vmatrix} b_1 & c_1 & d_1 \\ b_4 & c_4 & d_4 \\ b_3 & c_3 & d_3 \end{vmatrix} + a_3 \begin{vmatrix} b_1 & c_1 & d_1 \\ b_2 & c_2 & d_2 \\ b_4 & c_4 & d_4 \end{vmatrix} + a_4 \begin{vmatrix} b_1 & c_1 & d_1 \\ b_3 & c_3 & d_3 \\ b_2 & c_2 & d_2 \end{vmatrix}.$$

Now it is true, in general, that the coefficients A_1, A_2, A_3, &c., are determinants of the order $n - 1$. For, suppose Δ to be written as follows :—

$$\Delta = a_1 A_1 + a_2 A_2 + a_3 A_3 + \ldots + a_n A_n.$$

In effecting all the permutations of the suffixes $1, 2, 3, \ldots n$, suppose first 1 to remain in the leading place, as in the example referred to, we then obtain $1 . 2 . 3 \ldots (n - 1)$ terms which have a_1 as a factor, and

$$a_1 A_1 = a_1 \Sigma \pm b_2 c_3 \ldots l_n\,;$$

hence

$$A_1 = \Sigma \pm b_2 c_3 \ldots l_n = \begin{vmatrix} b_2 & c_2 & \ldots & l_2 \\ b_3 & c_3 & \ldots & l_3 \\ . & . & . & . \\ b_n & c_n & \ldots & l_n \end{vmatrix};$$

and this determinant is the minor corresponding to the constituent a_1, or $A_1 = \Delta_{a_1}$.

To find the value of A_2, we bring a_2 into the leading place by one displacement of rows. This changes the sign of Δ, so that we obtain

$$A_2 = - \Delta_{a_2}\,;$$

i.e. $A_2 =$ the minor corresponding to a_2 with its sign changed. Again, bringing a_3 to the leading place by two displacements, we have

$$A_3 = \Delta_{a_3}\,;$$

and so on.

Thus we conclude that, in general,

$$\Delta = a_1 \Delta_{a_1} - a_2 \Delta_{a_2} + a_3 \Delta_{a_3} - a_4 \Delta_{a_4} + \&c.$$

Similarly, we can expand Δ in terms of the constituents of any other column, or any row. For example,

$$\Delta = a_1 \Delta_{a_1} - b_1 \Delta_{b_1} + c_1 \Delta_{c_1} - \&c.$$

If it be required to obtain the proper sign to be attached to the minor which multiplies any constituent in the expanded form, we have only to consider how many displacements would bring that constituent to the leading place. For example, suppose the determinant $(a_1 b_2 c_3 d_4 e_5)$ is expanded in terms of its fourth column, and that it is required to find what sign is to be attached to $d_3 \Delta_{d_3}$. Here two displacements upwards, and afterwards three to the left, will bring d_3 to the leading place ; hence the sign is negative. This rule may be stated simply as follows :— *Proceed from a_1 to the constituent in question along the top row, and down the column containing the constituent; the number of letters passed over before reaching the constituent will decide the sign to be attached to the minor.* In the example just given ; beginning at a_1 we count a_1, b_1, c_1, d_1, d_2, i. e. five ; and this number being odd, the required sign is negative ; if the number were even the sign would be positive.

EXAMPLES.

1.
$$\begin{vmatrix} a_1 & b_1 & c_1 \\ a_2 & b_2 & c_2 \\ a_3 & b_3 & c_3 \end{vmatrix} = a_1 \begin{vmatrix} b_2 & c_2 \\ b_3 & c_3 \end{vmatrix} - a_2 \begin{vmatrix} b_1 & c_1 \\ b_3 & c_3 \end{vmatrix} + a_3 \begin{vmatrix} b_1 & c_1 \\ b_2 & c_2 \end{vmatrix}.$$

$$= a_1 b_2 c_3 - a_1 b_3 c_2 - a_2 b_1 c_3 + a_2 b_3 c_1 + a_3 b_1 c_2 - a_3 b_2 c_1.$$

(Compare (2), Art. 107.)

2.
$$\begin{vmatrix} a & h & g \\ h & b & f \\ g & f & c \end{vmatrix} = a \begin{vmatrix} b & f \\ f & c \end{vmatrix} - h \begin{vmatrix} h & g \\ f & c \end{vmatrix} + g \begin{vmatrix} h & g \\ b & f \end{vmatrix}$$

$$= abc + 2fgh - af^2 - bg^2 - ch^2.$$

3. Expand the determinant of the fourth order in terms of the constituents of the fourth row.

$$\Delta = - a_4 \Delta_{a_4} + b_4 \Delta_{b_4} - c_4 \Delta_{c_4} + d_4 \Delta_{d_4}$$

$$= - a_4 \begin{vmatrix} b_1 & c_1 & d_1 \\ b_2 & c_2 & d_2 \\ b_3 & c_3 & d_3 \end{vmatrix} + b_4 \begin{vmatrix} a_1 & c_1 & d_1 \\ a_2 & c_2 & d_2 \\ a_3 & c_3 & d_3 \end{vmatrix} - c_4 \begin{vmatrix} a_1 & b_1 & d_1 \\ a_2 & b_2 & d_2 \\ a_3 & b_3 & d_3 \end{vmatrix} + d_4 \begin{vmatrix} a_1 & b_1 & c_1 \\ a_2 & b_2 & c_2 \\ a_3 & b_3 & c_3 \end{vmatrix}.$$

When the determinants of the third order are expanded this will give the expression of Ex. 3, Art. 108, as the student will easily verify.

4.
$$\begin{vmatrix} 3 & 2 & 4 \\ 7 & 6 & 1 \\ 5 & 3 & 8 \end{vmatrix} = 3 \begin{vmatrix} 6 & 1 \\ 3 & 8 \end{vmatrix} - 7 \begin{vmatrix} 2 & 4 \\ 3 & 8 \end{vmatrix} + 5 \begin{vmatrix} 2 & 4 \\ 6 & 1 \end{vmatrix}$$

$$= 3 (48 - 3) - 7 (16 - 12) + 5 (2 - 24).$$

$$= - 3.$$

5. Find the value of the determinant

$$\Delta = \begin{vmatrix} 8 & 7 & 2 & 20 \\ 3 & 1 & 4 & 7 \\ 5 & 0 & 11 & 0 \\ 8 & 1 & 0 & 6 \end{vmatrix}.$$

It is evidently convenient to expand this in terms of the third row, since two of the constituents in that row vanish.

$$\Delta = 5 \begin{vmatrix} 7 & 2 & 20 \\ 1 & 4 & 7 \\ 1 & 0 & 6 \end{vmatrix} + 11 \begin{vmatrix} 8 & 7 & 20 \\ 3 & 1 & 7 \\ 8 & 1 & 6 \end{vmatrix} ;$$

and expanding the two determinants of the 3rd. order, we find $\Delta = 2188$.

6. Expand

$$\begin{vmatrix} 1 & \alpha & -\beta \\ -\alpha & 1 & \gamma \\ \beta & -\gamma & 1 \end{vmatrix}.$$

Ans. $1 + \alpha^2 + \beta^2 + \gamma^2$.

7. Expand

$$\begin{vmatrix} 0 & c & b & d \\ c & 0 & a & e \\ b & a & 0 & f \\ d & e & f & 0 \end{vmatrix}.$$

Ans. $a^2 d^2 + b^2 e^2 + c^2 f^2 - 2 bcef - 2 cafd - 2 abde$.

8. Prove

$$\begin{vmatrix} 1 & \alpha & \beta & \gamma \\ -\alpha & 1 & \gamma' & -\beta' \\ -\beta & -\gamma' & 1 & \alpha' \\ -\gamma & \beta' & -\alpha' & 1 \end{vmatrix} = 1 + \alpha^2 + \beta^2 + \gamma^2 + (\alpha\alpha' + \beta\beta' + \gamma\gamma')^2.$$

9. Expand

$$\begin{vmatrix} -a & b & c & d \\ b & -a & d & c \\ c & d & -a & b \\ d & c & b & -a \end{vmatrix}.$$

Ans. $a^4 + b^4 + c^4 + d^4 - 2b^2c^2 - 2c^2a^2 - 2a^2b^2 - 2a^2d^2 - 2b^2d^2 - 2c^2d^2 - 8abcd.$

10. Prove the following identity, and expand the determinants :—

$$\begin{vmatrix} 0 & 1 & 1 & 1 \\ 1 & 0 & z^2 & y^2 \\ 1 & z^2 & 0 & x^2 \\ 1 & y^2 & x^2 & 0 \end{vmatrix} = \begin{vmatrix} 0 & x & y & z \\ x & 0 & z & y \\ y & z & 0 & x \\ z & y & x & 0 \end{vmatrix}.$$

Ans. $x^4 + y^4 + z^4 - 2y^2z^2 - 2z^2x^2 - 2x^2y^2.$

11. Show that if the first two columns of a determinant of the third order be written after the third, as follows :—

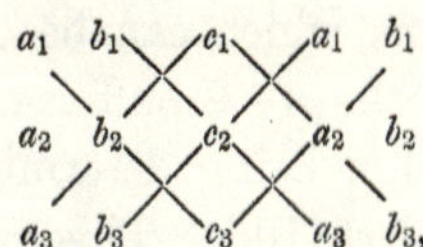

then the expansion of the determinant may be obtained by writing with positive signs the three products which lie in the diagonal lines descending from left to right, and with negative signs the three products which lie in the diagonal lines ascending from left to right.

This rule, due to M. Sarrus, is practically convenient in expanding a determinant of the third order.

12. Find the value of the determinant

$$\Delta \equiv \begin{vmatrix} a & h & g & \lambda \\ h & b & f & \mu \\ g & f & c & \nu \\ \lambda & \mu & \nu & 0 \end{vmatrix}.$$

Expand first in terms of the last row or last column, and then each of the determinants of the third order in terms of λ, μ, ν.

Ans. $-\Delta = (bc - f^2)\lambda^2 + (ca - g^2)\mu^2 + (ab - h^2)\nu^2 + 2(gh - af)\mu\nu + 2(hf - bg)\nu\lambda + 2(fg - ch)\lambda\mu.$

115. Development of Determinants continued.—
The expansion explained in the preceding Article is included
in a more general mode of development given by Laplace.
In place of expanding the determinant as a linear function
of the constituents of any line, we expand it as a linear function of the minors comprised in any number of lines.

Consider, for example, the first two columns (a, b) of any
determinant; and let all possible determinants of the second
order $(a_p b_q)$, obtained by taking any two rows of these two
columns, be formed. Let the minor formed by suppressing
the a_p and b_q lines be represented by $\Delta_{p, q}$; then the determinant can be expanded in the form $\Sigma \pm (a_p b_q) \Delta_{p, q}$, where
each term is the product of two complementary determinants
(see Art. 113). To prove this, we observe that every term of
the determinant must contain one constituent from the column
a and one from the column b. Suppose a term to contain the
factor $a_p b_q$, there must then (interchanging p and q) be another
term containing the factor $-a_q b_p$; hence, the determinant can
be expanded in the form $\Sigma (a_p b_q) A_{p, q}$; and $A_{p, q}$ is plainly
the sum of all the terms which can be obtained by permuting
in every possible way the $n-2$ suffixes of the letters c, d, e,
&c., viz., $\pm \Delta_{p, q}$, the sign being determined in any particular
instance by the rule of Art. 108. This reasoning can easily be
extended to the case where any number p of columns are taken,
and all possible minors formed by taking p rows of these
columns. Each minor is then multiplied by the *complementary*
minor, and the determinant expressed as the sum of all such
products with their proper signs.

Examples.

1. Expand the determinant $(a_1 b_2 c_3 d_4)$ in terms of the minors of the second order
formed from the first two columns.

Employing the bracket notation, we can write down the result as follows :—

$$(a_1 b_2)(c_3 d_4) - (a_1 b_3)(c_2 d_4) + (a_1 b_4)(c_2 d_3) + (a_2 b_3)(c_1 d_4) - (a_2 b_4)(c_1 d_3) + (a_3 b_4)(c_1 d_2);$$

the signs being determined by the method of Art. 108; for since the letters within
the brackets represent the leading terms of the several determinants, the signs may

be determined as if the brackets were expunged, and the terms were then elements of the determinant.

2. Expand similarly the determinant $(a_1 b_2 c_3 d_4 e_5)$.

$$Ans. \quad (a_1 b_2)(c_3 d_4 e_5) - (a_1 b_3)(c_2 d_4 e_5) + (a_1 b_4)(c_2 d_3 e_5) - (a_1 b_5)(c_2 d_3 e_4)$$

$$+ (a_2 b_3)(c_1 d_4 e_5) - (a_2 b_4)(c_1 d_3 e_5) + (a_2 b_5)(c_1 d_3 e_4) + (a_3 b_4)(c_1 d_2 e_5)$$

$$- (a_3 b_5)(c_1 d_2 e_4) + (a_4 b_5)(c_1 d_2 e_3).$$

3. Prove the identity

$$\begin{vmatrix} a_1 & b_1 & c_1 & d_1 & e_1 & f_1 \\ a_2 & b_2 & c_2 & d_2 & e_2 & f_2 \\ a_3 & b_3 & c_3 & d_3 & e_3 & f_3 \\ 0 & 0 & 0 & \alpha_1 & \beta_1 & \gamma_1 \\ 0 & 0 & 0 & \alpha_2 & \beta_2 & \gamma_2 \\ 0 & 0 & 0 & \alpha_3 & \beta_3 & \gamma_3 \end{vmatrix} = \begin{vmatrix} a_1 & b_1 & c_1 \\ a_2 & b_2 & c_2 \\ a_3 & b_3 & c_3 \end{vmatrix} \begin{vmatrix} \alpha_1 & \beta_1 & \gamma_1 \\ \alpha_2 & \beta_2 & \gamma_2 \\ \alpha_3 & \beta_3 & \gamma_3 \end{vmatrix}.$$

This appears by expanding the determinant in terms of the minors formed from the first three columns, for it is evident that all these minors vanish (having one row of ciphers) except one, viz. $(a_1 b_2 c_3)$.

In general it appears in the same way that if a determinant of the $2m^{th}$ order contains in any position a square of m^2 ciphers, it can be expressed as the product of two determinants of the mth order.

4. Expand the determinant

$$\begin{vmatrix} a & h & g & \lambda & \lambda' \\ h & b & f & \mu & \mu' \\ g & f & c & \nu & \nu' \\ \lambda & \mu & \nu & 0 & 0 \\ \lambda' & \mu' & \nu' & 0 & 0 \end{vmatrix}$$

in powers of α, β, γ, where

$$\alpha \equiv \mu\nu' - \mu'\nu, \quad \beta \equiv \nu\lambda' - \nu'\lambda, \quad \gamma \equiv \lambda\mu' - \lambda'\mu.$$

$$Ans. \quad a\alpha^2 + b\beta^2 + c\gamma^2 + 2f\beta\gamma + 2g\gamma\alpha + 2h\alpha\beta.$$

116. Addition of Determinants. Prop. V.—*If every constituent in any row or column can be resolved into the sum of two others, the determinant can be resolved into the sum of two others.*

Suppose the constituents of the first column to be $a_1 + a_1,$

$a_2 + a_2$, $a_3 + a_3$, &c.　Substituting these in the expansion of Art. 114, we have

$$\Delta = (a_1 + a_1)\,A_1 + (a_2 + a_2)\,A_2 + (a_3 + a_3)\,A_3 + \&c.,$$

$$= a_1 A_1 + a_2 A_2 + a_3 A_3 + \,.\,. \,\&c. + a_1 A_1 + a_2 A_2 + a_3 A_3 + \&c.\,;$$

or,

$$
\begin{vmatrix}
a_1 + a_1 & b_1 & c_1 .. \\
a_2 + a_2 & b_2 & c_2 .. \\
a_3 + a_3 & b_3 & c_3 .. \\
.\;\;.\;\;.\;\;.\;\;.\;\;.\;\;.
\end{vmatrix}
=
\begin{vmatrix}
a_1 & b_1 & c_1 .. \\
a_2 & b_2 & c_2 .. \\
a_3 & b_3 & c_3 .. \\
.\;\;.\;\;.\;\;.\;\;.\;\;.
\end{vmatrix}
+
\begin{vmatrix}
a_1 & b_1 & c_1 .. \\
a_2 & b_2 & c_2 .. \\
a_3 & b_3 & c_3 .. \\
.\;\;.\;\;.\;\;.\;\;.\;\;.
\end{vmatrix},
$$

which proves the proposition.

If a second column consists of the sum of two others, it is easily seen, by first resolving with reference to one column, and afterwards with reference to the other, that the determinant can be resolved into the sum of four others.　For example, the determinant

$$
\begin{vmatrix}
a_1 + a_1 & b_1 + \beta_1 & c_1 \\
a_2 + a_2 & b_2 + \beta_2 & c_2 \\
a_3 + a_3 & b_3 + \beta_3 & c_3
\end{vmatrix}
$$

is (using the notation of Art. 113) equal to the sum of the four determinants

$$(a_1 b_2 c_3) + (a_1 b_2 c_3) + (a_1 \beta_2 c_3) + (a_1 \beta_2 c_3).$$

Similarly it follows that if each of the constituents of one column consists of the algebraical sum of any number of terms, the determinant can be resolved into a corresponding number of determinants.　For example

$$
\begin{vmatrix}
a_1 - a_1 + a'_1 & b_1 & c_1 \\
a_2 - a_2 + a'_2 & b_2 & c_2 \\
a_3 - a_3 + a'_3 & b_3 & c_3
\end{vmatrix}
=
\begin{vmatrix}
a_1 & b_1 & c_1 \\
a_2 & b_2 & c_2 \\
a_3 & b_3 & c_3
\end{vmatrix}
-
\begin{vmatrix}
a_1 & b_1 & c_1 \\
a_2 & b_2 & c_2 \\
a_3 & b_3 & c_3
\end{vmatrix}
+
\begin{vmatrix}
a'_1 & b_1 & c_1 \\
a'_2 & b_2 & c_2 \\
a'_3 & b_3 & c_3
\end{vmatrix}.
$$

And, in general, if one column consist of the algebraic sum of

m others, a second column of the sum of n others, a third of the sum of p others, &c., the determinant can be resolved into the sum of $mnp \ldots$, &c., others.

The results here established in the case of the columns are also of course true when rows can be similarly resolved into sums.

117. PROP. VI.—*If the constituents of one line are equal to the sums of the corresponding constituents of the other lines multiplied by constant factors, the determinant vanishes.*

For it can then be resolved into the sum of a number of determinants which separately vanish. For example,

$$
\begin{vmatrix} ma_1 + nb_1 & a_1 & b_1 \\ ma_2 + nb_2 & a_2 & b_2 \\ ma_3 + nb_3 & a_3 & b_3 \end{vmatrix} = m \begin{vmatrix} a_1 & a_1 & b_1 \\ a_2 & a_2 & b_2 \\ a_3 & a_3 & b_3 \end{vmatrix} + n \begin{vmatrix} b_1 & a_1 & b_1 \\ b_2 & a_2 & b_2 \\ b_3 & a_3 & b_3 \end{vmatrix},
$$

and each of the latter determinants vanishes (Art. 110).

118. PROP. VII.—*A determinant is unchanged when to each constituent of any row or column are added those of several other rows or columns, multiplied respectively by constant factors.*

For when the determinant is resolved into the sum of others, as in Art. 116, the determinants in which the added lines occur all vanish, since each of them must, when the constant factor is removed, contain two identical lines. Thus, for example,

$$
\begin{vmatrix} a_1 & b_1 & c_1 \\ a_2 & b_2 & c_2 \\ a_3 & b_3 & c_3 \end{vmatrix} = \begin{vmatrix} a_1 + mb_1 + nc_1 & b_1 & c_1 \\ a_2 + mb_2 + nc_2 & b_2 & c_2 \\ a_3 + mb_3 + nc_3 & b_3 & c_3 \end{vmatrix};
$$

for when the second determinant is expressed as the sum of three others, the two arising from the added columns vanish identically (Art. 117).

The proposition of the present Article supplies in practice one of the most useful properties in the evaluation of determinants.

EXAMPLES.

1. Show that the following determinant vanishes :—

$$\begin{vmatrix} \beta + \gamma & \alpha & 1 \\ \gamma + \alpha & \beta & 1 \\ \alpha + \beta & \gamma & 1 \end{vmatrix}.$$

Adding the constituents of the second column to those of the first, we can take out $\alpha + \beta + \gamma$ as a factor, and two columns then become identical.

2. Find the value of the determinant

$$\begin{vmatrix} 1 & 2 & 4 \\ 2 & 3 & 7 \\ 3 & 4 & 10 \end{vmatrix}.$$

Subtracting the constituents of the first column from those of the second, and three times the constituents of the first column from those of the third, we obtain

$$\begin{vmatrix} 1 & 1 & 1 \\ 2 & 1 & 1 \\ 3 & 1 & 1 \end{vmatrix},$$

which vanishes identically.

3.
$$\begin{vmatrix} -1 & 1 & 1 & 1 \\ 1 & -1 & 1 & 1 \\ 1 & 1 & -1 & 1 \\ 1 & 1 & 1 & -1 \end{vmatrix} = \begin{vmatrix} -1 & 1 & 1 & 1 \\ 0 & 0 & 2 & 2 \\ 0 & 2 & 0 & 2 \\ 0 & 2 & 2 & 0 \end{vmatrix} = - \begin{vmatrix} 0 & 2 & 2 \\ 2 & 0 & 2 \\ 2 & 2 & 0 \end{vmatrix} = -16.$$

Here the first transformation is obtained by adding in succession the constituents of the first row to those of the second, third, and fourth.

4.
$$\begin{vmatrix} 7 & 11 & 4 \\ 13 & 15 & 10 \\ 3 & 9 & 6 \end{vmatrix} = 3\begin{vmatrix} 7 & 11 & 4 \\ 13 & 15 & 10 \\ 1 & 3 & 2 \end{vmatrix} = 3\begin{vmatrix} 7 & -10 & -10 \\ 13 & -24 & -16 \\ 1 & 0 & 0 \end{vmatrix} = 3\begin{vmatrix} 10 & 10 \\ 24 & 16 \end{vmatrix}$$

$$= 30\,(16 - 24) = -240.$$

Here the second transformation is obtained by subtracting three times the first column from the second, and twice the first from the third. In examples of this kind attempts should be made to reduce to zero all the constituents except one in some row or column, in which case the calculation reduces to that of a determinant of lower order. This can always be done by reducing any one line to units, as in Ex. 7, Art. 112; but in general it can be effected more readily by direct additions or subtractions, as in the present instance.

5.
$$\begin{vmatrix} 7 & -2 & 0 & 5 \\ -2 & 6 & -2 & 2 \\ 0 & -2 & 5 & 3 \\ 5 & 2 & 3 & 4 \end{vmatrix} = \begin{vmatrix} 7 & -2 & 0 & 5 \\ 19 & 0 & -2 & 17 \\ -7 & 0 & 5 & -2 \\ 12 & 0 & 3 & 9 \end{vmatrix} = 2\begin{vmatrix} 19 & -2 & 17 \\ -7 & 5 & -2 \\ 12 & 3 & 9 \end{vmatrix}.$$

The first transformation is obtained by adding to the second row three times the first, subtracting the first from the third row, and adding the first to the fourth row. The reduced determinant is easily calculated by subtracting four times the second column from the first, and three times the second column from the third. Thus

$$2\begin{vmatrix} 19 & -2 & 17 \\ -7 & 5 & -2 \\ 12 & 3 & 9 \end{vmatrix} = 2\begin{vmatrix} 27 & -2 & 23 \\ -27 & 5 & -17 \\ 0 & 3 & 0 \end{vmatrix} = -6\begin{vmatrix} 27 & 23 \\ -27 & -17 \end{vmatrix} = -972.$$

6. Calculate the determinant

$$\Delta \equiv \begin{vmatrix} 1 & 15 & 14 & 4 \\ 12 & 6 & 7 & 9 \\ 8 & 10 & 11 & 5 \\ 13 & 3 & 2 & 16 \end{vmatrix}.$$

The first sixteen natural numbers are arranged here in what is called a " magic square," *i.e.* the sum of all the figures in any row or in any column is constant. In general for a square of the first n^2 numbers this sum is $\frac{1}{2}n(n^2+1)$. Determinants of this kind can be at once reduced one degree. Here adding the last three columns to the first, and subtracting the last row from each of the others, we have

$$\Delta = 34\begin{vmatrix} 1 & 15 & 14 & 4 \\ 1 & 6 & 7 & 9 \\ 1 & 10 & 11 & 5 \\ 1 & 3 & 2 & 16 \end{vmatrix} = 34\begin{vmatrix} 0 & 12 & 12 & -12 \\ 0 & 3 & 5 & -7 \\ 0 & 7 & 9 & -11 \\ 1 & 3 & 2 & 16 \end{vmatrix} = -34 \times 12\begin{vmatrix} 1 & 1 & -1 \\ 3 & 5 & -7 \\ 7 & 9 & -11 \end{vmatrix};$$

and subtracting the second row from the last row, it is evident that the reduced determinant vanishes; hence $\Delta = 0$.

7. Calculate the determinant formed by the first nine natural numbers arranged in a magic square:

$$\begin{vmatrix} 4 & 9 & 2 \\ 3 & 5 & 7 \\ 8 & 1 & 6 \end{vmatrix}.$$

Ans. 360.

8. Calculate the determinant formed by the first twenty-five natural numbers arranged in a magic square :

$$
\begin{vmatrix}
10 & 18 & 1 & 14 & 22 \\
4 & 12 & 25 & 8 & 16 \\
23 & 6 & 19 & 2 & 15 \\
17 & 5 & 13 & 21 & 9 \\
11 & 24 & 7 & 20 & 3
\end{vmatrix} .
$$

Ans. $-4680000.$

9. Expand, by the method of the present Article, the determinant of Ex. 10, Art. 114.

$$
\Delta = \begin{vmatrix}
0 & 1 & 1 & 1 \\
1 & 0 & z^2 & y^2 \\
1 & z^2 & 0 & x^2 \\
1 & y^2 & x^2 & 0
\end{vmatrix}
= \begin{vmatrix}
0 & 1 & 0 & 0 \\
1 & 0 & z^2 & y^2 \\
1 & z^2 & -z^2 & x^2-z^2 \\
1 & y^2 & x^2-y^2 & -y^2
\end{vmatrix}
= - \begin{vmatrix}
1 & z^2 & y^2 \\
1 & -z^2 & x^2-z^2 \\
1 & x^2-y^2 & -y^2
\end{vmatrix} .
$$

Here, to obtain the second determinant we subtract the second column from each of the following ones. In the reduced determinant, subtracting the first row from each of the following, we find

$$
\Delta = - \begin{vmatrix}
1 & z^2 & y^2 \\
0 & -2z^2 & x^2-z^2-y^2 \\
0 & x^2-y^2-z^2 & -2y^2
\end{vmatrix}
= - \begin{vmatrix}
2z^2 & y^2+z^2-x^2 \\
y^2+z^2-x^2 & 2y^2
\end{vmatrix}
$$

$$
= (y^2 + z^2 - x^2)^2 - 4y^2 z^2
$$
$$
= (y^2 + z^2 - x^2 + 2yz)(y^2 + z^2 - x^2 - 2yz)
$$
$$
= \{(y + z)^2 - x^2\} \{(y - z)^2 - x^2\}
$$
$$
= - (x + y + z)(y + z - x)(z + x - y)(x + y - z).
$$

10. Prove the identity

$$
\Delta \equiv \begin{vmatrix}
(b + c)^2 & a^2 & a^2 \\
b^2 & (c + a)^2 & b^2 \\
c^2 & c^2 & (a + b)^2
\end{vmatrix} = 2abc(a + b + c)^3.
$$

Subtracting the last column from each of the others, $(a + b + c)^2$ may be taken out as a factor. Calling the remaining determinant Δ', and subtracting in it the sum of the first two rows from the last, we have

$$
\Delta' = \begin{vmatrix}
b+c-a & 0 & a^2 \\
0 & c+a-b & b^2 \\
c-a-b & c-a-b & (a+b)^2
\end{vmatrix}
= \begin{vmatrix}
b+c-a & 0 & a^2 \\
0 & c+a-b & b^2 \\
-2b & -2a & 2ab
\end{vmatrix}
$$

$$
= \frac{1}{ab} \begin{vmatrix}
a(b+c-a) & 0 & a^2 \\
0 & b(c+a-b) & b^2 \\
-2ab & -2ab & 2ab
\end{vmatrix} .
$$

Adding the last column to each of the others, we obtain

$$\Delta' = \frac{1}{ab} \begin{vmatrix} a(b+c) & a^2 & a^2 \\ b^2 & b(c+a) & b^2 \\ 0 & 0 & 2ab \end{vmatrix} = 2 \begin{vmatrix} a(b+c) & a^2 \\ b^2 & b(c+a) \end{vmatrix} = 2ab \begin{vmatrix} b+c & a \\ b & c+a \end{vmatrix}$$

$$= 2abc(a+b+c).$$

Hence,

$$\Delta = \Delta'(a+b+c)^2 = 2abc(a+b+c)^3.$$

11. Prove the identity

$$\begin{vmatrix} 1 & 1 & 1 \\ \alpha & \beta & \gamma \\ \alpha^3 & \beta^3 & \gamma^3 \end{vmatrix} \equiv (\beta - \gamma)(\gamma - \alpha)(\alpha - \beta)(\alpha + \beta + \gamma).$$

Subtracting the first column from each of the others, $\beta - \alpha$ and $\gamma - \alpha$ become factors. In the reduced determinant, subtract the first row multiplied by α^2 from the second row.

12. Resolve into simple factors the determinant

$$\Delta \equiv \begin{vmatrix} 1 & 1 & 1 & 1 \\ \alpha & \beta & \gamma & \delta \\ \alpha^2 & \beta^2 & \gamma^2 & \delta^2 \\ \alpha^4 & \beta^4 & \gamma^4 & \delta^4 \end{vmatrix}.$$

Proceeding as in Ex. 11, we easily find that $(\beta - \alpha)(\gamma - \alpha)(\delta - \alpha)$ is a factor, and that the reduced determinant is

$$\begin{vmatrix} 1 & 1 & 1 \\ \beta + \alpha & \gamma + \alpha & \delta + \alpha \\ \beta^3 + \beta^2\alpha + \beta\alpha^2 + \alpha^3 & \gamma^3 + \gamma^2\alpha + \gamma\alpha^2 + \alpha^3 & \delta^3 + \delta^2\alpha + \delta\alpha^2 + \alpha^3 \end{vmatrix}.$$

Subtracting the first column from each of the others, $(\gamma - \beta)(\delta - \beta)$ comes out as a factor, and the remaining factor is easily found to be $(\delta - \gamma)(\alpha + \beta + \gamma + \delta)$. Hence, finally,

$$\Delta = -(\beta - \gamma)(\alpha - \delta)(\gamma - \alpha)(\beta - \delta)(\alpha - \beta)(\gamma - \delta)(\alpha + \beta + \gamma + \delta).$$

119. Multiplication of Determinants. Prop. VIII.— *The product of two determinants of any order is itself a determinant of the same order.*

We shall prove this for two determinants of the third order. The student will observe, from the nature of the proof, that it is equally applicable in general. We propose to show that the product of the two determinants $(a_1 b_2 c_3)$, $(a_1 \beta_2 \gamma_3)$ is

$$\begin{vmatrix} a_1 a_1 + b_1 \beta_1 + c_1 \gamma_1 & a_1 a_2 + b_1 \beta_2 + c_1 \gamma_2 & a_1 a_3 + b_1 \beta_3 + c_1 \gamma_3 \\ a_2 a_1 + b_2 \beta_1 + c_2 \gamma_1 & a_2 a_2 + b_2 \beta_2 + c_2 \gamma_2 & a_2 a_3 + b_2 \beta_3 + c_2 \gamma_3 \\ a_3 a_1 + b_3 \beta_1 + c_3 \gamma_1 & a_3 a_2 + b_3 \beta_2 + c_3 \gamma_2 & a_3 a_3 + b_3 \beta_3 + c_3 \gamma_3 \end{vmatrix} ;$$

whose constituents are the sums of the products of the constituents in any row of $(a_1 b_2 c_3)$ by the corresponding constituents in any row of $(a_1 \beta_2 \gamma_3)$.

Since each column consists of the sum of three terms, this determinant can be expanded into the sum of 27 others (Art. 116). Now it will be observed that when any one of these is written down, a common factor can be taken off each column ; and that several of the partial determinants will, when these factors are removed, have two (or more) columns identical. The determinants which do not vanish in this way can be easily selected. Taking, for example, the first vertical line of the first column; this would give a vanishing determinant if we were to take along with it the first line of the second column. We take then the second line of the second column, and along with these two we must take the third line of the third column to obtain a determinant which does not vanish. Retaining still the first line of the first column, we may take the third line of the second column along with the second line of the third column. Taking out the common factors of the columns, we write down these two determinants as follows :—

$$a_1 \beta_2 \gamma_3 \begin{vmatrix} a_1 & b_1 & c_1 \\ a_2 & b_2 & c_2 \\ a_3 & b_3 & c_3 \end{vmatrix} + a_1 \gamma_2 \beta_3 \begin{vmatrix} a_1 & b_1 & c_1 \\ a_2 & b_2 & c_2 \\ a_3 & b_3 & c_3 \end{vmatrix} .$$

Taking in turn each of the other lines of the first column, we obtain four other determinants which do not vanish. Thus there are in all six terms ; and it is plain that $(a_1 b_2 c_3)$ is a factor in each of these. Taking out this factor, there remains the sum of six terms—

$$a_1\beta_2\gamma_3 - a_1\beta_3\gamma_2 - a_2\beta_1\gamma_3 + a_3\beta_1\gamma_2 + a_2\beta_3\gamma_1 - a_3\beta_2\gamma_1,$$

and this is the determinant $(a_1\beta_2\gamma_3)$. We have thus proved that the determinant above written is the product of the two given determinants.

In either of the given determinants the rows may be written in place of columns; hence, the product may be written in several different forms as a determinant; but these will, of course, give the same value when expanded.

Cor.—The square of a determinant is a symmetrical determinant.

Two constituents of a determinant are said to be *conjugate* when one is situated in the rows in the same place relatively to the leading constituent as the other is in the columns; and a *symmetrical* determinant is one in which the conjugate constituents are equal to one another. The student will find examples of such determinants on referring to Examples 2, 9, 10 of Art. 114, and Example 4, Art. 115. From these definitions the corollary follows immediately, by putting the several constituents of $(a_1\beta_2\gamma_3)$ equal to those of $(a_1b_2c_3)$ in the above expression for the product of these two determinants.

120. Multiplication of Determinants continued.—Another mode of proof of the proposition of the last Article, expressing as a determinant the product of two given determinants of the same order, may be derived from Laplace's mode of development already explained.

In the first of the following examples this method is applied to determine the product of two determinants of the third order. The student will have no difficulty in extending the proof to determinants of any order by the aid of the principles established in Art. 115.

EXAMPLES.

1. The product of the two determinants $(a_1 b_2 c_3)$, $(a_1 \beta_2 \gamma_3)$ is (see Ex. 3, Art. 115) plainly equal to the determinant

$$
\begin{vmatrix}
a_1 & b_1 & c_1 & 0 & 0 & 0 \\
a_2 & b_2 & c_2 & 0 & 0 & 0 \\
a_3 & b_3 & c_3 & 0 & 0 & 0 \\
-1 & 0 & 0 & \alpha_1 & \alpha_2 & \alpha_3 \\
0 & -1 & 0 & \beta_1 & \beta_2 & \beta_3 \\
0 & 0 & -1 & \gamma_1 & \gamma_2 & \gamma_3
\end{vmatrix}
$$

In this determinant add to the fourth column the sum of the first multiplied by α_1, the second by β_1, and the third by γ_1; add to the fifth column the sum of the first multiplied by α_2, the second by β_2, and the third by γ_2; and add to the sixth column the sum of the first multiplied by α_3, the second by β_3, and the third by γ_3. The determinant becomes then

$$
\begin{vmatrix}
a_1 & b_1 & c_1 & a_1\alpha_1 + b_1\beta_1 + c_1\gamma_1 & a_1\alpha_2 + b_1\beta_2 + c_1\gamma_2 & a_1\alpha_3 + b_1\beta_3 + c_1\gamma_3 \\
a_2 & b_2 & c_2 & a_2\alpha_1 + b_2\beta_1 + c_2\gamma_1 & a_2\alpha_2 + b_2\beta_2 + c_2\gamma_2 & a_2\alpha_3 + b_2\beta_3 + c_2\gamma_3 \\
a_3 & b_3 & c_3 & a_3\alpha_1 + b_3\beta_1 + c_3\gamma_1 & a_3\alpha_2 + b_3\beta_2 + c_3\gamma_2 & a_3\alpha_3 + b_3\beta_3 + c_3\gamma_3 \\
-1 & 0 & 0 & 0 & 0 & 0 \\
0 & -1 & 0 & 0 & 0 & 0 \\
0 & 0 & -1 & 0 & 0 & 0
\end{vmatrix}
$$

And this is, by Art. 115, equal to the product (with the proper sign) of the determinant

$$
\begin{vmatrix}
-1 & 0 & 0 \\
0 & -1 & 0 \\
0 & 0 & -1
\end{vmatrix}
\quad \text{(which is equal to} -1\text{),}
$$

by the complementary minor, which is the same determinant as that obtained in the preceding Article. That the sign to be attached to the product is negative is easily seen by moving down the first three rows till the diagonals of the two minors in question form the diagonal of the determinant itself. The student will have no difficulty in observing that, in the general case, the number of such displacements is odd when the order of the given determinants is odd, and even when it is even; so that the sign to be placed before the product-determinant of Art. 119 is always positive.

2. Show that the product of the two determinants

$$\begin{vmatrix} a + ib & c + id \\ -c + id & a - ib \end{vmatrix}, \quad \begin{vmatrix} a' - ib' & c' - id' \\ -c' - id' & a' + ib' \end{vmatrix},$$

where $i = \sqrt{-1}$, may be written in the form

$$\begin{vmatrix} D - iC & B - iA \\ -B - iA & D + iC \end{vmatrix};$$

where

$$A \equiv bc' - b'c + ad' - a'd, \quad B \equiv ca' - c'a + bd' - b'd,$$

$$C \equiv ab' - a'b + cd' - c'd, \quad D \equiv aa' + bb' + cc' + dd';$$

hence prove Euler's theorem

$$(a^2 + b^2 + c^2 + d^2)(a'^2 + b'^2 + c'^2 + d'^2)$$

$$\equiv (aa' + bb' + cc' + dd')^2 + (bc' - b'c + ad' - a'd)^2$$

$$+ (ca' - c'a + bd' - b'd)^2 + (ab' - a'b + cd' - c'd)^2,$$

viz., *the product of two sums each of four squares can be expressed as the sum of four squares.*

3. Prove the following expression for the square of a determinant of the third order :—

$$2 \begin{vmatrix} a & b & c \\ a' & b' & c' \\ a'' & b'' & c'' \end{vmatrix}^2 = \begin{vmatrix} 2(ac - b^2) & ac' + a'c - 2bb' & ac'' + a''c - 2bb'' \\ ac' + a'c - 2bb' & 2(a'c' - b'^2) & a'c'' + a''c' - 2b'b'' \\ ac'' + a''c - 2bb'' & a'c'' + a''c' - 2b'b'' & 2(a''c'' - b''^2) \end{vmatrix}.$$

This appears by multiplying the two determinants

$$\begin{vmatrix} a & b & c \\ a' & b' & c' \\ a'' & b'' & c'' \end{vmatrix}, \quad \begin{vmatrix} c & -2b & a \\ c' & -2b' & a' \\ c'' & -2b'' & a'' \end{vmatrix},$$

which differ only by the factor 2.

4. Show that two determinants of different orders may be multiplied together.

For their orders may be made equal; since the order of any determinant can be increased by adding any number of columns and the same number of rows consisting of units in the diagonal, and all the rest zero constituents. For example,

$$\begin{vmatrix} a_1 & b_1 \\ a_2 & b_2 \end{vmatrix} \quad \text{may be written} \quad \begin{vmatrix} 1 & 0 & 0 & 0 \\ 0 & 1 & 0 & 0 \\ 0 & 0 & a_1 & b_1 \\ 0 & 0 & a_2 & b_2 \end{vmatrix},$$

the only effect of the added constituents being to multiply the determinant by unity. More generally, one set of added constituents (*i.e.* those either to the right or the left of the diagonal) might be taken to be any quantities whatever, the remaining set being ciphers. Thus $(a_1 b_2)$ might be written

$$\begin{vmatrix} 1 & \alpha & \beta & \gamma \\ 0 & 1 & \delta & \epsilon \\ 0 & 0 & a_1 & b_1 \\ 0 & 0 & a_2 & b_2 \end{vmatrix};$$

for, expanding this according to the left-hand columns in succession, it is plain that the added constituents α, β, γ, δ, ϵ will not enter into the result.

121. Rectangular Arrays.—Arrays in which the number of rows is not equal to the number of columns may be called *rectangular*. These do not themselves represent any definite function; but if two such arrays of the same dimensions are given, a determinant can be derived from them by the process of Art. 119, whose value we proceed to investigate.

(1). *When the number of columns exceeds the number of rows.*

Take, for example, the two rectangular arrays,

$$\left. \begin{array}{cccc} a_1 & b_1 & c_1 & d_1 \\ a_2 & b_2 & c_2 & d_2 \end{array} \right\} (1), \qquad \left. \begin{array}{cccc} a_1 & \beta_1 & \gamma_1 & \delta_1 \\ a_2 & \beta_2 & \gamma_2 & \delta_2 \end{array} \right\} (2);$$

and, performing on these a process similar to that employed in multiplying two determinants, we obtain the determinant

$$\begin{vmatrix} a_1 a_1 + b_1 \beta_1 + c_1 \gamma_1 + d_1 \delta_1 & a_1 a_2 + b_1 \beta_2 + c_1 \gamma_2 + d_1 \delta_2 \\ a_2 a_1 + b_2 \beta_1 + c_2 \gamma_1 + d_2 \delta_1 & a_2 a_2 + b_2 \beta_2 + c_2 \gamma_2 + d_2 \delta_2 \end{vmatrix}$$

The value of this is easily found to be

$$(a_1 b_2)(a_1 \beta_2) + (a_1 c_2)(a_1 \gamma_2) + (a_1 d_2)(a_1 \delta_2) + (b_1 c_2)(\beta_1 \gamma_2)$$
$$+ (b_1 d_2)(\beta_1 \delta_2) + (c_1 d_2)(\gamma_1 \delta_2),$$

i. e. *the sum of the products of all possible determinants which can be formed from one array (by taking a number of columns equal to the number of rows) multiplied by the corresponding determinants formed from the other array.*

The student will have no difficulty in extending this proof to any two arrays of the kind here treated.

(2). *When the number of rows exceeds the number of columns.*

Take, for example, the two arrays,

$$\left.\begin{array}{cc} a_1 & b_1 \\ a_2 & b_2 \\ a_3 & b_3 \end{array}\right\} \ (1), \qquad \left.\begin{array}{cc} a_1 & \beta_1 \\ a_2 & \beta_2 \\ a_3 & \beta_3 \end{array}\right\} \ (2).$$

Performing the process of multiplication, we obtain the determinant

$$\begin{vmatrix} a_1 a_1 + b_1 \beta_1 & a_1 a_2 + b_1 \beta_2 & a_1 a_3 + b_1 \beta_3 \\ a_2 a_1 + b_2 \beta_1 & a_2 a_2 + b_2 \beta_2 & a_2 a_3 + b_2 \beta_3 \\ a_3 a_1 + b_3 \beta_1 & a_3 a_2 + b_3 \beta_2 & a_3 a_3 + b_3 \beta_3 \end{vmatrix}.$$

It will be observed that this determinant is the same as would arise if a column of ciphers were added to each of the given arrays and the determinants so formed then multiplied.

Hence the determinant vanishes, since it is the product of two factors, each equal to zero.

It readily appears that a similar proof applies in general. It is only necessary to add to each array columns of ciphers so as to make the number of columns equal to the number of rows, and then multiply the two determinants.

EXAMPLES.

1. From the two arrays

$$\left.\begin{array}{ccc} 1 & 1 & 1 \\ \alpha & \beta & \gamma \end{array}\right\} \ (1), \qquad \left.\begin{array}{ccc} 1 & 1 & 1 \\ \alpha & \beta & \gamma \end{array}\right\} \ (2),$$

prove

$$\begin{vmatrix} 3 & \alpha + \beta + \gamma \\ \alpha + \beta + \gamma & \alpha^2 + \beta^2 + \gamma^2 \end{vmatrix} = (\alpha - \beta)^2 + (\alpha - \gamma)^2 + (\beta - \gamma)^2.$$

2. From the two arrays

$$\left.\begin{array}{ccc} a & b & c \\ a' & b' & c' \end{array}\right\} \ (1), \qquad \left.\begin{array}{ccc} c & -2b & a \\ c' & -2b' & a' \end{array}\right\} \ (2),$$

prove

$$4\,(ac - b^2)\,(a'c' - b'^2) - (ac' + a'c - 2bb')^2 \equiv 4\,(bc' - b'c)\,(ab' - a'b) - (ac' - a'c)^2.$$

3. By squaring the array

$$\left. \begin{array}{ccc} a & b & c \\ a' & b' & c' \end{array} \right\} ,$$

prove

$$(a^2 + b^2 + c^2)(a'^2 + b'^2 + c'^2) = (aa' + bb' + cc')^2 + (bc' - b'c)^2 + (ca' - c'a)^2 + (ab' - a'b)^2.$$

4. Verify, by squaring the array

$$\left. \begin{array}{cccc} a & b & c & d \\ a' & b' & c' & d' \end{array} \right\} ,$$

the result of **Ex. 2, Art. 120.**

5. Prove the determinant identity

$$\begin{vmatrix} (a_1 - b_1)^2 & (a_1 - b_2)^2 & (a_1 - b_3)^2 & (a_1 - b_4)^2 \\ (a_2 - b_1)^2 & (a_2 - b_2)^2 & (a_2 - b_3)^2 & (a_2 - b_4)^2 \\ (a_3 - b_1)^2 & (a_3 - b_2)^2 & (a_3 - b_3)^2 & (a_3 - b_4)^2 \\ (a_4 - b_1)^2 & (a_4 - b_2)^2 & (a_4 - b_3)^2 & (a_4 - b_4)^2 \end{vmatrix} \equiv 0.$$

This can be proved by multiplying the two arrays

$$\left. \begin{array}{ccc} a_1^2 & a_1 & 1 \\ a_2^2 & a_2 & 1 \\ a_3^2 & a_3 & 1 \\ a_4^2 & a_4 & 1 \end{array} \right\} \ (1), \qquad \left. \begin{array}{ccc} 1 & -2b_1 & b_1^2 \\ 1 & -2b_2 & b_2^2 \\ 1 & -2b_3 & b_3^2 \\ 1 & -2b_4 & b_4^2 \end{array} \right\} \ (2).$$

122. Solution of a System of Linear Equations.—
We have seen in Art. 114 that a determinant may be expanded
as a linear homogeneous function of the constituents in any row
or column, the coefficient of any constituent being the corresponding minor with its proper sign. We have, for example,

$$\Delta = a_1 A_1 + a_2 A_2 + a_3 A_3 + \&c.$$

Now, the coefficients A_1, A_2, &c., are connected with the constituents of the other columns by $n - 1$ identical relations, viz.,

$$b_1 A_1 + b_2 A_2 + b_3 A_3 + \&c. = 0,$$

$$c_1 A_1 + c_2 A_2 + c_3 A_3 + \&c. = 0, \ \&c.,$$

for any one of these is what the determinant becomes when the constituents of the corresponding column are substituted for a_1, a_2, a_3, &c., and must therefore vanish.

By the aid of these relations we can write down the solution of a system of linear equations. The following application to the case of three variables is sufficient to explain the general process. Let the equations be

$$a_1 x + b_1 y + c_1 z = m_1,$$
$$a_2 x + b_2 y + c_2 z = m_2,$$
$$a_3 x + b_3 y + c_3 z = m_3.$$

Multiply the first equation by A_1, the second by A_2, and the third by A_3; and add. The coefficients of y and z vanish, in virtue of the relations above proved; and we obtain

$$(a_1 A_1 + a_2 A_2 + a_3 A_3)x = m_1 A_1 + m_2 A_2 + m_3 A_3,$$

or

$$\Delta\, x = \begin{vmatrix} m_1 & b_1 & c_1 \\ m_2 & b_2 & c_2 \\ m_3 & b_3 & c_3 \end{vmatrix},$$

where Δ represents the determinant formed from the nine constituents, a_1, b_1, c_1, &c.

Similarly, multiplying by B_1, B_2, B_3, we obtain

$$(b_1 B_1 + b_2 B_2 + b_3 B_3)y = m_1 B_1 + m_2 B_2 + m_3 B_3,$$

$$\Delta\, y = \begin{vmatrix} a_1 & m_1 & c_1 \\ a_2 & m_2 & c_2 \\ a_3 & m_3 & c_3 \end{vmatrix},$$

where the determinant on the right-hand side is what Δ becomes when m_1, m_2, m_3 are substituted for the constituents of the second column. Similarly, we obtain for z

$$\Delta\, z = \begin{vmatrix} a_1 & b_1 & m_1 \\ a_2 & b_2 & m_2 \\ a_3 & b_3 & m_3 \end{vmatrix}.$$

These values may be written more compactly, as follows :—

$$\Delta x = (m_1 b_2 c_3), \quad \Delta y = (a_1 m_2 c_3), \quad \Delta z = (a_1 b_2 m_3).$$

In general, the values of x, y, z, &c., may be written as follows :—

$$x = \frac{(m_1 b_2 c_3 \ldots l_n)}{(a_1 b_2 c_3 \ldots l_n)}, \quad y = \frac{(a_1 m_2 b_3 \ldots l_n)}{(a_1 b_2 c_3 \ldots l_n)}, \quad z = \frac{(a_1 b_2 m_3 \ldots l_n)}{(a_1 b_2 c_3 \ldots l_n)}, \text{ &c.,}$$

where, to obtain the value of any variable, the given quantities m_1, m_2, &c., on the right-hand side of the given equations are to be substituted in Δ for the coefficients of the variable in question, and the determinant so formed to be divided by Δ.

123. Linear Homogeneous Equations.—When $n - 1$ linear homogeneous equations between n variables are given, the ratios of the variables can be determined by bringing any one of them to the right-hand side of the equations, and solving as in the previous article ; or we can determine these ratios more conveniently, as follows. We take the particular case of three equations between four variables, which will be sufficient to illustrate the general process :

$$\left. \begin{aligned} a_1 x + b_1 y + c_1 z + d_1 w = 0, \\ a_2 x + b_2 y + c_2 z + d_2 w = 0, \\ a_3 x + b_3 y + c_3 z + d_3 w = 0. \end{aligned} \right\} \cdot \qquad (1)$$

To these we can add a fourth equation whose coefficients are undetermined, viz.,

$$a_4 x + b_4 y + c_4 z + d_4 w = \lambda. \qquad (2)$$

Calling $(a_1 b_2 c_3 d_4)$ as usual Δ, and solving from these four equations by the method of the last Article, we obtain, since $m_1 = 0$, $m_2 = 0$, $m_3 = 0$, $m_4 = \lambda$, the following values :—

$$\Delta x = \lambda A_4, \quad \Delta y = \lambda B_4, \quad \Delta z = \lambda C_4, \quad \Delta w = \lambda D_4,$$

or,

$$\frac{x}{A_4} = \frac{y}{B_4} = \frac{z}{C_4} = \frac{w}{D_4} = \frac{\lambda}{\Delta}. \qquad (3)$$

The first three of these equations express the ratios of the four variables in terms of the coefficients in the three given equations. And in general, *the variables are proportional to the coefficients in the expansion of* Δ *of the constituents of the* n^{th} *row supposed added to the* $n - 1$ *rows resulting from the given equations.*

We can now express the condition that n linear homogeneous equations should be consistent with one another; for example, that the equation (2) should, when $\lambda = 0$, be consistent with the equations (1). We have only to substitute in (2) the ratios derived from (1), when we obtain

$$a_4 A_4 + b_4 B_4 + c_4 C_4 + d_4 D_4 = 0,$$

or

$$\Delta = 0.$$

The same thing appears from the equations (3), for if $\lambda = 0$, and if the variables do not all vanish, Δ must vanish.

What has been proved may be expressed as follows:—*The result of eliminating the variables between n linear homogeneous equations in n variables is the vanishing of the determinant formed by the coefficients of the given equations.*

124. **Reciprocal Determinants.** — The quantities $A_1, B_1, C_1 \ldots A_2$, &c., which occur in the expansion of a determinant (*i.e.* the first minors, with their proper signs), may be called *inverse constituents ;* and the determinant formed with them the inverse or *reciprocal determinant.* We proceed to prove certain useful relations connecting the two determinants.

(1). *To express the reciprocal in terms of the given determinant.* Let the reciprocal of Δ be denoted by Δ', and multiply the two determinants

$$\Delta = \begin{vmatrix} a_1 & b_1 & c_1 \\ a_2 & b_2 & c_2 \\ a_3 & b_3 & c_3 \end{vmatrix}, \quad \Delta' = \begin{vmatrix} A_1 & B_1 & C_1 \\ A_2 & B_2 & C_2 \\ A_3 & B_3 & C_3 \end{vmatrix}.$$

All the constituents of the resulting determinant except

those in the diagonal vanish (see Art. 122) ; and the result is

$$\Delta\Delta' = \begin{vmatrix} \Delta & 0 & 0 \\ 0 & \Delta & 0 \\ 0 & 0 & \Delta \end{vmatrix} = \Delta^3 ;$$

whence

$$\Delta' = \Delta^2.$$

The process here employed is equally applicable in general ; giving $\Delta\Delta' = \Delta^n$, or $\Delta' = \Delta^{n-1}$. Hence *the reciprocal determinant is equal to the $(n-1)^{th}$ power of the given determinant.*

(2). *To express any minor of the reciprocal determinant in terms of the original constituents.*

We take, for example, the determinant of the fourth order, and proceed to express the first minors of its reciprocal. Multiplying the two determinants on the left-hand side of the following equation, and employing the identical equations of Art. 122, we obtain

$$\begin{vmatrix} a_1 & b_1 & c_1 & d_1 \\ a_2 & b_2 & c_2 & d_2 \\ a_3 & b_3 & c_3 & d_3 \\ a_4 & b_4 & c_4 & d_4 \end{vmatrix} \begin{vmatrix} 1 & 0 & 0 & 0 \\ A_2 & B_2 & C_2 & D_2 \\ A_3 & B_3 & C_3 & D_3 \\ A_4 & B_4 & C_4 & D_4 \end{vmatrix} = \begin{vmatrix} a_1 & 0 & 0 & 0 \\ a_2 & \Delta & 0 & 0 \\ a_3 & 0 & \Delta & 0 \\ a_4 & 0 & 0 & \Delta \end{vmatrix} ;$$

whence

$$\Delta \begin{vmatrix} B_2 & C_2 & D_2 \\ B_3 & C_3 & D_3 \\ B_4 & C_4 & D_4 \end{vmatrix} = a_1 \Delta^3 ,$$

or

$$(B_2\,C_3\,D_4) = a_1 \Delta^2 ,$$

thus expressing the first minor of Δ' complementary to A_1.

Again, to express the second minors of Δ', we have, by an exactly similar process,

$$\begin{vmatrix} a_1 & b_1 & c_1 & d_1 \\ a_2 & b_2 & c_2 & d_2 \\ a_3 & b_3 & c_3 & d_3 \\ a_4 & b_4 & c_4 & d_4 \end{vmatrix} \begin{vmatrix} 1 & 0 & 0 & 0 \\ 0 & 1 & 0 & 0 \\ A_3 & B_3 & C_3 & D_3 \\ A_4 & B_4 & C_4 & D_4 \end{vmatrix} = \begin{vmatrix} a_1 & b_1 & 0 & 0 \\ a_2 & b_2 & 0 & 0 \\ a_3 & b_3 & \Delta & 0 \\ a_4 & b_4 & 0 & \Delta \end{vmatrix} ;$$

whence

$$\Delta \begin{vmatrix} C_3 & D_3 \\ C_4 & D_4 \end{vmatrix} = \begin{vmatrix} a_1 & b_1 \\ a_2 & b_2 \end{vmatrix} \Delta^2 ,$$

or

$$(C_3 D_4) = (a_1 b_2) \Delta .$$

The general theorem of which these are particular cases can be proved in a similar manner, and may be expressed as follows :—*A minor of the order m formed out of the inverse constituents is equal to the complementary of the corresponding minor of the original determinant Δ multiplied by the $(m-1)^{th}$ power of Δ.*

125. Skew Symmetrical Determinants.—We have seen in Art. 119 that a symmetrical determinant is one in which the conjugate constituents are equal to one another.

A *skew symmetrical* determinant is one in which each constituent is equal to its conjugate with sign changed ; and including the constituents in the leading diagonal, it follows that all these vanish since each is its own conjugate.

In a symmetrical determinant the minor corresponding to any constituent is equal to the minor corresponding to its conjugate ; for, since one occupies in the rows the same position as the other in the columns, the two minors will differ only by an interchange of rows and columns.

It easily appears also that, in the case of a skew symmetrical determinant of the nth order, the minor corresponding to any constituent differs from that corresponding to its conjugate by an interchange of rows for columns, and by the signs of all the constituents. Hence the two minors are equal to one another when their order is even, *i.e.*, when n is odd; and they are equal with contrary signs when n is even.

The following are useful properties of skew symmetrical determinants :—

(1). *A skew symmetrical determinant of odd order vanishes.*— For any skew symmetrical determinant is unaltered by changing the columns into rows, and then changing the signs of all the rows ; but when n is odd this process ought to change the sign of Δ ; hence Δ must, in this case, vanish.   For example,

$$\begin{vmatrix} 0 & h & g \\ -h & 0 & f \\ -g & -f & 0 \end{vmatrix} = 0.$$

(2). *A skew symmetrical determinant of even order is a perfect square.*

This Proposition follows from the principles established in Art. 124.

Take, for example, the determinant of the fourth order

$$\Delta \equiv \begin{vmatrix} 0 & a & b & c \\ -a & 0 & f & e \\ -b & -f & 0 & d \\ -c & -e & -d & 0 \end{vmatrix} ;$$

and let the inverse constituents forming its reciprocal be denoted by A_1, B_1, ... A_2, &c.   We have then, by (2), Art. 124,

$$A_1 B_2 - A_2 B_1 = \Delta \begin{vmatrix} 0 & d \\ -d & 0 \end{vmatrix} = d^2 \Delta.$$

Now A_1, and B_2, being skew symmetrical determinants of odd order, vanish ; and $A_2 = - B_1$, since these are conjugate minors ; hence $d^2 \Delta = A_2^2$, which proves that Δ is a perfect square. Similarly, for the determinant of the sixth order, it is proved that the product of Δ by a skew determinant of the fourth order is a perfect square ; and since the latter determinant has been just proved to be a perfect square, it follows that Δ is also.   By an exactly similar process, assuming the truth of the Proposition for the determinant of the sixth order, it follows for one of the eighth ; and so on.

EXAMPLES.

1.
$$\begin{vmatrix} 0 & a & b & c \\ -a & 0 & f & e \\ -b & -f & 0 & d \\ -c & -e & -d & 0 \end{vmatrix} \equiv (ad - be + cf)^2.$$

In the above proof Δ is multiplied by a square factor; but this must divide out in every instance, since Δ is plainly not fractional.

2. The square of any determinant of even order can be expressed as a skew symmetrical determinant.

The following method of proof is applicable in general.

The square of $(a_1 b_2 c_3 d_4)$ is obtained by multiplying the two following determinants:—

$$\begin{vmatrix} a_1 & b_1 & c_1 & d_1 \\ a_2 & b_2 & c_2 & d_2 \\ a_3 & b_3 & c_3 & d_3 \\ a_4 & b_4 & c_4 & d_4 \end{vmatrix}, \quad \begin{vmatrix} -b_1 & a_1 & -d_1 & c_1 \\ -b_2 & a_2 & -d_2 & c_2 \\ -b_3 & a_3 & -d_3 & c_3 \\ -b_4 & a_4 & -d_4 & c_4 \end{vmatrix};$$

and the product of these is

$$\begin{vmatrix} 0, & -(a_1 b_2) - (c_1 d_2), & -(a_1 b_3) - (c_1 d_3), & -(a_1 b_4) - (c_1 d_4), \\ (a_1 b_2) + (c_1 d_2), & 0, & -(a_2 b_3) - (c_2 d_3), & -(a_2 b_4) - (c_2 d_4), \\ (a_1 b_3) + (c_1 d_3), & (a_2 b_3) + (c_2 d_3), & 0, & -(a_3 b_4) - (c_3 d_4), \\ (a_1 b_4) + (c_1 d_4), & (a_2 b_4) + (c_2 d_4), & (a_3 b_4) + (c_3 d_4), & 0, \end{vmatrix},$$

which is a skew symmetrical determinant.

MISCELLANEOUS EXAMPLES.

1. Prove that

$$\begin{vmatrix} a_0 & a_1 & a_2 \\ a_1 & a_2 & a_3 \\ a_2 & a_3 & a_4 \end{vmatrix} = J,$$

where J has the same signification as in Art. 38.

2. Prove that

$$\begin{vmatrix} \beta + \gamma & \gamma + \alpha & \alpha + \beta \\ \beta' + \gamma' & \gamma' + \alpha' & \alpha' + \beta' \\ \beta'' + \gamma'' & \gamma'' + \alpha'' & \alpha'' + \beta'' \end{vmatrix} = 2 \begin{vmatrix} \alpha & \beta & \gamma \\ \alpha' & \beta' & \gamma' \\ \alpha'' & \beta'' & \gamma'' \end{vmatrix}.$$

3. Prove that

$$\begin{vmatrix} \beta\gamma & \beta\gamma' + \beta'\gamma & \beta'\gamma' \\ \gamma\alpha & \gamma\alpha' + \gamma'\alpha & \gamma'\alpha' \\ \alpha\beta & \alpha\beta' + \alpha'\beta & \alpha'\beta' \end{vmatrix} = (\beta\gamma')\,(\gamma\alpha')\,(\alpha\beta'),$$

where the factors on the right-hand side are determinants of the second order.

Dividing the rows by $\beta'\gamma'$, $\gamma'\alpha'$, $\alpha'\beta'$; and putting $\lambda = \dfrac{\alpha}{\alpha'}$, $\mu = \dfrac{\beta}{\beta'}$, $\nu = \dfrac{\gamma}{\gamma'}$, the determinant (omitting a factor) reduces to the form

$$\begin{vmatrix} 1 & \mu + \nu & \mu\nu \\ 1 & \nu + \lambda & \nu\lambda \\ 1 & \lambda + \mu & \lambda\mu \end{vmatrix} = \begin{vmatrix} 1 & -\lambda & \mu\nu \\ 1 & -\mu & \nu\lambda \\ 1 & -\nu & \lambda\mu \end{vmatrix} = -(\mu - \nu)(\nu - \lambda)(\lambda - \mu), \ \&c.$$

4. Prove that

$$\begin{vmatrix} a & b & c & d \\ a' & b' & c' & d' \\ \dfrac{a^2}{a'} & \dfrac{b^2}{b'} & \dfrac{c^2}{c'} & \dfrac{d^2}{d'} \\ \dfrac{a'^2}{a} & \dfrac{b'^2}{b} & \dfrac{c'^2}{c} & \dfrac{d'^2}{d} \end{vmatrix} = \dfrac{-(bc')\,(ad')\,(ca')\,(bd')\,(ab')\,(cd')}{a\,b\,c\,d\,a'\,b'\,c'\,d'}.$$

Multiplying the columns by $\dfrac{a'}{a^2}$, $\dfrac{b'}{b^2}$, $\dfrac{c'}{c^2}$, $\dfrac{d'}{d^2}$, the determinant reduces to the form treated in Ex. 10, Art. 112.

5. Prove that

$$\begin{vmatrix} \beta^2\gamma^2 + \alpha^2\delta^2 & \beta\gamma + \alpha\delta & 1 \\ \gamma^2\alpha^2 + \beta^2\delta^2 & \gamma\alpha + \beta\delta & 1 \\ \alpha^2\beta^2 + \gamma^2\delta^2 & \alpha\beta + \gamma\delta & 1 \end{vmatrix} = (\beta - \gamma)(\alpha - \delta)(\gamma - \alpha)(\beta - \delta)(\alpha - \beta)(\gamma - \delta).$$

Add the last column multiplied by $2\alpha\beta\gamma\delta$ to the first. The determinant becomes then of the form of Ex. 9, Art. 112.

6. Prove that

$$\begin{vmatrix} (\beta + \gamma - \alpha - \delta)^4 & (\beta + \gamma - \alpha - \delta)^2 & 1 \\ (\gamma + \alpha - \beta - \delta)^4 & (\gamma + \alpha - \beta - \delta)^2 & 1 \\ (\alpha + \beta - \gamma - \delta)^4 & (\alpha + \beta - \gamma - \delta)^2 & 1 \end{vmatrix} = 64\,(\beta - \gamma)(\alpha - \delta)(\gamma - \alpha)(\beta - \delta)(\alpha - \beta)(\gamma - \delta).$$

7. Prove that

$$\begin{vmatrix} a & b & ax + b \\ b & c & bx + c \\ ax + b & bx + c & 0 \end{vmatrix} = -(ac - b^2)(ax^2 + 2bx + c).$$

Subtract from the third row the second row plus the first multiplied by x.

8. Prove similarly that

$$
\begin{vmatrix}
a & b & c & ax^2 + 2bx + c \\
b & c & d & bx^2 + 2cx + d \\
c & d & e & cx^2 + 2dx + e \\
ax^2 + 2bx + c & bx^2 + 2cx + d & cx^2 + 2dx + e & 0
\end{vmatrix}
$$

$$
\equiv -
\begin{vmatrix}
a & b & c \\
b & c & d \\
c & d & e
\end{vmatrix}
(ax^4 + 4bx^3 + 6cx^2 + 4dx + e).
$$

9. Given

$$
f_1(x) = a_1 x^3 + 3b_1 x^2 + 3c_1 x + d_1,
$$
$$
f_2(x) = a_2 x^3 + 3b_2 x^2 + 3c_2 x + d_2,
$$
$$
f_3(x) = a_3 x^3 + 3b_3 x^2 + 3c_3 x + d_3 ;
$$

prove the identity

$$
\begin{vmatrix}
f_1(x) & f_1'(x) & f_1''(x) \\
f_2(x) & f_2'(x) & f_2''(x) \\
f_3(x) & f_3'(x) & f_3''(x)
\end{vmatrix}
\equiv - 18
\begin{vmatrix}
1 & -x & x^2 & -x^3 \\
a_1 & b_1 & c_1 & d_1 \\
a_2 & b_2 & c_2 & d_2 \\
a_3 & b_3 & c_3 & d_3
\end{vmatrix} .
$$

The first determinant reduces easily (omitting a factor) to the following :—

$$
\begin{vmatrix}
a_1 x + b_1 & b_1 x + c_1 & c_1 x + d_1 \\
a_2 x + b_2 & b_2 x + c_2 & c_2 x + d_2 \\
a_3 x + b_3 & b_3 x + c_3 & c_3 x + d_3
\end{vmatrix} .
$$

An artifice which is often useful in simplifying a determinant is to increase its order by bordering it with rows or columns whose constituents are ciphers, except those in the diagonal, which are units. The value of the determinant remains unaltered (see Ex. 4, Art. 120). The determinant last written is plainly equal to

$$
\begin{vmatrix}
1 & 0 & 0 & 0 \\
a_1 & a_1 x + b_1 & b_1 x + c_1 & c_1 x + d_1 \\
a_2 & a_2 x + b_2 & b_2 x + c_2 & c_2 x + d_2 \\
a_3 & a_3 x + b_3 & b_3 x + c_3 & c_3 x + d_3
\end{vmatrix} .
$$

Subtracting from the second column the first multiplied by x; subtracting then from the third the new second column multiplied by x; and, finally, from the fourth the new third column multiplied by x, we have the result above stated.

10. Show that the determinant

$$\begin{vmatrix} \lambda x^2 + cy^2 + bz^2 - 1 & (\lambda - c)\,xy & (\lambda - b)\,xz \\ (\lambda - c)\,xy & \lambda y^2 + az^2 + cx^2 - 1 & (\lambda - a)\,yz \\ (\lambda - b)\,xz & (\lambda - a)\,yz & \lambda z^2 + bx^2 + cy^2 - 1 \end{vmatrix}$$

contains $\lambda(x^2 + y^2 + z^2) - 1$ as a factor, and that the remaining factor is independent of λ.

Border the determinant, as in Ex. 9, with a first column whose constituents are $1, \lambda x, \lambda y, \lambda z$; and with a first row whose constituents are $1, 0, 0, 0$. Subtract then x times the first column from the second, y times the first column from the third, and z times the first column from the fourth. In the determinant thus altered subtract from the first row x times the second, plus y times the third, plus z times the fourth; and the result follows.

11. Expand the determinant

$$\begin{vmatrix} a_1 + x & b_1 & c_1 & d_1 \\ a_2 & b_2 + x & c_2 & d_2 \\ a_3 & b_3 & c_3 + x & d_3 \\ a_4 & b_4 & c_4 & d_4 + x \end{vmatrix}.$$

Ans. $(a_1 b_2 c_3 d_4) + \{(b_2 c_3 d_4) + (a_1 c_3 d_4) + (a_1 b_2 d_4) + (a_1 b_2 c_3)\}\, x + \{(b_2 c_3) + (a_1 d_4)$
$+ (a_1 c_3) + (b_2 d_4) + (a_1 b_2) + (c_3 d_4)\}\, x^2 + (a_1 + b_2 + c_3 + d_4)\, x^3 + x^4.$

12. Prove the identities

$$\begin{vmatrix} 1 & \alpha & \alpha' & \alpha\alpha' \\ 1 & \beta & \beta' & \beta\beta' \\ 1 & \gamma & \gamma' & \gamma\gamma' \\ 1 & \delta & \delta' & \delta\delta' \end{vmatrix} \equiv \begin{vmatrix} B & C \\ B' & C' \end{vmatrix} \equiv \begin{vmatrix} C & A \\ C' & A' \end{vmatrix} \equiv \begin{vmatrix} A & B \\ A' & B' \end{vmatrix},$$

where

$$A = (\beta - \gamma)(\alpha - \delta), \qquad B = (\gamma - \alpha)(\beta - \delta), \qquad C = (\alpha - \beta)(\gamma - \delta),$$
$$A' = (\beta' - \gamma')(\alpha' - \delta'), \quad B' = (\gamma' - \alpha')(\beta' - \delta'), \quad C' = (\alpha' - \beta')(\gamma' - \delta').$$

Expanding the first determinant in terms of the minors formed from the first two columns (see Art. 115), we easily prove that it is equal to

$$A(\beta'\gamma' + \alpha'\delta') + B(\gamma'\alpha' + \beta'\delta') + C(\alpha'\beta' + \gamma'\delta');$$

and employing the identical equation $A + B + C \equiv 0$, along with the relations of Ex. 18, Art. 27, the result follows.

13. Prove that the determinant of Ex. 12 is equal to

$$\begin{vmatrix} 1 & \beta\gamma + \alpha\delta & \beta'\gamma' + \alpha'\delta' \\ 1 & \gamma\alpha + \beta\delta & \gamma'\alpha' + \beta'\delta' \\ 1 & \alpha\beta + \gamma\delta & \alpha'\beta' + \gamma'\delta' \end{vmatrix}.$$

This follows at once from the relations of Ex. 18, Art. 27. If α', β', γ', δ' be put equal to α^m, β^m, γ^m, δ^m in the result, we obtain an identity which includes Ex. 5, p. 258, as a particular case.

14. Prove the equation

$$\begin{vmatrix} x & a_1 & a_2 & a_3 & 1 \\ a & x & b_1 & b_2 & 1 \\ a & \beta & x & c_1 & 1 \\ a & \beta & \gamma & x & 1 \\ a & \beta & \gamma & \delta & 1 \end{vmatrix} = (x - \alpha)(x - \beta)(x - \gamma)(x - \delta) \ ;$$

a_1, a_2, a_3, b_1, b_2, c_1 being any quantities.

This follows by subtracting α times the last column from the first, β times the last from the second, &c. The student will have no difficulty in writing down the corresponding determinant of the $(n + 1)^{th}$ degree which is equal to the polynomial $f(x)$ whose roots are α_1, α_2, α_3, ... α_n.

15. Resolve into factors the determinant

$$\Delta \equiv \begin{vmatrix} (\alpha - \alpha')^2 & (\alpha - \beta')^2 & (\alpha - \gamma')^2 \\ (\beta - \alpha')^2 & (\beta - \beta')^2 & (\beta - \gamma')^2 \\ (\gamma - \alpha')^2 & (\gamma - \beta')^2 & (\gamma - \gamma')^2 \end{vmatrix} .$$

Here
$$\Delta = \begin{vmatrix} \alpha^2 & \alpha & 1 \\ \beta^2 & \beta & 1 \\ \gamma^2 & \gamma & 1 \end{vmatrix} \times \begin{vmatrix} 1 & -2\alpha' & \alpha'^2 \\ 1 & -2\beta' & \beta'^2 \\ 1 & -2\gamma' & \gamma'^2 \end{vmatrix} ;$$

and these two determinants may be resolved as in Ex. 9, Art. 112.

16. Resolve into factors the determinant

$$\Delta \equiv \begin{vmatrix} (\alpha - \alpha')^3 & (\alpha - \beta')^3 & (\alpha - \gamma')^3 \\ (\beta - \alpha')^3 & (\beta - \beta')^3 & (\beta - \gamma')^3 \\ (\gamma - \alpha')^3 & (\gamma - \beta')^3 & (\gamma - \gamma')^3 \end{vmatrix} .$$

Multiplying the two rectangular arrays

$$\left. \begin{array}{cccc} \alpha^3 & \alpha^2 & \alpha & 1 \\ \beta^3 & \beta^2 & \beta & 1 \\ \gamma^3 & \gamma^2 & \gamma & 1 \end{array} \right\} (1), \qquad \left. \begin{array}{cccc} 1 & -3\alpha' & 3\alpha'^2 & -\alpha'^3 \\ 1 & -3\beta' & 3\beta'^2 & -\beta'^3 \\ 1 & -3\gamma' & 3\gamma'^2 & -\gamma'^3 \end{array} \right\} (2),$$

Δ becomes equal to the sum of four terms, from each of which we can take out as a factor the product of the two determinants

$$\begin{vmatrix} 1 & \alpha & \alpha^2 \\ 1 & \beta & \beta^2 \\ 1 & \gamma & \gamma^2 \end{vmatrix} , \qquad \begin{vmatrix} 1 & \alpha' & \alpha'^2 \\ 1 & \beta' & \beta'^2 \\ 1 & \gamma' & \gamma'^2 \end{vmatrix} .$$

The remaining factor is

$$3\{3\alpha\beta\gamma - \Sigma\beta\gamma\,\Sigma\alpha' + \Sigma\beta'\gamma'\,\Sigma\alpha - 3\alpha'\beta'\gamma'\},$$

which can be written also in the form

$$3\{(\alpha-\alpha')(\beta-\beta')(\gamma-\gamma') + (\alpha-\beta')(\beta-\gamma')(\gamma-\alpha') + (\alpha-\gamma')(\beta-\alpha')(\gamma-\beta')\}.$$

17. Prove that

$$\begin{vmatrix} L & L' \\ M & M' \end{vmatrix} = (\omega^2 - \omega) \begin{vmatrix} \alpha & \beta & \gamma \\ \alpha' & \beta' & \gamma' \\ 1 & 1 & 1 \end{vmatrix},$$

where L and M have the values of Art. 59.

This appears easily by multiplying

$$\begin{vmatrix} \alpha & \beta & \gamma \\ \alpha' & \beta' & \gamma' \\ 1 & 1 & 1 \end{vmatrix} \text{ by } \begin{vmatrix} 1 & 0 & 0 \\ 1 & \omega & \omega^2 \\ 1 & \omega^2 & \omega \end{vmatrix}.$$

18. Prove by determinants the following relation :—

$$9(\beta-\gamma)^2(\alpha-\delta)^2(\gamma-\alpha)^2(\beta-\delta)^2(\alpha-\beta)^2(\gamma-\delta)^2$$
$$= \{(\beta-\gamma)^3(\alpha-\delta)^3 + (\gamma-\alpha)^3(\beta-\delta)^3 + (\alpha-\beta)^3(\gamma-\delta)^3\}^2.$$

This appears from the equation

$$\begin{vmatrix} \alpha^3 & \alpha^2 & \alpha & 1 \\ \beta^3 & \beta^2 & \beta & 1 \\ \gamma^3 & \gamma^2 & \gamma & 1 \\ \delta^3 & \delta^2 & \delta & 1 \end{vmatrix} \times \begin{vmatrix} 1 & -3\alpha & 3\alpha^2 & -\alpha^3 \\ 1 & -3\beta & 3\beta^2 & -\beta^3 \\ 1 & -3\gamma & 3\gamma^2 & -\gamma^3 \\ 1 & -3\delta & 3\delta^2 & -\delta^3 \end{vmatrix} = \begin{vmatrix} 0 & (\alpha-\beta)^3 & (\alpha-\gamma)^3 & (\alpha-\delta)^3 \\ (\beta-\alpha)^3 & 0 & (\beta-\gamma)^3 & (\beta-\delta)^3 \\ (\gamma-\alpha)^3 & (\gamma-\beta)^3 & 0 & (\gamma-\delta)^3 \\ (\delta-\alpha)^3 & (\delta-\beta)^3 & (\delta-\gamma)^3 & 0 \end{vmatrix},$$

on applying the expansion of a skew determinant of the fourth order given in Ex. 1, Art. 125.

19. Prove the expansion

$$\begin{vmatrix} 1+a_1 & 1 & 1 & 1 \\ 1 & 1+a_2 & 1 & 1 \\ 1 & 1 & 1+a_3 & 1 \\ 1 & 1 & 1 & 1+a_4 \end{vmatrix} = a_1 a_2 a_3 a_4 \left\{ 1 + \frac{1}{a_1} + \frac{1}{a_2} + \frac{1}{a_3} + \frac{1}{a_4} \right\}.$$

This is easily proved by subtracting the first column from each of the others, and then expanding the determinant as a linear function of the constituents of the first column. It will be apparent from the nature of the proof that the value of the similar determinant of the nth order is $a_1 a_2 a_3 \ldots a_n \left\{ 1 + \Sigma\dfrac{1}{a_1} \right\}.$

20. Prove the relation

$$\begin{vmatrix} \alpha & x & x & x \\ x & \beta & x & x \\ x & x & \gamma & x \\ x & x & x & \delta \end{vmatrix} = f(x) - xf'(x),$$

where $\qquad f(x) \equiv (x - \alpha)(x - \beta)(x - \gamma)(x - \delta).$

This can be derived from the previous example, or proved independently in a similar way. As in the last example the determinant of this form of the nth degree can be similarly expressed.

21. Prove that the equation in λ

$$\phi(\lambda) \equiv \begin{vmatrix} a - \lambda & h & g \\ h & b - \lambda & f \\ g & f & c - \lambda \end{vmatrix} = 0$$

has its roots all real. (Cf. Ex. 5, p. 164.)

This determinant is formed by subtracting the variable from each of the diagonal terms of a symmetrical determinant. The following method of proof, due to Prof. Sylvester, is equally applicable to the general case. Let $\phi(\lambda)$ be multiplied by $\phi(-\lambda)$. The result is

$$\begin{vmatrix} A - \lambda^2 & H & G \\ H & B - \lambda^2 & F \\ G & F & C - \lambda^2 \end{vmatrix},$$

where $\quad A = a^2 + h^2 + g^2, \qquad B = b^2 + h^2 + f^2, \qquad C = c^2 + f^2 + g^2,$

$\quad F = (b + c)f + gh, \qquad G = (c + a)g + hf, \qquad H = (a + b)h + fg.$

This determinant, when expanded, may be written as follows:—

$$\lambda^6 - L\,\lambda^4 + M\,\lambda^2 - N = 0,$$

where L, M, N are all essentially positive; for N is plainly the square of $\phi(0)$, and L, M are easily found to be

$$L = a^2 + b^2 + c^2 + 2(f^2 + g^2 + h^2),$$

$$M = (bc - f^2)^2 + (ca - g^2)^2 + (ab - h^2)^2 + 2(af - gh)^2 + 2(bg - hf)^2 + 2(ch - fg)^2.$$

The equation $\phi(\lambda) = 0$, therefore, cannot have a root of the form $\beta\sqrt{-1}$, for this would lead to a negative root of the above equation in λ^2, and no such root can exist since the signs are alternately positive and negative. It appears further, by substituting in the above proof $a - \alpha$, $b - \alpha$, $c - \alpha$ for a, b, c, respectively, that λ cannot have a value of the form $\alpha + \beta\sqrt{-1}$.

22. Show that the most general values of x, y, z, w which satisfy the two homogeneous equations

$$ax + by + cz + dw = 0, \qquad a'x + b'y + c'z + d'w = 0$$

may be expressed symmetrically in terms of two indeterminates X, Y, in the form

$$(ab')(ac')(ad')x = aX + a'Y,$$

$$(ba')(bc')(bd')y = bX + b'Y, \ \&c.$$

This can be proved by adding to the two given equations the two following :—

$$\frac{a^2}{a'}x + \frac{b^2}{b'}y + \frac{c^2}{c'}z + \frac{d^2}{d'}w = \lambda, \quad \frac{a'^2}{a}x + \frac{b'^2}{b}y + \frac{c'^2}{c}z + \frac{d'^2}{d}w = \mu,$$

where λ, μ are indeterminate quantities; by then solving for x, y, z, w, as in Art· 122, and reducing the determinants as in Ex. 4, p. 258; and finally making $X = a'b'c'd'\lambda$, $Y = -abcd\mu$.

23. When a polynomial U is divided by another U' of lower dimensions, the coefficients of the quotient, and of the remainder, can be expressed as determinants in terms of the coefficients of U and U'.

The method employed in the following particular case is equally applicable in general. Let U be of the 5th, and U' of the 3rd degree; the quotient and remainder can then be represented as follows :—

$$Q \equiv q_0 x^2 + q_1 x + q_2, \qquad R \equiv r_0 x^2 + r_1 x + r_2.$$

Also, let

$$U \equiv a_0 x^5 + a_1 x^4 + a_2 x^3 + a_3 x^2 + a_4 x + a_5, \quad U' \equiv a_0' x^3 + a_1' x^2 + a_2' x + a_3'.$$

From the identity

$$U \equiv QU' + R$$

we have the following equations :—

$$a_0 = q_0 a_0',$$
$$a_1 = q_0 a_1' + q_1 a_0',$$
$$a_2 = q_0 a_2' + q_1 a_1' + q_2 a_0',$$
$$a_3 = q_0 a_3' + q_1 a_2' + q_2 a_1' + r_0,$$
$$a_4 = \qquad\ \ q_1 a_3' + q_2 a_2' + r_1,$$
$$a_5 = \qquad\qquad\ q_2 a_3' + r_2.$$

Solving by Art. 122; q_0, q_1, q_2 are expressed as determinants by means of the first three of these equations; and taking the first three with each of the others in succession, we determine r_0, r_1, r_2. For example, to find r_0 we have from the first four equations

$$\begin{vmatrix} a_0' & 0 & 0 & -a_0 \\ a_1' & a_0' & 0 & -a_1 \\ a_2' & a_1' & a_0' & -a_2 \\ a_3' & a_2' & a_1' & -a_3 + r_0 \end{vmatrix} = 0, \text{ or } a_0'^3 r_0 = \begin{vmatrix} a_0' & 0 & 0 & a_0 \\ a_1' & a_0' & 0 & a_1 \\ a_2' & a_1' & a_0' & a_2 \\ a_3' & a_2' & a_1' & a_3 \end{vmatrix}.$$

24. Each of the coefficients of any equation can be expressed in terms of the roots as the quotient of two determinants.

The student will easily extend to any degree the following application to the equation of the third degree.

From Ex. 10, Art. 112, we have

$$\begin{vmatrix} x^3 & x^2 & x & 1 \\ \alpha^3 & \alpha^2 & \alpha & 1 \\ \beta^3 & \beta^2 & \beta & 1 \\ \gamma^3 & \gamma^2 & \gamma & 1 \end{vmatrix} \equiv -(\beta-\gamma)(\gamma-\alpha)(\alpha-\beta)(x-\alpha)(x-\beta)(x-\gamma).$$

Expanding the determinant, this identity can be written

$$\begin{vmatrix} \alpha^2 & \alpha & 1 \\ \beta^2 & \beta & 1 \\ \gamma^2 & \gamma & 1 \end{vmatrix} x^3 - \begin{vmatrix} \alpha^3 & \alpha & 1 \\ \beta^3 & \beta & 1 \\ \gamma^3 & \gamma & 1 \end{vmatrix} x^2 + \begin{vmatrix} \alpha^3 & \alpha^2 & 1 \\ \beta^3 & \beta^2 & 1 \\ \gamma^3 & \gamma^2 & 1 \end{vmatrix} x - \begin{vmatrix} \alpha^3 & \alpha^2 & \alpha \\ \beta^3 & \beta^2 & \beta \\ \gamma^3 & \gamma^2 & \gamma \end{vmatrix}$$

$$= \begin{vmatrix} \alpha^2 & \alpha & 1 \\ \beta^2 & \beta & 1 \\ \gamma^2 & \gamma & 1 \end{vmatrix} \{x^3 - p_1 x^2 + p_2 x - p_3\},$$

from which the above proposition follows; p_1, p_2, p_3 being the coefficients of the equation whose roots are α, β, γ.

25. To express as a determinant the reducing cubic in the solution of a biquadratic.

Employing Descartes' method, and substituting from equations (1), Art. 64, in the identity

$$\begin{vmatrix} 1 & 1 & 0 \\ p & p' & 0 \\ q & q' & 0 \end{vmatrix} \begin{vmatrix} 1 & 1 & 0 \\ p' & p & 0 \\ q' & q & 0 \end{vmatrix} \equiv \begin{vmatrix} 2 & p+p' & q+q' \\ p+p' & 2pp' & pq'+p'q \\ q+q' & pq'+p'q & 2qq' \end{vmatrix} = 0,$$

we find the equation

$$\begin{vmatrix} a & b & c+2a\phi \\ b & c-a\phi & d \\ c+2a\phi & d & e \end{vmatrix} = 0,$$

a cubic for ϕ, which is easily seen to be identical with the cubic

$$4a^3\phi^3 - Ia\phi + J = 0$$

of Art. 64.

CHAPTER XII.

126. Newton's Theorem on the Sums of the Powers of the Roots.—We now resume the discussion of symmetric functions of the roots of an equation, of which a short account has been previously given (see Art. 27) ; and proceed to prove certain general propositions relating to these functions.

PROP. I.—*The sums of the similar powers of the roots of an equation can be expressed rationally in terms of the coefficients.*

Let the equation be

$$f(x) = x^n + p_1 x^{n-1} + p_2 x^{n-2} + \ldots + p_n,$$
$$= (x - a_1)(x - a_2)(x - a_3) \ldots (x - a_n).$$

We proceed to calculate Σa^2, Σa^3, $\ldots \Sigma a^m$; or, adopting the usual notation, s_2, s_3, $\ldots s_m$, in terms of the coefficients $p_1, p_2, \ldots p_n$.

We have, by Art. 72,

$$f'(x) = \frac{f(x)}{x - a_1} + \frac{f(x)}{x - a_2} + \ldots + \frac{f(x)}{x - a_n}$$
$$= nx^{n-1} + (n-1)p_1 x^{n-2} + (n-2)p_2 x^{n-3} + \ldots + 2p_{n-2}x + p_{n-1} ;$$

and we find, dividing by the method of Art. 8,

$$\frac{f(x)}{x - a} =
\begin{array}{l|l|l|l}
x^{n-1} + a & x^{n-2} + a^2 & x^{n-3} + a^3 & x^{n-4} + \ldots + a^{n-1} \\
\quad + p_1 & \quad + p_1 a & \quad + p_1 a^2 & \quad\quad + p_1 a^{n-2} \\
 & \quad + p_2 & \quad + p_2 a & \quad\quad + p_2 a^{n-3} \\
 & & \quad + p_3 & \quad\quad + \ldots \\
 & & & \quad\quad + p_{n-2}\, a \\
 & & & \quad\quad + p_{n-1}.
\end{array}$$

If in this equation we replace successively a by each of the quantities $a_1, a_2, \ldots a_n$, and put $s_p = \Sigma a^p = a_1{}^p + a_2{}^p + a_3{}^p + \ldots + a_n{}^p$, we have, by adding all these results, the following value for $f'(x)$:—

$$
\begin{array}{l|l|l|l}
f'(x) = nx^{n-1} + s_1 & x^{n-2} + s_2 & x^{n-3} + s_3 & x^{n-4} + \ldots + s_{n-1} \\[4pt]
\qquad\quad + np_1 & \quad + p_1 s_1 & \quad + p_1 s_2 & \quad\quad + p_1 s_{n-2} \\[4pt]
& \quad + np_2 & \quad + p_2 s_1 & \quad\quad + p_2 s_{n-3} \\[4pt]
& & \quad + np_3 & \quad\quad\ \ \cdot\ \ \cdot\ \ \cdot \\[4pt]
& & & \quad\quad + p_{n-2} s_1 \\[4pt]
& & & \quad\quad + np_{n-1} ;
\end{array}
$$

whence, comparing this value of $f'(x)$ with the former, we have the following relations :—

$$s_1 + p_1 = 0,$$

$$s_2 + p_1 s_1 + 2p_2 = 0,$$

$$s_3 + p_1 s_2 + p_2 s_1 + 3p_3 = 0,$$

$$s_4 + p_1 s_3 + p_2 s_2 + p_3 s_1 + 4p_4 = 0,$$

$$\cdot \qquad \cdot \qquad \cdot \qquad \cdot \qquad \cdot \qquad \cdot \qquad \cdot$$

$$s_{n-1} + p_1 s_{n-2} + p_2 s_{n-3} + \ldots + p_{n-2} s_1 + (n-1)p_{n-1} = 0.$$

The first equation determines s_1 in terms of $p_1, p_2, \ldots p_n$; the second s_2; the third s_3; and so on, until s_{n-1} is determined. We find in this way

$$s_1 = -p_1, \quad s_2 = p_1{}^2 - 2p_2, \quad s_3 = -p_1{}^3 + 3p_1 p_2 - 3p_3,$$

$$s_4 = p_1{}^4 - 4p_1{}^2 p_2 + 4p_1 p_3 + 2p_2{}^2 - 4p_4,$$

$$s_5 = -p_1{}^5 + 5p_1{}^3 p_2 - 5p_1{}^2 p_3 - 5(p_2{}^2 - p_4)p_1 + 5(p_2 p_3 - p_5) ; \ \ \&c.$$

Having shown how $s_1, s_2, s_3, \ldots s_{n-1}$ can be calculated in terms of the coefficients, we proceed now to extend our results to the sums of all positive powers of the roots, viz., $s_n, s_{n+1} \ldots s_m$. For this purpose we have

$$x^{m-n} f(x) = x^m + p_1 x^{m-1} + p_2 x^{m-2} + \ldots + p_n x^{m-n}.$$

Replacing, in this identity, x by the roots $a_1, a_2, \ldots a_n$, in succession, and adding, we have

$$s_m + p_1 s_{m-1} + p_2 s_{m-2} + \ldots + p_n s_{m-n} = 0.$$

Now, giving m the values n, $n+1$, $n+2$, &c., successively, and observing that $s_0 = n$, the last equation gives

$$s_n \quad + p_1 s_{n-1} + p_2 s_{n-2} + \ldots + n p_n \quad = 0,$$

$$s_{n+1} + p_1 s_n \quad + p_2 s_{n-1} + \ldots + \quad p_n s_1 = 0,$$

$$s_{n+2} + p_1 s_{n+1} + p_2 s_n \quad + \ldots + \quad p_n s_2 = 0, \text{ &c.}$$

Hence the sums of all positive powers of the roots may be expressed by integral functions of the coefficients. And by transforming the equation into one whose roots are the reciprocals of $a_1, a_2, a_3, \ldots a_n$, and applying the above formulas, we may express similarly all negative powers of the roots.

127. PROP. II.—*Every rational symmetric function of the roots of an algebraic equation can be expressed rationally in terms of the coefficients.*

It is sufficient to prove this theorem for integral symmetric functions, since fractional symmetric functions can be reduced to a single fraction whose numerator and denominator are integral symmetric functions. Every integral function of $a_1, a_2, \ldots a_n$ is the sum of a number of terms of the form $N a_1^p a_2^q a_3^r \ldots$, where N is a numerical constant; but if this function is symmetrical we can write it under the form $N \Sigma a_1^p a_2^q a_3^r \ldots$, all the terms being of the same type. Therefore, if we prove that this quantity can be expressed rationally in terms of the coefficients, the theorem will be demonstrated. We shall first establish the following value of the symmetric function $\Sigma a_1^p a_2^q$:—

$$\Sigma a_1^p a_2^q = s_p s_q - s_{p+q}. \qquad (1)$$

To prove this, multiply together s_p and s_q, where

$$s_p = a_1^p + a_2^p + a_3^p + \ldots + a_n^p,$$

$$s_q = a_1^q + a_2^q + a_3^q + \ldots + a_n^q \; ;$$

whence

$$s_p s_q = a_1^{p+q} + a_2^{p+q} + \ldots a_n^{p+q} + a_1^p a_2^q + a_1^q a_2^p + \&c.,$$

or

$$s_p s_q = s_{p+q} + \Sigma a_1^p a_2^q,$$

which expresses the double function $\Sigma a_1^p a_2^q$ in terms of the single functions $s_p,\, s_q,\, s_{p+q}$ in the form above written.

We proceed now to prove a similar expression for the triple function, *i. e.*,

$$\Sigma a_1^p a_2^q a_3^r = s_p s_q s_r - s_{q+r} s_p - s_{r+p} s_q - s_{p+q} s_r + 2 s_{p+q+r}. \tag{2}$$

Multiplying together $\Sigma a_1^p a_2^q$ and s_r, where

$$\Sigma a_1^p a_2^q = a_1^p a_2^q + a_1^q a_2^p + a_1^p a_3^q + \ldots$$

$$s_r = a_1^r + a_2^r + a_3^r + \ldots + a_n^r,$$

the result will consist of three different parts, viz., terms of the form $\Sigma a_1^{p+r} a_2^q$, $\Sigma a_1^{q+r} a_2^p$, and $\Sigma a_1^p a_2^q a_3^r$.

Hence

$$s_r \Sigma a_1^p a_2^q = \Sigma a_1^{p+r} a_2^q + \Sigma a_1^{q+r} a_2^p + \Sigma a_1^p a_2^q a_3^r,$$

a formula connecting double and triple symmetric functions of $a_1,\, a_2,\, a_3,\, \ldots a_n$.

But, by (1),

$$\Sigma a_1^{p+r} a_2^q = s_{p+r} s_q - s_{p+q+r},$$

$$\Sigma a_1^{q+r} a_2^p = s_{q+r} s_p - s_{p+q+r},$$

$$\Sigma a_1^p a_2^q = s_p s_q - s_{p+q}.$$

Substituting these values, we find the triple function $\Sigma a_1^p a_2^q a_3^r$ expressed as above in terms of single functions in the series $s_1,\, s_2,\, s_3,\, \&c.$

In the same manner the quadruple function $\Sigma a_1^p a_2^q a_3^r a_4^s$

can be made to depend on the triple function $\Sigma a_1{}^p a_2{}^q a_3{}^r$, and ultimately on $s_1,\ s_2,\ s_3$, &c.; and so on. Whence, finally, every rational symmetric function of the roots may be expressed in terms of the coefficients, since, by Prop. I., $s_1,\ s_2,\ s_3$, &c., can be so expressed.

The formulas (1) and (2) require to be modified when any of the exponents become equal.

Thus, if $p = q$, $a_1{}^p a_2{}^q = a_2{}^p a_1{}^q$, and the terms in (1) become equal two and two; therefore $\Sigma a_1{}^p a_2{}^q = 2\Sigma a_1{}^p a_2{}^p$; whence

$$\Sigma a_1{}^p a_2{}^p = \tfrac{1}{2}\left(s_p{}^2 - s_{2p}\right).$$

Similarly, if $p = q = r$ in $\Sigma a_1{}^p a_2{}^q a_3{}^r$, the six terms obtained by interchanging the roots in $a_1{}^p a_2{}^q a_3{}^r$ become all equal; hence

$$\Sigma a_1{}^p a_2{}^p a_3{}^p = \frac{1}{2 \cdot 3}\left(s_p{}^3 - 3s_p s_{2p} + 2s_{3p}\right).$$

And, in general, if t exponents become equal, each term is repeated $1 \cdot 2 \cdot 3 \ldots t$ times.

EXAMPLES.

1. Prove

$$\Sigma a_1{}^p a_2{}^q a_3{}^r a_4{}^s = s_p s_q s_r s_s - \Sigma s_p s_q s_{r+s} + 2\Sigma s_p s_{q+r+s} + \Sigma s_{p+q} s_{r+s} - 6 s_{p+q+r+s}$$

where Σ on the right signifies the sum of all similar terms found by combining the suffixes $p,\ q,\ r,\ s$.

2. Prove

$$24 \Sigma a_1{}^m a_2{}^m a_3{}^m a_4{}^m = s_m{}^4 - 6 s_m{}^2 s_{2m} + 8 s_m s_{3m} + 3 s_{2m}{}^2 - 6 s_{4m}.$$

128. PROP. III.—*The value of s_r, expressed in terms of $p_1,\ p_2,\ \ldots\ p_n$, is the coefficient of y^r in the expansion by ascending powers of y of* $-r \log y^n f\!\left(\dfrac{1}{y}\right).$

Since

$$x^n + p_1 x^{n-1} + p_2 x^{n-2} + \ldots + p_n \equiv (x - a_1)(x - a_2) \ldots (x - a_n),$$

putting $\dfrac{1}{y}$ for x in this identical equation, we find

$$1 + p_1 y + p_2 y^2 + p_3 y^3 + \ldots + p_n y^n \equiv (1 - a_1 y)(1 - a_2 y) \ldots (1 - a_n y).$$

Now, taking the Napierian logarithms of both sides

$$p_1 y + p_2 \ \Big|\ y^2 + p_3 \ \Big|\ y^3 + p_4 \ \Big|\ y^4 + p_5 \ \Big|\ y^5 + \&\text{c.} \ldots + P_r y^r + \&\text{c.}$$

$$-\frac{1}{2} p_1{}^2 \ \Big|\ \begin{array}{l} -p_1 p_2 \\[4pt] +\dfrac{1}{3} p_1{}^3 \end{array} \ \Big|\ \begin{array}{l} -p_1 p_3 \\[4pt] -\dfrac{1}{2} p_2{}^2 \\[4pt] +p_1{}^2 p_2 \\[4pt] -\dfrac{1}{4} p_1{}^4 \end{array} \ \Big|\ \begin{array}{l} -p_2 p_3 \\[4pt] +p_1 p_2{}^2 \\[4pt] +p_1{}^2 p_3 \\[4pt] -p_1 p_4 \\[4pt] -p_1{}^3 p_2 \\[4pt] +\dfrac{1}{5} p_1{}^5 \end{array}$$

$$\equiv -y s_1 - \frac{1}{2} y^2 s_2 - \frac{1}{3} y^3 s_3 - \ldots - \frac{1}{r} y^r s_r - \&\text{c.}$$

Therefore, equating coefficients of y^r in both expansions,

$$s_r = -r P_r,$$

where P_r is the coefficient of y^r in $\log y^n f\!\left(\dfrac{1}{y}\right)$.

Remark.—From the above identical equation it may be seen that s_r (r less than n) involves only the coefficients $p_1, p_2, p_3, \ldots p_r$; and, therefore, $p_{r+1}, p_{r+2}, \ldots p_n$ may be made to vanish without affecting the form of the expression of s_r in terms of the coefficients.

129. *To express the coefficients in terms of the sums of the powers of the roots.*

Since

$$1 + p_1 y + p_2 y^2 + \ldots + p_n y^n \equiv (1 - a_1 y)(1 - a_2 y) \ldots (1 - a_n y),$$

we have

$$\log (1 + p_1 y + p_2 y^2 + \ldots + p_n y^n) \equiv -y s_1 - \frac{1}{2} y^2 s_2 - \ldots - \frac{1}{r} y^r s_r - \ldots;$$

and, therefore,

$$1 + p_1 y + p_2 y^2 + \ldots + p_n y^n \equiv e^{-y s_1 - \frac{1}{2} y^2 s_2 - \frac{1}{3} y^3 s_3 - \cdots},$$

which becomes by expansion

$$
\left| 1 - s_1 y - \frac{1}{2}\, s_2 + \frac{1}{1 \cdot 2}\, s_1^{\,2} \right|
\quad
\left| y^2 - \frac{1}{3}\, s_3 + \frac{1}{1 \cdot 2}\, s_1 s_2 - \frac{1}{1 \cdot 2 \cdot 3}\, s_1^{\,3} \right|
\quad
\left| y^3 - \frac{1}{4}\, s_4 + \frac{1}{3}\, s_1 s_3 - \frac{1}{4}\, s_1^{\,2} s_2 + \frac{1}{2 \cdot 4}\, s_2^{\,2} + \frac{1}{2 \cdot 3 \cdot 4}\, s_1^{\,4} \right|
\quad
\left| y^4 - \ldots \right.
$$

Now, comparing the coefficients of the different powers of y, we obtain values for $p_1,\, p_2,\, p_3,\, \ldots p_n$, in terms of $s_1,\, s_2,\, \ldots s_n$; and it may be seen that p_r involves only $s_1,\, s_2,\, \ldots s_r$ in its expression.

Remark.—It is important for the student to observe that it is a perfectly definite problem to express any symmetric function of the roots in terms of the coefficients, or to express the coefficients in terms of the sums of the powers of the roots; there being only one solution in each case.

We add some examples depending on the principles established in the preceding propositions.

Examples.

1. Determine the value of

$$
\phi(a_1) + \phi(a_2) + \ldots + \phi(a_n),
$$

where $a_1,\, a_2,\, a_3,\, \ldots a_n$ are the roots of $f(x) = 0$, and $\phi(x)$ is any rational and integral function of x.

We have

$$
\frac{f'(x)}{f(x)} = \frac{1}{x - a_1} + \frac{1}{x - a_2} + \ldots + \frac{1}{x - a_n},
$$

and

$$
\frac{f'(x)\,\phi(x)}{f(x)} = \frac{\phi(x)}{x - a_1} + \frac{\phi(x)}{x - a_2} + \ldots + \frac{\phi(x)}{x - a_n}.
$$

Performing the division, and retaining only the remainders on both sides of this equation, we have

$$\frac{R_0 x^{n-1} + R_1 x^{n-2} + \ldots + R_{n-1}}{f(x)} = \frac{\phi(a_1)}{x - a_1} + \frac{\phi(a_2)}{x - a_2} + \ldots + \frac{\phi(a_n)}{x - a_n};$$

whence

$$R_0 x^{n-1} + R_1 x^{n-2} + \ldots + R_{n-1} = \Sigma \phi(a_1)(x - a_2)(x - a_3) \ldots (x - a_n);$$

and, comparing the coefficients of x^{n-1} on both sides of this equation,

$$R_0 = \Sigma \phi(a_1).$$

2. Prove that s_p is the coefficient of $\dfrac{1}{x^{p+1}}$ in the quotient of the division of $f'(x)$ by $f(x)$ arranged according to negative powers of x.

3. Prove that s_{-p} is the coefficient (with sign changed) of x^{p-1} in the same quotient arranged according to positive powers of x.

4. If the degree of $\phi(x)$ does not exceed $n - 2$, prove

$$\Sigma_1^n \frac{\phi(a_r)}{f'(a_r)} = 0,$$

where Σ_1^n denotes the sum obtained by giving r all values from 1 to n inclusive.

We have, by partial fractions,

$$\frac{\phi(x)}{f(x)} = \frac{A_1}{x - a_1} + \frac{A_2}{x - a_2} + \ldots + \frac{A_n}{x - a_n};$$

and, multiplying across by $f(x)$, and putting x equal to a_1, a_2, $\ldots$ in succession,

$$\frac{\phi(x)}{f(x)} = \frac{\phi(a_1)}{f'(a_1)} \frac{1}{x - a_1} + \frac{\phi(a_2)}{f'(a_2)} \frac{1}{x - a_2} + \ldots + \frac{\phi(a_n)}{f'(a_n)} \frac{1}{x - a_n};$$

whence

$$\frac{x \phi(x)}{f(x)} = \Sigma_1^n \frac{\phi(a_r)}{f'(a_r)} \left(1 + \frac{a_r}{x} + \frac{a_r^2}{x^2} + \ldots \right).$$

When $\phi(x)$ is of the degree $n - 2$; expressing the first side of the equation in terms of $\dfrac{1}{x}$, it becomes $\dfrac{1}{x} \cdot \dfrac{\phi\left(\dfrac{1}{x}\right)}{f\left(\dfrac{1}{x}\right)}$, and there is no term without $\dfrac{1}{x}$ as a multiplier. We have then, comparing coefficients,

$$\Sigma_1^n \frac{\phi(a_r)}{f'(a_r)} = 0.$$

And therefore, as ϕ may be any rational and integral function of degree not higher than $n - 2$,

$$\Sigma \frac{a^{n-2}}{f'(a)} = 0, \quad \Sigma \frac{a^{n-3}}{f'(a)} = 0, \ldots \Sigma \frac{a}{f'(a)} = 0, \quad \Sigma \frac{1}{f'(a)} = 0.$$

5. Prove that the sum of all the homogeneous products Π_r, of the rth degree of the quantities $a_1, a_2, \ldots a_n,$ is equal to

$$\Sigma \frac{a^{n+r-1}}{f'(a)}.$$

We have, putting $y = \frac{1}{x}$,

$$\frac{x^n}{f(x)} = \frac{1}{(1-a_1 y)(1-a_2 y)\ldots(1-a_n y)}$$

$$= (1 + a_1 y + a_1^2 y^2 + \ldots)(1 + a_2 y + a_2^2 y^2 + \ldots)\ldots(1 + a_n y + a_n^2 y^2 + \ldots)$$

$$= 1 + \Pi_1 y + \Pi_2 y^2 + \ldots + \Pi_r y^r + \ldots$$

Also

$$\frac{x^{n-1}}{f(x)} = \Sigma \frac{a^{n-1}}{f'(a)} \frac{1}{x-a},$$

and therefore

$$\frac{x^n}{f(x)} = \Sigma \frac{a^{n-1}}{f'(a)} \frac{1}{1-ay} = \Sigma \frac{a^{n+r-1}}{f'(a)} y^r;$$

whence, comparing coefficients of y^r in these two expansions,

$$\Pi_r = \Sigma \frac{a^{n+r-1}}{f'(a)}.$$

130. Definitions. Theorem.—The *weight* of any symmetric function of the roots is the degree in *all* the roots of any term in the function. For example, the weight of $\Sigma a\beta^2\gamma^3$ is six.

The *order* of any symmetric function of the roots is the highest degree in which each root enters the function. For example, the order of $\Sigma a\beta^2\gamma^3$ is three.

It has been proved (see Art. 28), that the weight of any symmetric function of the roots, when expressed by the coefficients $a_0, a_1, a_2, \ldots a_n,$ is the same as the sum of the suffixes of each term in the expression. We now prove another important theorem, viz. :

If any symmetric function be expressed in terms of the coefficients $p_1, p_2, \ldots p_n,$ the degree in the coefficients is the same as the order of the symmetric function. For example, $\Sigma a^2\beta^2 = p_2^2 - 2p_1 p_3 + 2p_4,$ no term being of higher degree than the second in the coefficients, and the order of the symmetric function being two.

The student may easily satisfy himself in general of the truth of this theorem by observing that in the equations (2) of Art. 23, the value of each coefficient in terms of the roots contains each root in the first power only; hence the highest degree in the coefficients will be the same as the highest degree of the corresponding symmetric function in any individual root. We add the following more rigorous proof, as it is in accordance with the proofs of certain general propositions to be given subsequently.

Replace the coefficients $p_1, p_2, \ldots p_n$ by $\dfrac{a_1}{a_0}, \dfrac{a_2}{a_0}, \ldots \dfrac{a_n}{a_0}$.

Now, if $\phi(a_1, a_2, \ldots a_n)$ denote any rational and integral symmetric function of the roots, we have

$$a_0{}^\varpi \phi(a_1, a_2, \ldots a_n) = F(a_0, a_1, a_2, \ldots a_n),$$

where ϖ is the degree of the coefficients in $F(a_0, a_1, a_2, \ldots a_n)$, which is a function integral and homogeneous in the coefficients, and not divisible by a_0.

We require now to show that ϖ is the order of ϕ. For this purpose change the roots into their reciprocals, and, therefore, $a_0, a_1, \ldots a_n$ into $a_n, a_{n-1}, \ldots a_0$. Whence

$$a_n{}^\varpi \phi\left(\frac{1}{a_1}, \frac{1}{a_2}, \ldots \frac{1}{a_n}\right) = F(a_n, a_{n-1}, a_{n-2}, \ldots a_0); \qquad (1)$$

also

$$\phi\left(\frac{1}{a_1}, \frac{1}{a_2}, \ldots \frac{1}{a_n}\right) = \frac{\psi(a_1, a_2, a_3 \ldots a_n)}{(a_1 a_2 a_3 \ldots a_n)^p},$$

where p is the order of ϕ, and ψ an integral function not divisible by the product of all the roots; $(a_1 a_2 a_3 \ldots a_n)^p$ being the lowest common denominator of all the terms. Substituting in (1), we have

$$a_0{}^p \psi(a_1, a_2, \ldots a_n) = \pm\, a_n{}^{p-\varpi} F(a_n, a_{n-1}, \ldots a_0);$$

wherefore $p = \varpi$; for if not, F would be divisible by $a_n{}^{\varpi-p}$, which is contrary to our hypothesis.

131. Calculation of Symmetric Functions of the Roots.—Any rational symmetric function can be calculated by the method of Art. 127. In practice, however, other methods are usually more convenient, as will appear from the examples given at the end of the present Article, and from the following Articles, in which we shall give certain general propositions which in many cases facilitate the calculation of symmetric functions.

The number of terms in any symmetric function of the roots is easily determined. For example, the number of terms in $\Sigma a_1^3 a_2^2 a_3$ of the equation of the nth degree is $n(n-1)(n-2)$, this being the number of permutations of n things taken three together. If the exponents of the roots in any term be not all different, the number of terms will be reduced. Thus, $\Sigma a^2 \beta \gamma$ for a biquadratic consists of 12 terms only (see Ex. 6, p. 48), and not of 24, since the two permutations $a\beta\gamma$, $a\gamma\beta$ give only one distinct term, viz., $a^2\beta\gamma$, in $\Sigma a^2\beta\gamma$. The student acquainted with the theory of permutations will have no difficulty in effecting these reductions in any particular case. When two exponents of roots are equal, the number obtained on the supposition that they are all unequal is to be divided by $1 . 2$; when three become equal this number is to be divided by $1 . 2 . 3$; and so on. In general, the number of terms in $\Sigma a_1^p a_2^q a_3^r \ldots$ of the equation of the nth degree, each term containing m roots, and ν of the indices being equal, is

$$\frac{n(n-1)(n-2)\ldots(n-m+1)}{1 . 2 . 3 \ldots \nu}.$$

When the highest power in which any one root enters into the symmetric function is small, *i.e.*, when the order of the function (see Art. 130) is low, the methods already illustrated in Art. 27 may be employed with advantage for the calculation of the symmetric function of the roots in terms of the coefficients.

It is important to observe that when any symmetric function whose degree in all the roots, *i.e.*, its weight, is n, is calculated in terms of the coefficients $p_1, p_2 \ldots p_n$ for the equation of the

n^{th} degree, its value for an equation of any higher degree (the numerical coefficients being all equal to unity) is precisely the same; for it is plain that no coefficient beyond p_n can enter into this value, and the equations of Art. 126, by means of which the calculation can be supposed to be made, have precisely the same form for an equation of the nth degree as for equations of all higher degrees. It is also evident that the value of the same symmetric function for an equation of a degree m (lower than n) is obtained by putting $p_{m+1}, p_{m+2}, \ldots p_n$ all equal to zero in the calculated value for an equation of the nth degree, since the equation of lower degree can be derived from that of the nth by putting the coefficients beyond p_m equal to zero, and the corresponding symmetric function reduces similarly by putting the roots $a_{m+1}, a_{m+2}, \ldots a_n$ each equal to zero.

EXAMPLES.

1. Calculate $\Sigma a_1^2 a_2 a_3$ of the roots of the equation

$$x^n + p_1 x^{n-1} + p_2 x^{n-2} + \ldots + p_{n-1}x + p_n = 0.$$

Multiply together the equations

$$\Sigma a_1 = -p_1,$$

$$\Sigma a_1 a_2 a_3 = -p_3.$$

In the product the term $a_1^2 a_2 a_3$ occurs only once; the term $a_1 a_2 a_3 a_4$ occurs four times, arising from the product of a_1 by $a_2 a_3 a_4$, of a_2 by $a_1 a_3 a_4$, of a_3 by $a_1 a_2 a_4$, and of a_4 by $a_1 a_2 a_3$. Hence

$$\Sigma a_1^2 a_2 a_3 + 4\Sigma a_1 a_2 a_3 a_4 = p_1 p_3 \,;$$

therefore

$$\Sigma a_1^2 a_2 a_3 = p_1 p_3 - 4p_4. \qquad \text{(Compare Ex. 6, Art. 27.)}$$

If the calculation were conducted by the method of Art. 127, we should have

$$\Sigma a_1^2 a_2 a_3 = \tfrac{1}{2} s_2 s_1^2 - s_1 s_3 - \tfrac{1}{2} s_2^2 + s_4,$$

which leads, on substituting the values of Art. 126, to the same result; but it is evident that in this case the former process is much more simple, since the values of s_1, s_2, &c., introduce a number of terms which destroy one another.

2. Calculate $\Sigma a_1^2 a_2^2$ for the general equation.

Squaring

$$\Sigma a_1 a_2 = p_2,$$

we have

$$\Sigma a_1^2 a_2^2 + 2\Sigma a_1^2 a_2 a_3 + 6\Sigma a_1 a_2 a_3 a_4 = p_2^2.$$

In squaring it is evident that the term $\alpha_1 \alpha_2 \alpha_3 \alpha_4$ will arise from the product of $\alpha_1 \alpha_2$ by $\alpha_3 \alpha_4$, or of $\alpha_1 \alpha_3$ by $\alpha_2 \alpha_4$, or of $\alpha_1 \alpha_4$ by $\alpha_2 \alpha_3$; hence the coefficient of $\alpha_1 \alpha_2 \alpha_3 \alpha_4$ in the result is 6, since each of these occurs twice in the square. The result differs from the similar equation of Ex. 8, Art. 27, only in having Σ before the term $\alpha_1 \alpha_2 \alpha_3 \alpha_4$. Hence the result

$$\Sigma\alpha_1{}^2\alpha_2{}^2 = p_2{}^2 - 2p_1 p_3 + 2p_4.$$

3. Calculate $\Sigma\alpha_1{}^3\alpha_2$ for the general equation.

We have, as in Ex. 9, Art. 27,

$$\Sigma\alpha_1{}^2\, \Sigma\alpha_1\alpha_2 = \Sigma\alpha_1{}^3\alpha_2 + \Sigma\alpha_1{}^2\alpha_2\alpha_3.$$

Hence, employing previous results,

$$\Sigma\alpha_1{}^3\alpha_2 = p_1{}^2 p_2 - 2p_2{}^2 - p_1 p_3 + 4p_4.$$

4. Calculate $\Sigma\alpha_1{}^2\alpha_2{}^2\alpha_3$ for the general equation.

The result will be the same as if the calculation were made for the equation of the 5th degree.

To obtain the symmetric function we multiply together $\Sigma\alpha_1\alpha_2$ and $\Sigma\alpha_1\alpha_2\alpha_3$; and consider what types of terms, involving the five roots $\alpha_1, \alpha_2, \alpha_3, \alpha_4, \alpha_5$, can result. The term $\alpha_1{}^2\alpha_2{}^2\alpha_3$ will occur only once in the product, since it can only arise by multiplying $\alpha_1\alpha_2$ by $\alpha_1\alpha_2\alpha_3$. Terms of the type $\alpha_1{}^2\alpha_2\alpha_3\alpha_4$ will occur, each three times; since $\alpha_1{}^2\alpha_2\alpha_3\alpha_4$ will arise from the product of $\alpha_1\alpha_2$ by $\alpha_1\alpha_3\alpha_4$, of $\alpha_1\alpha_3$ by $\alpha_1\alpha_2\alpha_4$, or of $\alpha_1\alpha_4$ by $\alpha_1\alpha_2\alpha_3$; and it cannot arise in any other way. The term $\alpha_1\alpha_2\alpha_3\alpha_4\alpha_5$ will occur ten times in the product, since it will arise from the product of any pair by the other three roots, and there are ten combinations in pairs of the five roots. We have then, for the general equation,

$$\Sigma\alpha_1\alpha_2\, \Sigma\alpha_1\alpha_2\alpha_3 = \Sigma\alpha_1{}^2\alpha_2{}^2\alpha_3 + 3\Sigma\alpha_1{}^2\alpha_2\alpha_3\alpha_4 + 10\,\Sigma\alpha_1\alpha_2\alpha_3\alpha_4\alpha_5.$$

[We can verify this equation when $n = 5$, just as in Ex. 9, Art. 27; for the product of two factors, each consisting of 10 terms, will contain 100 terms. These are made up of the 30 terms contained in $\Sigma\alpha_1{}^2\alpha_2{}^2\alpha_3$, along with the 20 terms contained in $\Sigma\alpha_1{}^2\alpha_2\alpha_3\alpha_4$, each taken three times, and the term $\alpha_1\alpha_2\alpha_3\alpha_4\alpha_5$ taken 10 times.]

Thus the calculation of the required symmetric function involves that of $\Sigma\alpha_1{}^2\alpha_2\alpha_3\alpha_4$; for which we easily find

$$\Sigma\alpha_1\, \Sigma\alpha_1\alpha_2\alpha_3\alpha_4 = \Sigma\alpha_1{}^2\alpha_2\alpha_3\alpha_4 + 5\Sigma\alpha_1\alpha_2\alpha_3\alpha_4\alpha_5.$$

Hence, finally, we obtain

$$\Sigma\alpha_1{}^2\alpha_2{}^2\alpha_3 = -p_2 p_3 + 3p_1 p_4 - 5p_5.$$

The process of Art. 127 would involve the calculation of s_5; and many terms would be introduced through the values of s_1, s_2, &c., which disappear in the result.

5. Find the value of $\Sigma\alpha_1{}^2\alpha_2{}^2\alpha_3\alpha_4$ for the general equation.

We multiply together $\Sigma\alpha_1\alpha_2$ and $\Sigma\alpha_1\alpha_2\alpha_3\alpha_4$, and consider what types of terms can arise involving the six roots $\alpha_1 \alpha_2 \alpha_3 \alpha_4 \alpha_5 \alpha_6$. The term $\alpha_1{}^2\alpha_2{}^2\alpha_3\alpha_4$ can occur only once. Terms of the type $\alpha_1{}^2\alpha_2\alpha_3\alpha_4\alpha_5$ will each occur four times, this term arising from the product of $\alpha_1\alpha_2$ by $\alpha_1\alpha_3\alpha_4\alpha_5$, or of $\alpha_1\alpha_3$ by $\alpha_1\alpha_2\alpha_4\alpha_5$, or of $\alpha_1\alpha_4$ by $\alpha_1\alpha_2\alpha_3\alpha_5$, or of $\alpha_1\alpha_5$ by $\alpha_1\alpha_2\alpha_3\alpha_4$. The term $\alpha_1\alpha_2\alpha_3\alpha_4\alpha_5\alpha_6$ will occur 15 times, this being the number of combinations in pairs of the six roots.   Hence

$$\Sigma\alpha_1\alpha_2\,\Sigma\alpha_1\alpha_2\alpha_3\alpha_4 = \Sigma\alpha_1{}^2\alpha_2{}^2\alpha_3\alpha_4 + 4\Sigma\alpha_1{}^2\alpha_2\alpha_3\alpha_4\alpha_5 + 15\Sigma\alpha_1\alpha_2\alpha_3\alpha_4\alpha_5\alpha_6.$$

We have again, for the calculation of $\Sigma\alpha_1{}^2\alpha_2\alpha_3\alpha_4\alpha_5$,

$$\Sigma\alpha_1\,\Sigma\alpha_1\alpha_2\alpha_3\alpha_4\alpha_5 = \Sigma\alpha_1{}^2\alpha_2\alpha_3\alpha_4\alpha_5 + 6\Sigma\alpha_1\alpha_2\alpha_3\alpha_4\alpha_5\alpha_6.$$

Hence, finally,

$$\Sigma\alpha_1{}^2\alpha_2{}^2\alpha_3\alpha_4 = p_2\,p_4 - 4p_1\,p_5 + 9p_6.$$

6. Find the value of $\Sigma\alpha_1{}^2\alpha_2{}^2\alpha_3{}^2$ in terms of the coefficients of the general equation.

Here, squaring $\Sigma\alpha_1\alpha_2\alpha_3$, we have

$$\Sigma\alpha_1\alpha_2\alpha_3\,\Sigma\alpha_1\alpha_2\alpha_3 = \Sigma\alpha_1{}^2\alpha_2{}^2\alpha_3{}^2 + 2\Sigma\alpha_1{}^2\alpha_2{}^2\alpha_3\alpha_4 + 6\Sigma\alpha_1{}^2\alpha_2\alpha_3\alpha_4\alpha_5 + 20\Sigma\alpha_1\alpha_2\alpha_3\alpha_4\alpha_5\alpha_6,$$

from which we obtain

$$\Sigma\alpha_1{}^2\alpha_2{}^2\alpha_3{}^2 = p_3{}^2 - 2p_2p_4 + 2p_1p_5 - 2p_6.$$

132. Differential Equation between the Sums of the Powers of the Roots and the Coefficients of an Equation.—M. Brioschi has given the following differential equation connecting the coefficients and sums of powers of the roots :—

$$\frac{dp_{r+k}}{ds_r} = -\frac{1}{r}\,p_k.$$

To prove this we have, as in Art. 128,

$$\log\,(1+p_1y+p_2y^2+\ldots+p_ny^n) = -ys_1 - \frac{1}{2}\,y^2\,s_2 - \frac{1}{3}\,y^3\,s_3\ldots - \frac{1}{r}\,y^r s_r\ldots,$$

and differentiating,

$$\frac{d}{ds_r}\,(1+p_1y+p_2y^2+\ldots+p_ny^n) = -(1+p_1y+p_2y^2+\ldots+p_ny^n)\,\frac{y^r}{r}\,;$$

whence, comparing the coefficients of the different powers of y,

$$\frac{dp_q}{ds_r} = 0,\ \text{ when } q < r;$$

$$\frac{dp_r}{ds_r} = -\frac{1}{r},\quad \frac{dp_{r+k}}{ds_r} = -\frac{1}{r}p_k.$$

We can now express the result of differentiating with respect to s_r any function of the coefficients

$$F(p_1, p_2, p_3, \ldots p_n).$$

Since

$$\frac{dp_1}{ds_r}, \quad \frac{dp_2}{ds_r}, \ldots \frac{dp_{r-1}}{ds_r}$$

all vanish,

$$\frac{d}{ds_r} F(p_1, p_2, p_3, \ldots p_n) = \frac{dF}{dp_r}\frac{dp_r}{ds_r} + \frac{dF}{dp_{r+1}}\frac{dp_{r+1}}{ds_r} + \ldots + \frac{dF}{dp_n}\frac{dp_n}{ds_r},$$

and, applying the formula given above, this reduces to

$$-\frac{1}{r}\left(\frac{dF}{dp_r} + p_1\frac{dF}{dp_{r+1}} + p_2\frac{dF}{dp_{r+2}} + \ldots + p_{n-r}\frac{dF}{dp_n}\right).$$

By means of Brioschi's formula symmetric functions can often be calculated with great facility, as will appear from the following examples.

EXAMPLES.

1. Calculate the value of the symmetric function $\Sigma \alpha_1^2 \alpha_2^2 \alpha_3^2 \alpha_4^2$ of the roots of the equation

$$x^n + p_1 x^{n-1} + p_2 x^{n-2} + \ldots + p_n = 0.$$

Knowing the order and weight of any symmetric function, we can write down the literal part of its value in terms of the coefficients. Here Σ is of the second order, and its weight is eight; whence

$$\Sigma = t_0\, p_8 + t_1\, p_7\, p_1 + t_2\, p_6\, p_2 + t_3\, p_5\, p_3 + t_4\, p_4^2,$$

where t_0, t_1, t_2, &c., are numerical coefficients to be determined.

Terms such as $p_6 p_1^2$, $p_5 p_1 p_2$, $p_5 p_1^3$, &c., although of the right weight, are of too high an order, and therefore cannot enter into the expression for Σ. Again, Σ expressed in terms of the sums of the powers of the roots is of the form $F(s_2, s_4, s_6, s_8)$; for, in general, $\Sigma \alpha_1^p \alpha_2^q \alpha_3^r \ldots$, expressed in terms of the sums of the powers of the roots, is made up of terms such as s_p, s_{p+q}, s_{p+q+r}, $\ldots s_{kp}$, $\ldots$ all of which are sums of even powers when p, q, r, $\ldots$ are even; therefore in this case none but even sums of powers enter into the expression for Σ.

Also, since $\dfrac{d\Sigma}{ds_3} = 0$, and $\dfrac{d\Sigma}{ds_7} = 0$, we have, using the formula above given for $\dfrac{dF}{ds_r}$,

$$t_0\, p_5 + t_1\, p_1\, p_4 + t_2\, p_3\, p_2 + t_3(p_2\, p_3 + p_5) + 2t_4\, p_1\, p_4 = 0,$$

$$t_0\, p_1 + t_1\, p_1 = 0.$$

From these equations we infer

$$t_0 + t_1 = 0, \quad t_2 + t_3 = 0, \quad t_3 + t_0 = 0, \quad t_1 + 2t_4 = 0 \,;$$

but $t_4 = 1$, since for a quartic $\Sigma = p_4{}^2$; therefore

$$t_1 = -2, \quad t_0 = 2, \quad t_3 = -2, \quad t_2 = 2\,;$$

and, substituting these values of $\quad t_0,\ t_1,\ t_2,\ t_3,\ t_4,$

$$\Sigma \alpha_1{}^2 \alpha_2{}^2 \alpha_3{}^2 \alpha_4{}^2 = 2p_8 - 2p_7 p_1 + 2p_6 p_2 - 2p_5 p_3 + p_4{}^2.$$

2. Calculate $\Sigma \alpha_1{}^2 \alpha_2{}^2 \alpha_3{}^2$ for the same equation.

$$Ans. \ -2p_6 + 2p_1 p_5 - 2p_2 p_4 + p_3{}^2. \quad \text{(Compare Ex. 6, Art. 131).}$$

3. Calculate for the same equation the symmetric function $\Sigma \alpha_1{}^3 \alpha_2{}^2 \alpha_3$.
Here the weight is six, and the order three; hence

$$\Sigma \alpha_1{}^3 \alpha_2{}^2 \alpha_3 = t_0 \, p_6 + t_1 \, p_5 p_1 + t_2 p_4 p_2 + t_3 \, p_4 p_1{}^2 + t_4 p_3{}^2 + t_5 p_1 p_2 p_3 + t_6 p_2{}^3.$$

Also Σ, expressed in terms of $s_1,\ s_2,\ s_3,$ &c., is (see Art. 127),

$$s_1 s_2 s_3 - s_1 s_5 - s_3{}^2 - s_2 s_4 + 2 s_6.$$

Now, differentiating by means of Brioschi's equation these two values of Σ with regard to s_6, and comparing differential coefficients, we have

$$t_0 \frac{d p_6}{d s_6} = -\frac{t_0}{6} = 2, \quad \text{or} \quad t_0 = -12.$$

Differentiating with regard to s_5, we have

$$t_0 p_1 + t_1 p_1 = 5 s_1 = -5 p_1\,; \ \therefore \ t_1 = 7.$$

Differentiating with regard to s_4,

$$t_0 p^2 + t_1 p_1{}^2 + t_2 p_2 + t_3 p_1{}^2 = 4 s_2 = 4 \left(p_1{}^2 - 2p_2 \right);$$

whence
$$t_0 + t_2 = -8, \quad t_1 + t_3 = 4\,;$$

and hence
$$t_3 = -3, \quad t_2 = 4.$$

Again, $t_6 = 0$; for Σ vanishes if $n-2$ roots vanish. And we find t_4 and t_5 by taking the particular case when $n-3$ roots vanish; for in this case

$$\Sigma \alpha_1{}^3 \alpha_2{}^2 \alpha_3 = \alpha_1 \alpha_2 \alpha_3 \Sigma \alpha_1{}^2 \alpha_2 = -p_3 \left(-p_1 p_2 + 3 p_3 \right) = p_1 p_2 p_3 - 3 p_3{}^2,$$

and, therefore
$$t_4 = -3, \quad t_5 = 1\,;$$
whence, finally,

$$\Sigma \alpha_1{}^3 \alpha_2{}^2 \alpha_3 = -12 p_6 + 7 p_1 p_5 + 4 p_4 p_2 - 3 p_4 p_1{}^2 - 3 p_3{}^2 + p_1 p_2 p_3.$$

133. Derivation of new Symmetric Functions from a given one.—From any relation such as

$$a_0^{\varpi} \Sigma\phi(a_1, a_2, \ldots a_n) = F(a_0, a_1, a_2 \ldots a_n),$$

where ϕ is an integral function, of the order ϖ, of some or all of the roots of the equation

$$a_0 x^n + na_1 x^{n-1} + \frac{n(n-1)}{1 \cdot 2} a_2 x^{n-2} + \ldots + a_n = 0,$$

we may derive a number of other symmetric functions and their equivalents in terms of the coefficients.

For this purpose diminish each of the roots by any quantity x, and consequently change any coefficient a_r into U_r (see Art. 36). When this is done the original relation becomes

$$a_0^{\varpi} \Sigma\phi(a_1 - x, a_2 - x, \ldots a_n - x) = F(U_0, U_1, U_2, \ldots U_n),$$

and comparing the coefficients of the different powers of x on both sides of this equation, we have a number of symmetric functions of the roots expressed in terms of the coefficients as required. It should be observed, however, that this method leads to no new symmetric functions when the given function ϕ is a function of the differences of the roots.

134. Equation of Operation.—We now proceed to deduce an important equation of operation in the notation of the differential calculus, which may be applied to furnish the results of the last Article.

Let $\qquad a_0^{\varpi} \phi(a_1, a_2, \ldots a_n) = F(a_0, a_1, a_2, \ldots a_n),$

as in the last Article. Adopting the notation

$$-\delta = \frac{d}{da_1} + \frac{d}{da_2} + \ldots + \frac{d}{da_n},$$

$$D = a_0 \frac{d}{da_1} + 2a_1 \frac{d}{da_2} + 3a_2 \frac{d}{da_3} + \ldots + na_{n-1} \frac{d}{da_n},$$

we have the following equation of operation :—

$$\delta a_0^{\varpi} \phi(a_1, a_2, \ldots a_n) = DF(a_0, a_1, \ldots a_n).$$

To prove this, we have, as in Art. 133,

$$a_0^{\varpi}\,\phi(a_1 - x,\, a_2 - x,\, \ldots a_n - x) = F(U_0,\, U_1,\, U_2,\, \ldots U_n)\,;$$

and, by Taylor's theorem,

$$\phi(a_1 - x,\, a_2 - x,\, \ldots a_n - x) = \phi_0 + x\delta\,\phi_0 + \frac{x^2}{1\cdot 2}\,\delta^2\phi_0 + \ldots,$$

where

$$\phi_0 = \phi(a_1,\, a_2,\, \ldots a_n).$$

Again, omitting all powers of x higher than the first,

$F(U_0,\, U_1,\, \ldots U_n)$ becomes $F(a_0,\, a_1 + a_0 x,\, a_2 + 2a_1 x,\, \ldots a_n + n a_{n-1} x)$,

or, when expanded,

$$F_0 + x\left(a_0\frac{d}{da_1} + 2a_1\frac{d}{da_2} + \ldots n a_{n-1}\frac{d}{da_n}\right)F_0 + \&\mathrm{c.},$$

where

$$F_0 = F(a_0,\, a_1,\, \ldots a_n)\,;$$

whence, comparing coefficients of x in both expansions, we find the equation above written, viz.,

$$a_0^{\varpi}\,\delta\phi(a_1,\, a_2,\, \ldots a_n) = DF(a_0,\, a_1,\, \ldots a_n).$$

This equation shows that if a symmetric function be derived from ϕ by the operation δ, its value in terms of the coefficients may be derived from the corresponding value of ϕ by the operation D.

Again, since $\delta\phi$ and DF may take the place of ϕ and F in this equation, $a_0^{\varpi}\,\delta^2\phi$ becomes $D^2 F$, &c. It may be noticed, moreover, that if $\delta\phi_0$ vanishes, $\delta^2\phi_0$, $\delta^3\phi_0$, &c., all vanish; and thus that x disappears in the expansion of

$$\phi(a_1 - x,\, a_2 - x,\, \ldots a_n - x).$$

Now this can happen only when ϕ is a function of the differences of $a_1,\, a_2,\, \ldots\, a_n$: whence we conclude that if $a_0^{-\varpi}F(a_0,\, a_1,\, a_2,\, \ldots a_n)$ is the value in terms of the coefficients of a function of the differences of the roots, then

$$DF(a_0,\, a_1,\, a_2,\, \ldots a_n)$$

vanishes identically.

This identical relation is often sufficient to determine the numerical coefficients in a function of the differences expressed by the coefficients, when the order and weight are known. It is not sufficient for this purpose when there exist more than one function of the differences of the required order and weight. We add examples of functions of the differences determined in this way.

EXAMPLES.

1. Determine a function of the differences whose order and weight are both three.

Assume

$$\phi = A a_0^2 a_3 + B a_0 a_1 a_2 + C a_1^3,$$

these being the only three terms which satisfy the required conditions. It is evident from the form of D that the operation is performed by applying to the suffix of any coefficient a_r the same process as in ordinary differentiation is applied to the index. Thus $D a_r = r a_{r-1},$ and therefore,

$$D\phi = (3A + B) a_0^2 a_2 + (2B + 3C) a_1^2 a_0 \equiv 0.$$

Hence

$$3A + B = 0, \quad \text{and} \quad 2B + 3C = 0 ;$$

and putting $A = 1$, we have

$$B = -3, \quad \text{and} \quad C = 2 ;$$

whence, finally,

$$\phi = a_0^2 a_3 - 3 a_0 a_1 a_2 + 2 a_1^3 \equiv G. \qquad \text{(See Art. 37.)}$$

2. Determine a function of the differences whose degree in the coefficients is four, and whose weight is six.

Assume

$$\phi = A a_0^2 a_3^2 + B a_0 a_2^3 + C a_3 a_1^3 + D a_1^2 a_2^2 + E a_0 a_1 a_2 a_3,$$

whence

$$D\phi = (6A + E) a_0^2 a_2 a_3 + (6B + 3E + 2D) a_0 a_1 a_2^2 + (3C + 4D) a_1^3 a_2$$

$$+ (3C + 2E) a_0 a_1^2 a_3 \equiv 0.$$

Now let $A = 1$, whence $E = -6$; also $3C + 2E = 0$, giving $C = 4$; and $3C + 4D = 0$, giving $D = -3$; and from $6B + 3E + 2D = 0$, we have finally $B = 4$.

Hence

$$\phi = a_0^2 a_3^2 + 4 a_0 a_2^3 + 4 a_3 a_1^3 - 3 a_1^2 a_2^2 - 6 a_0 a_1 a_2 a_3.$$

Compare Art. 41, where the value of ϕ is given in terms of the roots.

135. Operation involving the Sums of the Powers of the Roots. Theorem.—If

$$\phi(a_1, a_2, a_3 \ldots a_n) = F(s_1, s_2, s_3, \ldots s_n) \qquad (1)$$

be any equation connecting a function of the sums of the powers with another symmetric function of the roots, we have then the differential equation

$$\frac{d\phi}{da_1} + \frac{d\phi}{da_2} + \frac{d\phi}{da_3} + \ldots + \frac{d\phi}{da_n} = s_0 \frac{dF}{ds_1} + 2s_1 \frac{dF}{ds_2} + 3s_2 \frac{dF}{ds_3} + \ldots + rs_{r-1} \frac{dF}{ds_r}.$$

For, let the roots be increased by h; and comparing the coefficients of h on both sides of the equation (1) when

$$s_1 + h s_0, \quad s_2 + 2h s_1, \quad \ldots \quad s_r + rh s_{r-1}, \quad \ldots$$

are substituted for $s_1, s_2, \ldots s_r$, we have the required relation.

Employing the results of the last Article, we have, therefore, the following equation of operation connecting the coefficients and the sums of the powers of the roots :—

$$- D = a_0^\varpi \left(s_0 \frac{d}{ds_1} + 2s_1 \frac{d}{ds_2} + 3s_2 \frac{d}{ds_3} + \ldots + rs_{r-1} \frac{d}{ds_r} \right) = a_0^\varpi D_s,$$

where D_s represents the result of substituting s for a in the operator D.

From this it follows that if $f(a_0, a_1, a_2, \ldots a_n)$ is a function of the differences, $f(s_0, s_1, s_2, \ldots s_n)$ is a function of the differences also ; for it is plain that when $Df(a_0, a_1, a_2, \ldots a_n) = 0$, $D_s f(s_0, s_1, s_2, \ldots s_n) = 0$, and therefore $Df(s_0, s_1, s_2, \ldots s_n) = 0$, since $D_s = - a_0^{-\varpi} D$.

EXAMPLES.

1. $a_0 a_4 - 4a_1 a_3 + 3a_2^2 \equiv I$ is a function of the differences, whence $s_0 s_4 - 4s_1 s_3 + 3s_2^2$ is also a function of the differences.

2. $\begin{vmatrix} a_0 & a_1 & a_2 \\ a_1 & a_2 & a_3 \\ a_2 & a_3 & a_4 \end{vmatrix} = J$, when similarly transformed, gives $\begin{vmatrix} s_0 & s_1 & s_2 \\ s_1 & s_2 & s_3 \\ s_2 & s_3 & s_4 \end{vmatrix}$, which is, therefore, a function of the differences.

MISCELLANEOUS EXAMPLES.

1. Prove, by squaring the determinant of Example 10, Art. 112, the following relation between the roots α, β, γ, δ of the biquadratic :—

$$\begin{vmatrix} s_0 & s_1 & s_2 & s_3 \\ s_1 & s_2 & s_3 & s_4 \\ s_2 & s_3 & s_4 & s_5 \\ s_3 & s_4 & s_5 & s_6 \end{vmatrix} = (\beta - \gamma)^2 (\alpha - \delta)^2 (\gamma - \alpha)^2 (\beta - \delta)^2 (\alpha - \beta)^2 (\gamma - \delta)^2.$$

The student will have no difficulty in writing down in general the corresponding determinant in terms of the sums of the powers of the roots which is equal to the product of the squares of the differences.

2. Prove, for the general equation,

$$\begin{vmatrix} s_0 & s_1 \\ s_1 & s_2 \end{vmatrix} = \Sigma (\alpha - \beta)^2.$$

This appears by squaring the array

$$\left. \begin{array}{ccccccccc} 1 & 1 & 1 & 1 & 1 & . & . & . \\ \alpha & \beta & \gamma & \delta & \epsilon & . & . & . \end{array} \right\} \quad \text{(See Art. 121.)}$$

3. Prove similarly, for the general equation,

$$\begin{vmatrix} s_0 & s_1 & s_2 \\ s_1 & s_2 & s_3 \\ s_2 & s_3 & s_4 \end{vmatrix} = \Sigma (\beta - \gamma)^2 (\gamma - \alpha)^2 (\alpha - \beta)^2.$$

By the process of Art. 121, a series of relations of this kind can be established; and when the number of rows in the array becomes equal to the degree of the equation, the value of the determinant is the product of the squares of the differences, as in Ex. 1. When the number of rows exceeds the degree of the equation the value of the corresponding determinant vanishes. For example, the value of the determinant of Ex. 1 is zero for equations of the second and third degrees.

4. Prove from the equations of Art. 126 that the sums of the powers can be expressed in terms of the coefficients, or *vice versâ*, in the form of determinants, as follows :—

$$s_2 = \begin{vmatrix} p_1 & 1 \\ 2p_2 & p_1 \end{vmatrix}, \quad s_3 = - \begin{vmatrix} p_1 & 1 & 0 \\ 2p_2 & p_1 & 1 \\ 3p_3 & p_2 & p_1 \end{vmatrix}, \quad s_4 = \begin{vmatrix} p_1 & 1 & 0 & 0 \\ 2p_2 & p_1 & 1 & 0 \\ 3p_3 & p_2 & p_1 & 1 \\ 4p_4 & p_3 & p_2 & p_1 \end{vmatrix}, \quad \&\text{c.}$$

$$2p_2 = \begin{vmatrix} s_1 & 1 \\ s_2 & s_1 \end{vmatrix}, \quad 6p_3 = -\begin{vmatrix} s_1 & 1 & 0 \\ s_2 & s_1 & 2 \\ s_3 & s_2 & s_1 \end{vmatrix}, \quad 24p_4 = \begin{vmatrix} s_1 & 1 & 0 & 0 \\ s_2 & s_1 & 2 & 0 \\ s_3 & s_2 & s_1 & 3 \\ s_4 & s_3 & s_2 & s_1 \end{vmatrix}, \quad \&c.$$

5. Resolve into factors the determinant

$$\begin{vmatrix} s_6 & s_5 & s_4 & s_3 & x^3 \\ s_5 & s_4 & s_3 & s_2 & x^2 \\ s_4 & s_3 & s_2 & s_1 & x \\ s_3 & s_2 & s_1 & s_0 & 1 \\ y^3 & y^2 & y & 1 & 0 \end{vmatrix},$$

where s_0, s_1, s_2, &c., are the sums of the powers of three quantities α, β, γ. This determinant is the product of the two determinants

$$\begin{vmatrix} \alpha^3 & \beta^3 & \gamma^3 & x^3 & 0 \\ \alpha^2 & \beta^2 & \gamma^2 & x^2 & 0 \\ \alpha & \beta & \gamma & x & 0 \\ 1 & 1 & 1 & 1 & 0 \\ 0 & 0 & 0 & 0 & 1 \end{vmatrix}, \quad \begin{vmatrix} \alpha^3 & \beta^3 & \gamma^3 & 0 & y^3 \\ \alpha^2 & \beta^2 & \gamma^2 & 0 & y^2 \\ \alpha & \beta & \gamma & 0 & y \\ 1 & 1 & 1 & 0 & 1 \\ 0 & 0 & 0 & 1 & 0 \end{vmatrix};$$

and each of the latter can be resolved into simple factors.

6. If α, β, γ, δ be the roots of the equation

$$a_0 x^4 + 4a_1 x^3 + 6a_2 x^2 + 4a_3 x + a_4 = 0,$$

calculate in terms of a_0, H, I, J the value of the symmetric function

$$a_0{}^6 \Sigma (3\alpha - \beta - \gamma - \delta)^2 (3\beta - \gamma - \delta - \alpha)^2 (3\gamma - \delta - \alpha - \beta)^2.$$

Here

$$a_0{}^6 \Sigma = 4^6 \Sigma z_1{}^2 z_2{}^2 z_3{}^2,$$

where z_1, z_2, z_3, z_4 are the roots of the equation

$$z^4 + 6Hz^2 + 4Gz + a_0{}^2 I - 3H^2 = 0. \qquad \text{(See Art. 38.)}$$

Hence, by Ex. 2, Art. 132,

$$\textit{Ans. } 4^7 \{ -7H^3 + a_0{}^2 HI - 4a_0{}^3 J \}.$$

7. Prove that

$$\Pi \equiv a_0{}^6 (\beta - \gamma)^2 (\gamma - \alpha)^2 (\alpha - \beta)^2 (\alpha - \delta)^2 (\beta - \delta)^2 (\gamma - \delta)^2 = lI^3 + mJ^2,$$

where

$$m = -27l.$$

The weight of this function of the roots is 12, and the order 6.

We now make use of a proposition which will be proved subsequently, namely, that any even, rational, and integral symmetric function of the roots, of the order ϖ, and involving the differences only of the roots, is, when multiplied by $a_0{}^\varpi$, a rational and *integral* function of a_0, H, I, J. (Compare Ex. 8, p. 119.)

Hence, expressing the function whose order is 6, and weight 12, in terms of a_0, H, I, J, it is easy to see from the table

	Order.	Weight.
J	3	6
I	2	4
H	2	2

that H cannot enter, for the terms of the sixth order, viz., H^3, H^2I, HI^2, ... are of too low weight. Therefore Π must be of the form

$$lI^3 + mJ^2,$$

where l and m are numerical coefficients.

Now put a_3 and a_4 equal to zero, and Π will vanish, since in that case the quartic will have equal roots; hence

$$0 = l(3a_2{}^2)^3 + m(-a_2{}^3)^2,$$

and
$$m = -27\,l.$$

8. Calculate the symmetric function of the roots of a biquadratic

$$\Sigma(\beta - \gamma)^2 (\gamma - \alpha)^2 (\alpha - \beta)^2.$$

Since the order of this symmetric function is four, and its weight six, we may assume

$$a_0{}^4 \Sigma(\beta - \gamma)^2 (\gamma - \alpha)^2 (\alpha - \beta)^2 = lHI + ma_0 J. \tag{1}$$

We proceed to determine l, m by taking two biquadratics whose roots are known, and calculating in each case the symmetric function by actually substituting the roots, and then comparing both sides of the equation when H, I, J are replaced by their values calculated from the numerical coefficients.

First we take the biquadratic equation

$$x^4 - x^2 = 0,$$

whose roots are $0, 0, 1, -1$; whence

$$\Sigma = 8, \quad H = -\frac{1}{6}, \quad I = \frac{3}{6^2}, \quad J = \frac{1}{6^3}.$$

Substituting in equation (1), we have

$$192 \times 9 = -3l + m.$$

Proceeding in the same way with the biquadratic equation

$$x^4 - 6x^2 + 5 = 0, \quad \text{whose roots are} \quad \pm\sqrt{5}, \ \pm 1,$$

we find

$$\Sigma = 768, \quad H = -1, \quad I = 8, \quad J = -4;$$

whence

$$-192 = 2l + m,$$

and

$$l = -2 \times 192, \quad m = 3 \times 192;$$

and, finally,

$$a_0^4 \, \Sigma = 192 \, (-2HI + 3a_0 J).$$

9. Calculate the determinant

$$\Delta \equiv \begin{vmatrix} s_0 & s_1 & s_2 \\ s_1 & s_2 & s_3 \\ s_2 & s_3 & s_4 \end{vmatrix}$$

in terms of the coefficients of a quartic.

This determinant is a function of the differences of the roots (see Ex. 2, Art. 135), whence we may remove the second term of the quartic before calculating it; and if the equation so transformed be

$$y^4 + P_2 y^2 + P_3 y + P_4 = 0,$$

$$\Delta = \begin{vmatrix} 4 & 0 & -2P_2 \\ 0 & -2P_2 & -3P_3 \\ -2P_2 & -3P_3 & 2P_2^2 - 4P_4 \end{vmatrix} = 4 \{ 8P_2 P_4 - 2P_2^3 - 9P_3^2 \};$$

but

$$a_0^2 P_2 = 6H, \quad a_0^3 P_3 = 4G, \quad a_0^4 P_4 = a_0^2 I - 3H^2.$$

Substituting for P_2, P_3, P_4 these values, we have

$$a_0^4 \, \Delta = 192 \, (-2HI + 3a_0 J):$$

the same result as in the preceding example. (Compare Ex. 3, p. 286.)

10. If α, β, γ, δ be the roots of the equation

$$a_0 x^4 + 4a_1 x^3 + 6a_2 x^2 + 4a_3 x + a_4 = 0,$$

express H_s, I_s, J_s, G_s of the equation

$$s_0 x^4 + 4s_1 x^3 + 6s_2 x^2 + 4s_3 x + s_4 \equiv \Sigma (x + \alpha)^4 = 0$$

in terms of H, I, J, G.

$$Ans. \quad \frac{H_s}{s_0^2} = -3 \, \frac{H}{a_0^2}, \quad \frac{I_s}{s_0^2} = \frac{48H^2 - a_0^2 I}{a_0^4}, \quad \frac{G_s}{s_0^3} = -3 \, \frac{G}{a_0^3},$$

and by the aid of the relations

$$G^2 + 4H^3 \equiv a_0^2 (HI - a_0 J), \quad G_s^2 + 4H_s^3 \equiv s_0^2 (H_s I_s - s_0 J_s),$$

$$J_s = \frac{192}{a_0^4} (3a_0 J - 2HI).$$

11. When p is even, prove that

$$\Sigma(\alpha_1 - \alpha_2)^p = s_0 s_p - p s_1 s_{p-1} + \tfrac{1}{2}p(p-1)s_2 s_{p-2} - \&c.$$

Since

$$\Sigma(x - \alpha)^p = n x^p - p s_1 x^{p-1} + \frac{p \cdot p - 1}{2} s_2 x^{p-2} - \&c. \ldots - p s_{p-1} x + s_p,$$

changing x successively into $\quad \alpha_1, \alpha_2, \alpha_3, \ldots \alpha_n, \quad$ and adding the results on both sides of the equations thus obtained, we find

$$2\Sigma(\alpha_1 - \alpha_2)^p = s_0 s_p - p s_1 s_{p-1} + \frac{p \cdot p - 1}{1 \cdot 2} s_2 s_{p-2} - \ldots - p s_1 s_{p-1} + s_0 s_p,$$

where all the terms on the right side of this equation are repeated except the middle term. Thus

$$\Sigma(\alpha_1 - \alpha_2)^4 = s_0 s_4 - 4 s_1 s_3 + 3 s_2^2, \qquad \text{(Compare Ex. 1, Art. 135.)}$$

$$\Sigma(\alpha_1 - \alpha_2)^6 = s_0 s_6 - 6 s_1 s_5 + 15 s_2 s_4 - 10 s_3^2, \&c.$$

12. Form the equation whose roots are $\quad \phi'(\alpha), \ \phi'(\beta), \ \phi'(\gamma), \ \phi'(\delta), \quad$ where $\alpha, \beta, \gamma, \delta$ are the roots of the equation

$$\phi(x) \equiv a_0 x^4 + 4a_1 x^3 + 6a_2 x^2 + 4a_3 x + a_4 = 0.$$

$$Ans. \ \ \phi'^4 + \frac{32\,G}{a_0^3}\phi'^3 + \frac{96\,(2HI - 3a_0 J)}{a_0^4}\phi'^2 + \frac{256\,(I^3 - 27 J^2)}{a_0^6} = 0.$$

13. If $$\Sigma(\alpha - \beta)^2(\beta - \gamma)^2(\gamma - \alpha)^2(x - \delta)^4,$$

when expanded, becomes]

$$K_0 x^4 + 4K_1 x^3 + 6K_2 x^2 + 4K_3 x + K_4;$$

prove that

$$\frac{K_0 \alpha\beta\gamma + K_1(\beta\gamma + \gamma\alpha + \alpha\beta) + K_2(\alpha + \beta + \gamma) + K_3}{(\beta - \gamma)(\gamma - \alpha)(\alpha - \beta)} = \frac{\pm 16\sqrt{\Delta}}{a_0^3},$$

where

$$\Delta = I^3 - 27 J^2.$$

14. Prove that

$$a_0^4 \Sigma(\beta + \gamma - \alpha - \delta)^2(\beta - \gamma)^2(\alpha - \delta)^2 = 192\,(3a_0 J - 2HI).$$

15. Prove that

$$a_0^6 \Sigma(\beta + \gamma - \alpha - \delta)^4(\beta - \gamma)^2(\alpha - \delta)^2 = 512\,(a_0^2 I^2 - 36\,a_0 HJ + 12 H^2 I).$$

CHAPTER XIII.

136. Definitions.—Being given a system of n equations, homogeneous between n variables, or non-homogeneous between $n - 1$ variables, if we combine these equations in such a manner as to eliminate the variables, and obtain an equation $R = 0$, containing only the coefficients of the equations; the quantity R is, when expressed in a rational and integral form, called their *Resultant* or *Eliminant*.

In what follows we shall be chiefly concerned with the discussion of two equations involving one unknown quantity only. In this case the equation $R = 0$ asserts that the two equations are consistent; that is, they are both satisfied by a common value of the variable. We now proceed to show how the elimination may be performed so as to obtain the quantity R, illustrating the different methods by simple examples.

Let it be required to eliminate x between the equations

$$ax^2 + 2bx + c = 0, \quad a'x^2 + 2b'x + c' = 0.$$

Solving these equations, and equating the values of x so obtained, the result of elimination appears in the irrational form

$$-\frac{b}{a} + \frac{\sqrt{b^2 - ac}}{a} = -\frac{b'}{a'} + \frac{\sqrt{b'^2 - a'c'}}{a'}.$$

Multiplying by aa', we obtain

$$ab' - a'b = a\sqrt{b'^2 - a'c'} - a'\sqrt{b^2 - ac}.$$

Squaring both sides, and dividing by aa' (for R does not vanish when a or a' vanishes), and then squaring again, we find

$$R = 4(ac - b^2)(a'c' - b'^2) - (ac' + a'c - 2bb')^2.$$

This method of forming the resultant is very limited in application, as it is not, in general, possible to express by an algebraic formula a root of an equation higher than the fourth degree. Other methods have consequently been devised for determining the resultant without first solving the equations. We proceed now to explain the method of elimination by symmetric functions of the roots of the equations.

137. **Elimination by Symmetric Functions.**—Let two algebraic equations of the mth and n^{th} degrees be

$$\phi(x) \equiv a_0 x^m + a_1 x^{m-1} + a_2 x^{m-2} + \ldots + a_m = 0,$$

$$\psi(x) \equiv b_0 x^n + b_1 x^{n-1} + b_2 x^{n-2} + \ldots + b_n = 0;$$

and let it be required to find the condition that these equations should have a common root. For this purpose let the roots of the equation $\phi(x) = 0$ be $a_1, a_2, \ldots a_m$. If the given equations have a common root it is *necessary* and *sufficient* that one of the quantities

$$\psi(a_1), \quad \psi(a_2), \ldots, \quad \psi(a_m)$$

should be zero, or, in other words, that the product

$$\psi(a_1)\,\psi(a_2)\,\psi(a_3) \ldots \psi(a_m)$$

should vanish. If, now, we transform this product into a rational and integral function of the coefficients, which is always possible as it is a symmetric function of the roots of the equation $\phi(x) = 0$, we shall have the resultant required: further, if $\beta_1, \beta_2, \ldots \beta_n$ be the roots of the equation $\psi(x) = 0$, we have

$$\psi(a_1) = b_0(a_1 - \beta_1)(a_1 - \beta_2) \ldots (a_1 - \beta_n),$$

$$\psi(a_2) = b_0(a_2 - \beta_1)(a_2 - \beta_2) \ldots (a_2 - \beta_n),$$

$$\cdots \qquad \cdots \qquad \cdots$$

$$\psi(a_m) = b_0(a_m - \beta_1)(a_m - \beta_2) \ldots (a_m - \beta_n).$$

If we change the signs of the mn factors, and multiply these equations, taking together the factors which are situated in the same column, we find

$$a_0^n\,\psi(a_1)\,\psi(a_2) \ldots \psi(a_m) = (-1)^{mn}\,b_0^m\,\phi(\beta_1)\,\phi(\beta_2) \ldots \phi(\beta_n).$$

We may therefore take

$$R =(-1)^{mn}\, b_0{}^m\, \phi(\beta_1)\, \phi(\beta_2)\ldots \phi(\beta_n)=a_0{}^n\, \psi(a_1)\, \psi(a_2)\ldots \psi(a_m), \quad (1)$$

for both these values of R are integral functions of the coefficients of $\phi(x)$ and $\psi(x)$, which vanish only when $\phi(x)$ and $\psi(x)$ have a common factor, and which become identical when they are expressed in terms of the coefficients.

138. **Properties of the Resultant.**—(1). *The order of the resultant of two equations in the coefficients is equal to the sum of the degrees of the equations, the coefficients of the first equation entering R in the degree of the second, and the coefficients of the second entering in the degree of the first.*

This appears by reviewing the two forms of R in (1), Art. 137; for in the first form $a_0, a_1, \ldots a_m$ enter in the nth degree, and in the second form $b_0, b_1, \ldots b_n$ enter in the mth degree. Also it may be seen that two terms, one selected from each form, are $(-1)^{mn}\, b_0{}^m a_m{}^n,\; a_0{}^n b_n{}^m$.

(2). *If the roots of both equations be multiplied by the same quantity ρ, the resultant is multiplied by ρ^{mn}.*

This is evident, since any one of the mn factors of the form $a_p - \beta_q$ becomes $\rho\,(a_p - \beta_q)$, and therefore ρ^{mn} divides the resultant. From this we may conclude that *the weight of the resultant is mn*, in which form this proposition is often stated.

(3). *If the roots of both equations be increased by the same quantity, the resultant of the equations so transformed is equal to the resultant of the original equations.*

For we have

$$\pm R = a_0{}^n\, b_0{}^m\, \Pi\,(a_p - \beta_q),$$

where Π signifies the continued product of the mn terms of the form $a_p - \beta_q$; and this is unaltered when a_p and β_q receive the same increment.

(4). *If the roots be changed into their reciprocals, the value of R obtained from the transformed equations remains unaltered, except in sign when mn is an odd number.*

Making this transformation in

$$R = a_0{}^n\, b_0{}^m\, \Pi\,(a_p - \beta_q),$$

we have

$$R' = a_m{}^n b_n{}^m (-1)^{mn} \frac{\Pi (a_p - \beta_q)}{(a_1 a_2 \ldots a_m)^n (\beta_1 \beta_2 \ldots \beta_n)^m} \ ;$$

but

$$a_1 a_2 \ldots a_m = (-1)^m \frac{a_m}{a_0}, \quad \beta_1 \beta_2 \ldots \beta_n = (-1)^n \frac{b_n}{b_0} \ ;$$

substituting, we obtain

$$R' = a_0{}^n b_0{}^m (-1)^{mn} \Pi (a_p - \beta_q) = (-1)^{mn} R.$$

From this it follows that in the resultant of two equations the coefficients with complementary suffixes of both equations, e.g. a_0, a_m ; a_1, a_{m-1}, &c., may be all interchanged without altering the value of the resultant.

(5). *If both equations be transformed by homographic transformation; that is, by substituting for x*

$$\frac{\lambda x + \mu}{\lambda' x + \mu'}$$

and each simple factor multiplied by $\lambda' x + \mu'$, *to render the new equations integral; then the new resultant* $R' = (\lambda \mu' - \lambda' \mu)^{mn} R.$

To prove this, we have

$$\phi(x) = a_0 (x - a_1)(x - a_2) \ldots (x - a_m),$$

$$\psi(x) = b_0 (x - \beta_1)(x - \beta_2) \ldots (x - \beta_n) \ ;$$

also

$$x - a_r \ \text{becomes} \ (\lambda - \lambda' a_r)\left(x - \frac{\mu' a_r - \mu}{\lambda - \lambda' a_r}\right),$$

$$x - \beta_r \quad ,, \quad (\lambda - \lambda' \beta_r)\left(x - \frac{\mu' \beta_r - \mu}{\lambda - \lambda' \beta_r}\right).$$

Multiplying together all the factors of each equation,

$$a_0 \ \text{becomes} \ a_0(\lambda - \lambda' a_1)(\lambda - \lambda' a_2) \ldots (\lambda - \lambda' a_m),$$

$$b_0 \quad ,, \quad b_0(\lambda - \lambda' \beta_1)(\lambda - \lambda' \beta_2) \ldots (\lambda - \lambda' \beta_n).$$

Also, since a_r, β_r are transformed into $\dfrac{\mu' a_r - \mu}{\lambda - \lambda' a_r}, \ \dfrac{\mu' \beta_r - \mu}{\lambda - \lambda' \beta_r},$

$$a_r - \beta_r \ \text{becomes} \ \frac{(\lambda \mu' - \lambda' \mu)(a_r - \beta_r)}{(\lambda - \lambda' a_r)(\lambda - \lambda' \beta_r)} \ ;$$

whence

$$a_0{}^n b_0{}^m \, \Pi \, (a_r - \beta_r) \text{ becomes } a_0{}^n b_0{}^m (\lambda\mu' - \lambda'\mu)^{mn} \Pi \, (a_r - \beta_r),$$

that is, the resultant calculated from the new forms of $\phi(x)$ and $\psi(x)$ is

$$(\lambda\mu' - \lambda'\mu)^{mn} R.$$

This proposition includes the three foregoing; and they are collectively equivalent to the present proposition.

139. **Euler's Method of Elimination.**—When two equations $\phi(x) = 0$, and $\psi(x) = 0$, of the mth and n^{th} degrees respectively, have any common root θ, we may assume

$$\phi(x) \equiv (x - \theta) \, \phi_1(x),$$

$$\psi(x) \equiv (x - \theta) \, \psi_1(x),$$

where

$$\phi_1(x) \equiv p_1 x^{m-1} + p_2 x^{m-2} + \ldots + p_m,$$

$$\psi_1(x) \equiv q_1 x^{n-1} + q_2 x^{n-2} + \ldots + q_n,$$

the coefficients being undetermined constants, as θ is not supposed to be known.

Whence we have

$$\phi(x) \, \psi_1(x) \equiv \psi(x) \, \phi_1(x),$$

an identical equation of the $(m + n - 1)^{th}$ degree. Now, equating the coefficients of the different powers of x on both sides of the equation, we have $m + n$ homogeneous equations of the first degree in the $m + n$ constants $p_1, p_2, \ldots p_m, q_1, q_2, \ldots q_n$; and by eliminating these constants by the method of Art. 123, we obtain the resultant of the two given equations in the form of a determinant.

EXAMPLE.

Suppose the two equations

$$ax^2 + bx + c = 0, \quad a_1 x^2 + b_1 x + c_1 = 0$$

to have a common root. We have identically

$$(q_1 x + q_2)(ax^2 + bx + c) \equiv (p_1 x + p_2)(a_1 x^2 + b_1 x + c_1),$$

or

$$(q_1 a - p_1 a_1) x^3 + (q_1 b + q_2 a - p_1 b_1 - p_2 a_1) x^2$$
$$+ (q_1 c + q_2 b - p_1 c_1 - p_2 b_1) x + q_2 c - p_2 c_1 = 0.$$

Equating to zero all the coefficients of this equation, we have the four homogeneous equations

$$q_1 a \qquad - p_1 a_1 \qquad = 0,$$
$$q_1 b + q_2 a - p_1 b_1 - p_2 a_1 = 0,$$
$$q_1 c + q_2 b - p_1 c_1 - p_2 b_1 = 0,$$
$$q_2 c \qquad - p_2 c_1 = 0;$$

and eliminating the constants p_1, p_2, q_1, q_2, we obtain the resultant as follows :

$$\begin{vmatrix} a & 0 & a_1 & 0 \\ b & a & b_1 & a_1 \\ c & b & c_1 & b_1 \\ 0 & c & 0 & c_1 \end{vmatrix} = 0.$$

The student can easily verify that this result is the same as that of Art. 136.

140. Sylvester's Dialytic Method of Elimination.— This method leads to the same determinants for resultants as the method of Euler just explained; but it has an advantage over Euler's method in point of generality, since it can often be applied to form the resultant of equations involving several variables.

Suppose we require the resultant of the two equations

$$\phi(x) \equiv a_0 x^m + a_1 x^{m-1} + a_2 x^{m-2} + \ldots + a_m = 0,$$
$$\psi(x) \equiv b_0 x^n + b_1 x^{n-1} + b_2 x^{n-2} + \ldots + b_n = 0,$$

we multiply the first by the successive powers of x,

$$x^{n-1}, \; x^{n-2}, \; \ldots \; x^2, \; x, \; x^0 \, ;$$

and the second by

$$x^{m-1}, \; x^{m-2}, \; \ldots \; x^2, \; x, \; x^0,$$

thus obtaining $m+n$ equations, the highest power of x being $m+n-1$. We have, consequently, equations enough from which to eliminate

$$x^{m+n-1}, \; x^{m+n-2}, \; \ldots \; x^2, \; x,$$

considered as distinct variables.

EXAMPLE.

In the case of two quadratic equations

$$ax^2 + bx + c = 0, \quad \text{and} \quad a_1 x^2 + b_1 x + c_1 = 0,$$

we have

$$ax^3 + bx^2 + cx \qquad = 0,$$
$$ax^2 + bx + c = 0,$$
$$a_1 x^3 + b_1 x^2 + c_1 x \qquad = 0,$$
$$a_1 x^2 + b_1 x + c_1 = 0;$$

from which, eliminating x^3, x^2, x, we get the same determinant as before, columns now replacing rows :

$$\begin{vmatrix} a & b & c & 0 \\ 0 & a & b & c \\ a_1 & b_1 & c_1 & 0 \\ 0 & a_1 & b_1 & c_1 \end{vmatrix}.$$

141. Bezout's Method of Elimination.—The general method will be most easily comprehended by applying it in the first instance to particular cases. We proceed to this application—(1), when the equations are of the same degree, and (2), when they are of different degrees.

(1). Let us take the two cubic equations

$$ax^3 + bx^2 + cx + d = 0, \quad a_1 x^3 + b_1 x^2 + c_1 x + d_1 = 0.$$

Multiplying these two equations successively by

$$a_1 \quad \text{and} \quad a,$$
$$a_1 x + b_1 \quad ,, \quad ax + b,$$
$$a_1 x^2 + b_1 x + c_1 \quad ,, \quad ax^2 + bx + c,$$

and subtracting each time the products so formed, we find the three following equations :—

$$(ab_1)x^2 \qquad + (ac_1)\,x \qquad + (ad_1) = 0,$$
$$(ac_1)\,x^2 + \{(ad_1) + (bc_1)\}x + (bd_1) = 0,$$
$$(ad_1)x^2 + \qquad + (bd_1)\,x \qquad + (cd_1) = 0.$$

By eliminating from these equations x^2, x, as distinct variables, the resultant is obtained in the form of a symmetrical determinant as follows:—

$$\begin{vmatrix} (ab_1) & (ac_1) & (ad_1) \\ (ac_1) & (ad_1) + (bc_1) & (bd_1) \\ (ad_1) & (bd_1) & (cd_1) \end{vmatrix}.$$

To render the law of formation of the resultant more apparent, the following mode of procedure is given:—

Let the two equations be

$$ax^4 + bx^3 + cx^2 + dx + e = 0,$$

$$a_1x^4 + b_1x^3 + c_1x^2 + d_1x + e_1 = 0;$$

whence, following Cauchy's mode of presenting Bezout's method, we have the system of equations

$$\frac{a}{a_1} = \frac{bx^3 + cx^2 + dx + e}{b_1x^3 + c_1x^2 + d_1x + e_1},$$

$$\frac{ax + b}{a_1x + b_1} = \frac{cx^2 + dx + e}{c_1x^2 + d_1x + e_1},$$

$$\frac{ax^2 + bx + c}{a_1x^2 + b_1x + c_1} = \frac{dx + e}{d_1x + e_1},$$

$$\frac{ax^3 + bx^2 + cx + d}{a_1x^3 + b_1x + c_1x + d_1} = \frac{e}{e_1},$$

which, when rendered integral, lead, on the elimination of x^3, x^2, x, to the following form for the resultant:—

$$\begin{vmatrix} (ab_1) & (ac_1) & (ad_1) & (ae_1) \\ (ac_1) & (ad_1) + (bc_1) & (ae_1) + (bd_1) & (be_1) \\ (ad_1) & (ae_1) + (bd_1) & (bc_1) + (cd_1) & (ce_1) \\ (ae_1) & (be_1) & (ce_1) & (de_1) \end{vmatrix}.$$

If, now, we consider the two symmetrical determinants,

$$
\begin{vmatrix}
(ab_1) & (ac_1) & (ad_1) & (ae_1) \\
(ac_1) & (ad_1) & (ae_1) & (be_1) \\
(ad_1) & (ae_1) & (be_1) & (ce_1) \\
(ae_1) & (be_1) & (ce_1) & (de_1)
\end{vmatrix}, \quad
\begin{vmatrix}
(bc_1) & (bd_1) \\
(bd_1) & (cd_1)
\end{vmatrix},
$$

the formation of which is at once apparent, we observe that R is obtained by adding the constituents of the second to the four central constituents of the first.

Similarly in the case of the two equations of the fifth degree

$$ax^5 + bx^4 + cx^3 + dx^2 + ex + f = 0,$$

$$a_1x^5 + b_1x^4 + c_1x^3 + d_1x^2 + e_1x + f_1 = 0,$$

the resultant is obtained from the three following determinants :—

$$
\begin{vmatrix}
(ab_1) & (ac_1) & (ad_1) & (ae_1) & (af_1) \\
(ac_1) & (ad_1) & (ae_1) & (af_1) & (bf_1) \\
(ad_1) & (ae_1) & (af_1) & (bf_1) & (cf_1) \\
(ae_1) & (af_1) & (bf_1) & (cf_1) & (df_1) \\
(af_1) & (bf_1) & (cf_1) & (df_1) & (ef_1)
\end{vmatrix}, \quad
\begin{vmatrix}
(bc_1) & (bd_1) & (be_1) \\
(bd_1) & (be_1) & (ce_1) \\
(be_1) & (ce_1) & (de_1)
\end{vmatrix}, \quad (cd_1),
$$

by adding the constituents of the second to the six central constituents of the first, and then adding the third to the central constituent of the determinant so formed. The student will have no difficulty in applying a similar process of superposition to the formation of the determinant in general.

(2). We take now the case of two equations of different dimensions, for example,

$$ax^4 + bx^3 + cx^2 + dx + e = 0, \quad a_1x^2 + b_1x + c_1 = 0.$$

Multiplying these equations successively by

$$a_1 \quad \text{and} \quad ax^2,$$

$$a_1x + b_1 \quad ,, \quad (ax + b)x^2,$$

and subtracting each time the products so formed, we find the two following equations :—

$$(ab_1)x^3 + (ac_1)x^2 - da_1 x - ea_1 = 0,$$

$$(ac_1)x^3 + \{(bc_1) - da_1\}x^2 - \{db_1 + ea_1\}x - eb_1 = 0.$$

If, now, we join to these the two equations

$$a_1 x^3 + b_1 x^2 + c_1 x = 0,$$

$$a_1 x^2 + b_1 x + c_1 = 0,$$

we shall have four equations by means of which we can eliminate x^3, x^2, x; whence we obtain the resultant in the form of a determinant as follows :—

$$
\begin{vmatrix}
(ab_1) & (ac_1) & da_1 & ea_1 \\
(ac_1) & (bc_1) - da_1 & db_1 + ea_1 & eb_1 \\
a_1 & b_1 & -c_1 & 0 \\
0 & a_1 & -b_1 & -c_1
\end{vmatrix}.
$$

This determinant involves the coefficients of the first equation in the second degree, and the coefficients of the second equation in the fourth degree, as it should do; whence no extraneous factor enters this form of the resultant.

We now proceed to the general case of two equations of the mth and n^{th} degrees.

Let the equations be

$$\phi(x) = a_0 x^m + a_1 x^{m-1} + a_2 x^{m-2} + \ldots + a_m = 0,$$

$$\psi(x) = b_0 x^n + b_1 x^{n-1} + b_2 x^{n-2} + \ldots + b_n = 0,$$

where $m > n$; and let the second equation be multiplied by x^{m-n}. We have then

$$b_0 x^m + b_1 x^{m-1} + b_2 x^{m-2} + \ldots + b_n x^{m-n} = 0,$$

an equation of the same degree as the first. This equation has, however, in addition to the n roots of $\psi(x) = 0$, $m - n$ zero roots ;

so that we must be on our guard lest the factor $a_m{}^{m-n}$ (i. e. the result of substituting these roots in $\phi(x)$) enter the form of the resultant obtained. From these two equations we derive, as in the above case (1), the following n equations :—

$$\frac{a_0}{b_0} = \frac{a_1 x^{m-1} + a_2 x^{m-2} + \ldots + a_m}{b_1 x^{m-1} + b_2 x^{m-2} + \ldots b_n x^{m-n}},$$

$$\frac{a_0 x + a_1}{b_0 x + b_1} = \frac{a_2 x^{m-2} + a_3 x^{m-3} + \ldots + a_m}{b_2 x^{m-2} + b_3 x^{m-3} + \ldots b_n x^{m-n}},$$

$$. \quad . \quad . \quad . \quad . \quad . \quad . \quad . \quad . \quad . \quad .$$

$$\frac{a_0 x^{n-1} + a_1 x^{n-2} + \ldots + a_{n-1}}{b_0 x^{n-1} + b_1 x^{n-2} + \ldots + b_{n-1}} = \frac{a_n x^{m-n} + a_{n-1} x^{m-n-1} + \ldots + a_m}{b_n x^{m-n}},$$

which, when rendered integral, are all of the $(m-1)^{\text{th}}$ degree ; whence, eliminating $x^{m-1}, x^{m-2}, \ldots x$ between these n and the $m - n$ equations,

$$b_0 x^{m-1} + b_1 x^{m-2} + b_2 x^{m-3} + \ldots \qquad\qquad = 0,$$

$$b_0 x^{m-2} + b_1 x^{m-3} + \ldots \qquad\qquad = 0,$$

$$. \quad . \quad . \quad . \quad . \quad . \quad .$$

$$b_0 x^n + b_1 x^{n-1} + \ldots + b_n = 0,$$

we obtain the resultant in the form of a determinant of the mth order, the coefficients of the first equation entering in the degree n, and the coefficients of the second equation entering in the degree m ; whence it appears that no extraneous factor can enter ; and that the resultant as obtained by this method has not been affected by the introduction of the zero roots.

Remark.—If R be the resultant of two equations, $\phi(x) = 0$, $\psi(x) = 0$, whose degrees are both equal to m, the resultant R' of the system

$$\lambda \phi(x) + \mu \psi(x) = 0, \quad \lambda' \phi(x) + \mu' \psi(x) = 0$$

is

$$(\lambda \mu' - \lambda' \mu)^m R ;$$

for each of the minors $(a_r b_s)$, which in Bezout's method constitute the determinant form of R, becomes in this case

$$\begin{vmatrix} \lambda a_r + \mu b_r, & \lambda' a_r + \mu' b_r \\ \lambda a_s + \mu b_s, & \lambda' a_s + \mu' b_s \end{vmatrix} = (\lambda \mu' - \lambda' \mu)(a_r b_s);$$

whence $R' = (\lambda \mu' - \lambda' \mu)^m R$, since R is a determinant of order m.

142. We conclude our account of the different methods of Elimination with one which is often employed, but which has the disadvantage, when applied to equations of higher degree than the second, of giving the resultant multiplied by extraneous factors. The process is virtually equivalent to that usually described as the method of the greatest common measure.

In forming by this method the resultant of the two quadratic equations

$$ax^2 + bx + c = 0, \quad a_1 x^2 + b_1 x + c_1 = 0,$$

we multiply these equations successively by

$$a_1 \quad \text{and} \quad a, \quad c_1 \quad \text{and} \quad c,$$

and subtract the products so formed. We thus find the two linear equations

$$(ab_1) x + (ac_1) = 0,$$

$$(ac_1) x + (bc_1) = 0;$$

from which, eliminating x, we have

$$(ac_1)^2 - (ab_1)(bc_1) = 0.$$

As the degree of this expression is four, and its weight four, it can contain no extraneous factor, and is a correct form for the resultant.

To form by the same process the resultant of the cubic equations

$$ax^3 + bx^2 + cx + d = 0, \quad a_1 x^3 + b_1 x^2 + c_1 x + d_1 = 0,$$

we multiply these equations successively by

$$a_1 \text{ and } a, \quad d_1 \text{ and } d,$$

and subtract each time the products so formed. We have then

$$(ab_1)\, x^2 + (ac_1)\, x + (ad_1) = 0,$$
$$(ad_1)\, x^2 + (bd_1)\, x + (cd_1) = 0.$$

(1)

Now, eliminating x between these two quadratics by means of the formula above obtained, we find for their resultant

$$\begin{vmatrix} (ab_1) & (ad_1) \\ (ad_1) & (cd_1) \end{vmatrix}^2 - \begin{vmatrix} (ab_1) & (ac_1) \\ (ad_1) & (bd_1) \end{vmatrix} \times \begin{vmatrix} (ac_1) & (ad_1) \\ (bd_1) & (cd_1) \end{vmatrix},$$

an expression whose degree is 8, and weight 12, in place of degree 6, and weight 9 ; whence it ought to be divisible by a factor whose degree is 2 and weight 3.   This factor must therefore be of the form $l\,(bc_1) + m\,(ad_1)$.   We proceed now to show that it is (ad_1); and to find the quotient when this factor is removed.

For this purpose, retaining only the terms which do not directly involve (ad_1), we have

$$(ab_1)(cd_1)\,\{(ab_1)(cd_1) + (ca_1)(bd_1)\},$$

which is divisible by (ad_1), since

$$(bc_1)(ad_1) + (ca_1)(bd_1) + (ab_1)(cd_1) \equiv 0,$$

this being only another form of the identical equation

$$(\beta - \gamma)(\alpha - \delta) + (\gamma - \alpha)(\beta - \delta) + (\alpha - \beta)(\gamma - \delta) \equiv 0.$$

Expanding the determinants, and dividing off by (ad_1), we find ultimately the quotient

$$(ad_1)^3 - 2\,(ab_1)(cd_1)(ad_1) + (bd_1)(ca_1)(ad_1)$$
$$+ (ca_1)^2(cd_1) + (ab_1)(bd_1)^2 - (ab_1)(bc_1)(cd_1),$$

which, being of the proper degree and weight, is the resultant of the two cubics.   (Compare the determinant form in Art. 141.)

If we proceed in a similar manner to form the resultant of two biquadratic equations, by reducing the process to an elimination between two cubic equations, we shall have to remove an extraneous factor of the fourth degree, which is the condition that these cubics should have a common factor when the biquadratics from which they are derived have not necessarily a common factor; and in general, if we seek by this method the resultant of two equations of the nth degree, eliminating between two equations of the $(n-1)^{th}$ degree, we shall have to remove an extraneous factor of the order $2n-4$. Therefore as a general method this is inferior to all the methods previously given; and it cannot be conveniently used except when, from the nature of the investigation, extraneous factors can be easily removed.

143. **Discriminants.**—The *discriminant* of any equation is the simplest function of the coefficients, in a rational and integral form, whose vanishing expresses the condition for equal roots. We have had examples of such functions in Arts. 41 and 68. We proceed to show that they come under eliminants as particular cases. If an equation $f(x) = 0$ has a double root, this root must occur once in the equation $f'(x) = 0$; and, subtracting $xf'(x)$ from $nf(x)$, the same root must occur in the equation

$$nf(x) - xf'(x) = 0.$$

This is an equation of the $(n-1)^{th}$ degree in x; and by eliminating x between it and the equation $f'(x) = 0$, which is also of the $(n-1)^{th}$ degree, we obtain a function of the coefficients whose vanishing expresses the condition for equal roots. The degree of this eliminant in the coefficients of $f(x)$ is $2(n-1)$; and its weight is $n(n-1)$, as may be seen by examining the specimen terms given in section (1), Art. 138. Expressed as a symmetric function of the roots of the given equation, the discriminant will be the product of all the differences in the lowest power which can be expressed in a rational form in terms of the coefficients. Now the product of the squares of the differences $\Pi(a_1 - a_2)^2$ can be so expressed; and since it is of the $2(n-1)^{th}$ degree in any one root, and of the $n(n-1)^{th}$ degree in all the roots, it follows that the discriminant multiplied by a numerical factor is equal to $a_0^{2(n-1)} \Pi(a_1 - a_2)^2$; and is, moreover, identical with the eliminant just obtained. (Compare (6), Art. 41, and Ex. 2, Art. 62.)

EXAMPLES.

1. Find the discriminant of

$$a_0 x^3 + 3a_1 x^2 + 3a_2 x + a_3 = 0.$$

We have here to find the eliminant of the two equations

$$a_0 x^2 + 2a_1 x + a_2 = 0,$$

$$a_1 x^2 + 2a_2 x + a_3 = 0.$$

This is, by Art. 136,

$$4 (a_0 a_2 - a_1^2)(a_1 a_3 - a_2^2) - (a_0 a_3 - a_1 a_2)^2 = 0 ;$$

or it may be written in the form of a determinant, as follows, by Art. 140 :—

$$\begin{vmatrix} a_0 & 2a_1 & a_2 & 0 \\ 0 & a_0 & 2a_1 & a_2 \\ a_1 & 2a_2 & a_3 & 0 \\ 0 & a_1 & 2a_2 & a_3 \end{vmatrix} = 0.$$

It can be easily verified that this value of the discriminant is the same as that already obtained in Art. 41.

2. Express as a determinant the discriminant of the biquadratic

$$a_0 x^4 + 4a_1 x^3 + 6a_2 x^2 + 4a_3 x + a_4 = 0.$$

We have here to eliminate x from the equations

$$a_0 x^3 + 3a_1 x^2 + 3a_2 x + a_3 = 0,$$

$$a_1 x^3 + 3a_2 x^2 + 3a_3 x + a_4 = 0.$$

By the method of Art. 140, the result is

$$\begin{vmatrix} a_0 & 3a_1 & 3a_2 & a_3 & 0 & 0 \\ 0 & a_0 & 3a_1 & 3a_2 & a_3 & 0 \\ 0 & 0 & a_0 & 3a_1 & 3a_2 & a_3 \\ a_1 & 3a_2 & 3a_3 & a_4 & 0 & 0 \\ 0 & a_1 & 3a_2 & 3a_3 & a_4 & 0 \\ 0 & 0 & a_1 & 3a_2 & 3a_3 & a_4 \end{vmatrix} = 0.$$

This must be the same as $I^3 - 27 J^2$ of Art. 68.

3. Express the discriminant of the quartic as a determinant by Bezout's method of elimination.

X

4. Prove by elimination that $J = 0$ is one condition for the equality of three roots of the biquadratic of Ex. 2.

Since the triple root must be a double root of

$$U_3 \equiv a_0 x^3 + 3a_1 x^2 + 3a_2 x + a_3 = 0,$$

and therefore a single root of $\quad a_1 x^2 + 2a_2 x + a_3 = 0 ;\quad$ and since it must also be a single root of

$$U_2 \equiv a_0 x^2 + 2a_1 x + a_2 = 0,$$

it follows from the identity

$$U_4 \equiv x^2 U_2 + 2x\,(a_1 x^2 + 2a_2 x + a_3) + a_2 x^2 + 2a_3 x + a_4$$

that the triple root must be a root common to the three equations

$$a_0 x^2 + 2a_1 x + a_2 = 0,$$

$$a_1 x^2 + 2a_2 x + a_3 = 0,$$

$$a_2 x^2 + 2a_3 x + a_4 = 0.$$

Hence the condition

$$\begin{vmatrix} a_0 & a_1 & a_2 \\ a_1 & a_2 & a_3 \\ a_2 & a_3 & a_4 \end{vmatrix} \equiv J = 0.$$

That the other condition for a triple root is $I = 0$ may be inferred from Ex. 7, p. 287 ; for when three roots are equal the discriminant must vanish, and it is of the form $l I^3 + m J^2$.

5. Prove that the discriminant of the product of two functions is the product of their discriminants multiplied by the square of their eliminant.

This appears by applying the results of Art. 137 and the present Article ; for the product of the squares of the differences of all the roots is made up of the product of the squares of the differences of the roots of each equation separately, and the square of the product of the differences formed by taking each root of one equation with all the roots of the other.

144. Determination of a Root common to two Equations.—If R be the resultant of two equations

$$U \equiv a_m x^m + a_{m-1} x^{m-1} + \ldots + a_0 = 0,$$

$$V \equiv b_n x^n + b_{n-1} x^{n-1} + \ldots + b_0 = 0,$$

and a any common root, then

$$a = \dfrac{\frac{dR}{da_1}}{\frac{dR}{da_0}} = \dfrac{\frac{dR}{da_2}}{\frac{dR}{da_1}} = \dfrac{\frac{dR}{da_3}}{\frac{dR}{da_2}} = \text{\&c.}$$

To prove this we substitute in R, for a_0 and b_0, $a_0 - U$ and $b_0 - V$, and obtain an identical equation connecting U, V which is satisfied by every value of x, and which is of the form

$$R = U\phi + V\psi ;$$

whence

$$\frac{dR}{da_p} = x^p \phi + U \frac{d\phi}{da_p} + V \frac{d\psi}{da_p},$$

$$\frac{dR}{da_{p+1}} = x^{p+1}\phi + U\frac{d\phi}{da_{p+1}} + V\frac{d\psi}{da_{p+1}} ;$$

and when a is a common root of the equations $U = 0$, and $V = 0$, we have, substituting this value for x in the preceding equations,

$$a \frac{dR}{da_p} = \frac{dR}{da_{p+1}},$$

which proves the proposition, by giving p the values 0, 1, 2, &c.

A double root of an equation can be determined in a similar manner by differentiating the discriminant Δ.

145. **Symmetric Functions of the Roots of two Equations.**—If it be required to calculate a symmetric function involving the roots $a_1, a_2, a_3, \ldots a_m,$ of the equation

$$\phi(x) = a_0 x^m + a_1 x^{m-1} + a_2 x^{m-2} + \ldots + a_m = 0, \qquad (1)$$

along with the roots $\beta_1, \beta_2, \beta_3, \ldots \beta_n,$ of the equation

$$\psi(x) = b_0 y^n + b_1 y^{n-1} + b_2 y^{n-2} + \ldots + b_n = 0, \qquad (2)$$

we proceed as follows :—

Assume a new variable t connected with x and y by the equation

$$t = \lambda x + \mu y ;$$

and let y be eliminated by means of this equation from (2). The result is an equation of the nth degree in x whose coefficients involve λ, μ, and t in the nth power. Now let x be eliminated by any of the preceding methods from this equation and (1). We obtain an equation of the mn^{th} degree in t, whose roots are the mn values of the expression $\lambda a + \mu \beta$.

If, now, it be required to calculate in terms of the coefficients of $\phi(x)$ and $\psi(x)$ any symmetric function such as $\Sigma a^p \beta^q$, we form the sum of the $(p+q)^{th}$ powers of the roots of the equation in t. We thus find the value of $\Sigma (\lambda a + \mu \beta)^{p+q}$ expressed in terms of the original coefficients and the several powers of λ and μ. The coefficient of $\lambda^p \mu^q$ in this expression will furnish the required value of $\Sigma a^p \beta^q$ in terms of the coefficients of $\phi(x)$ and $\psi(x)$.

MISCELLANEOUS EXAMPLES.

1. Eliminate x from the equations

$$ax^2 + bx + c = 0,$$
$$x^3 = 1.$$

Multiplying the first equation by x, we have, since $x^3 = 1$,

$$bx^2 + cx + a = 0;$$

and multiplying again by x, we have

$$cx^2 + ax + b = 0.$$

Eliminating x^2 and x linearly (see Art. 123) from these three equations, the result is expressed as a determinant

$$\begin{vmatrix} a & b & c \\ b & c & a \\ c & a & b \end{vmatrix} = 0.$$

2. Eliminate similarly x from the equations

$$ax^4 + bx^3 + cx^2 + dx + e = 0,$$
$$x^5 = 1.$$

The result is a determinant of the fifth order similar to that in the preceding example. An analogous process may be applied in general to two equations of this kind.

3. Apply the method of Art. 139 to find the conditions that the two cubics

$$\phi(x) \equiv ax^3 + bx^2 + cx + d = 0,$$
$$\psi(x) \equiv a'x^3 + b'x^2 + c'x + d' = 0$$

should have two common roots.

When this is the case, identical results must be obtained by multiplying $\phi(x)$ by the third factor of $\psi(x)$, and $\psi(x)$ by the third factor of $\phi(x)$. We have, therefore,

$$(\lambda'x + \mu')\,\phi(x) \equiv (\lambda x + \mu)\,\psi(x),$$

where λ, μ, λ', μ' are indeterminate quantities. This identity leads to the equations

$$\lambda'a \qquad\qquad - \lambda a' \qquad\quad = 0,$$
$$\lambda'b + \mu'a - \lambda b' - \mu a' = 0,$$
$$\lambda'c + \mu'b - \lambda c' - \mu b' = 0,$$
$$\lambda'd + \mu'c - \lambda d' - \mu c' = 0,$$
$$\mu'd \qquad\qquad - \mu d' = 0.$$

Eliminating λ', μ', λ, μ from every four of these, we obtain five determinants, whose vanishing expresses the required conditions. There is a convenient notation in use to express the result of eliminating from a number of equations of this kind. In the present instance the vanishing of the five determinants is expressed as follows :—

$$\left\|\begin{array}{ccccc} a & b & c & d & 0 \\ 0 & a & b & c & d \\ a' & b' & c' & d' & 0 \\ 0 & a' & b' & c' & d' \end{array}\right\| = 0,$$

the determinants being formed by omitting each column in turn.

4. Prove the identity

$$\begin{vmatrix} \alpha^2 & 2\alpha\beta & \beta^2 \\ \alpha\alpha' & \alpha\beta' + \alpha'\beta & \beta\beta' \\ \alpha'^2 & 2\alpha'\beta' & \beta'^2 \end{vmatrix} \equiv (\alpha\beta' - \alpha'\beta)^3.$$

This appears by eliminating x and y from the equations

$$\alpha x + \beta y = 0, \qquad \alpha'x + \beta'y = 0\,;$$

for from these equations we derive

$$(\alpha x + \beta y)^2 = 0, \quad (\alpha x + \beta y)(\alpha'x + \beta'y) = 0, \quad (\alpha'x + \beta'y)^2 = 0.$$

The determinant above written is the result of eliminating x^2, xy, and y^2 from the latter equations; and this result must be a power of the determinant derived by eliminating x, y from the linear equations.

5. Prove similarly

$$\begin{vmatrix} \alpha^3 & 3\alpha^2\beta & 3\alpha\beta^2 & \beta^3 \\ \alpha^2\alpha' & \alpha^2\beta' + 2\alpha\alpha'\beta & 2\alpha\beta\beta' + \alpha'\beta^2 & \beta^2\beta' \\ \alpha\alpha'^2 & \alpha'^2\beta + 2\alpha\alpha'\beta' & 2\alpha'\beta\beta' + \alpha\beta'^2 & \beta\beta'^2 \\ \alpha'^3 & 3\alpha'^2\beta' & 3\alpha'\beta'^2 & \beta'^3 \end{vmatrix} = (\alpha\beta' - \alpha'\beta)^6.$$

This appears by deriving from the linear equations the following equations of the third degree :—

$$(\alpha x + \beta y)^3 = 0, \quad (\alpha x + \beta y)^2 (\alpha' x + \beta' y) = 0, \quad \&c.,$$

and eliminating x^3, $x^2 y$, xy^2, y^3.

6. Prove the result of Ex. 12, p. 260, by eliminating the constants λ, μ, λ', μ' from four equations

$$\alpha' = \frac{\lambda \alpha + \mu}{\lambda' \alpha + \mu'}, \quad \beta' = \frac{\lambda \beta + \mu}{\lambda' \beta + \mu'}, \quad \&c.,$$

connecting the variables in homographic transformation (cf. Art. 39).

7. Given

$$U \equiv A u^2 + 2 B u v + C v^2,$$

$$V \equiv A' u^2 + 2 B' u v + C' v^2,$$

$$u \equiv a x^2 + 2 b x y + c y^2,$$

$$v \equiv a' x^2 + 2 b' x y + c' y^2 \,;$$

determine the resultant of U and V considered as functions of x, y.

Since

$$U = A (u - \alpha v)(u - \beta v),$$

$$V = A'(u - \alpha' v)(u - \beta' v),$$

if U and V vanish for common values of x, y, some pair of factors, as $u - \alpha v$ and $u - \alpha' v$ must vanish; whence forming the resultant of

$$u - \alpha v \quad \text{and} \quad u - \alpha' v,$$

and representing the resultant of u and v by $R(u, v)$, we have

$$R(u - \alpha v, \quad u - \alpha' v) = (\alpha - \alpha')^2 R(u, v) \,;$$

and multiplying all these resultants together, we find

$$R(U_x, \ V_x) = A^4 A'^4 (\alpha - \alpha')^2 (\beta - \beta')^2 (\alpha - \beta')^2 (\beta - \alpha')^2 \{ R(u, v) \}^4,$$

or

$$R(U_x, \ V_x) = \{ R(U, V) \}^2 \{ R(u, v) \}^4.$$

8. Prove that the equation whose roots are the differences of the roots of a given equation $f(x) = 0$ may be obtained by eliminating x from the equations

$$f(x) = 0, \quad f'(x) + f''(x) \frac{y}{1 \cdot 2} + f'''(x) \frac{y^2}{1 \cdot 2 \cdot 3} + \&c. = 0 \,;$$

and determine the degree of the equation in y (cf. Art. 67).

CHAPTER XIV.

146. Definitions.—In this and the following Chapters the notation

$$(a_0,\ a_1,\ a_2,\ \ldots\ a_n)\,(x,\ y)^n$$

will be employed to represent the quantic

$$a_0 x^n + n a_1 x^{n-1} y + \frac{n\,(n-1)}{1\,.\,2}\, a_2 x^{n-2} y^2 + \ldots + n a_{n-1} x y^{n-1} + a_n y^n,$$

which is a homogeneous function of x and y, written with binomial coefficients. If we put $y = 1$, this quantic becomes U_n of Art. 36.

Let ϕ be a rational, integral, and homogeneous symmetric function, of the order ϖ, of the roots $a_1, a_2, a_3, \ldots a_n$ of the equation $U_n \equiv (a_0,\ a_1,\ a_2,\ \ldots\ a_n)\,(x,\ 1)^n = 0$, this function involving only the differences of the roots; then if

$$\frac{1}{a_1 - x}, \quad \frac{1}{a_2 - x}, \quad \ldots \quad \frac{1}{a_n - x}$$

be substituted for $a_1, a_2 \ldots a_n$, respectively, the result multiplied by U_n^{ϖ} (to remove fractions) is a *covariant* of U_n if it involves the variable x, and an *invariant* if it does not involve x.

From this definition of an invariant we may infer at once that

$$a_0^{\varpi}\, \phi\,(a_1,\ a_2,\ a_3,\ \ldots\ a_n)$$

is an invariant of U_n, provided that ϕ is made up of a number of terms of the same type, each of which involves all the roots, and each root in the same degree ϖ.

These definitions may be extended to the case where ϕ (the function of differences) involves symmetrically the roots of several equations $U_p = 0$, $U_q = 0$, $U_r = 0$, &c., the roots of these equations entering ϕ in the orders $\varpi, \varpi', \varpi''$, &c..., respectively. We may substitute for each root a, $\dfrac{1}{a - x}$ as before, and remove fractions by the multiplier $U_p{}^\varpi U_q{}^{\varpi'} U_r{}^{\varpi''} \ldots$ &c. If the result involves the variable x, we obtain a covariant of the system of quantics U_p, U_q, U_r, &c. ; and if it does not, ϕ is an invariant of the system.

147. Formation of Covariants and Invariants.—We proceed now to show how the foregoing transformations may be conveniently effected, and covariants and invariants calculated in terms of the coefficients. With this object, let the symmetric function of the differences of the roots be expressed in terms of the coefficients as follows :—

$$a_0{}^\varpi \phi(a_1, a_2, \ldots a_n) = F(a_0, a_1, a_2, \ldots a_n).$$

Now, changing the roots into their reciprocals, and consequently a_0 into a_n, &c., a_r into a_{n-r}, &c. (that is, giving the suffixes their complementary values), we have

$$a_0{}^\varpi \psi(a_1, a_2, \ldots a_n) = F(a_n, a_{n-1}, \ldots a_0),$$

where ψ is an integral symmetric function of the roots, and F the corresponding value in terms of the coefficients. This function is called the *source** of the covariant derived therefrom.

Again, substituting $a_1 - x$, $a_2 - x$, $\ldots a_n - x$ for $a_1, a_2, \ldots a_n$, and consequently U_r, &c., for a_r, &c., we find

$$a_0{}^\varpi \psi(a_1 - x, a_2 - x, \ldots a_n - x) = F(U_n, U_{n-1}, \ldots U_1, U_0).$$

Thus, by two steps we derive a covariant from a function of the differences, and find at the same time its equivalent calculated in terms of the coefficients.

To illustrate this mode of procedure we take the example in the case of the cubic

$$a_0{}^2 \Sigma (a - \beta)^2 = 18(a_1{}^2 - a_0 a_2) ;$$

---

* This term was introduced by Mr. Roberts.

whence, changing the roots into their reciprocals, and a_0, a_1, a_2, a_3 into a_3, a_2, a_1, a_0, we have

$$a_0^2 \, \Sigma a^2 \, (\beta - \gamma)^2 = 18 \, (a_2^2 - a_3 a_1).$$

Again, changing a, β, γ into $a - x$, $\beta - x$, $\gamma - x$, and a_1, a_2, a_3 into U_1, U_2, U_3, respectively, we find

$$a_0^2 \, \Sigma \, (\beta - \gamma)^2 \, (x - a)^2 = 18 \, (U_2^2 - U_3 U_1).$$

The second member of this equation becomes when expanded

$$U_1 U_3 - U_2^2 \equiv (a_0 a_2 - a_1^2) x^2 + (a_0 a_3 - a_1 a_2) x + (a_1 a_3 - a_2^2).$$

This covariant is called the *Hessian* of U_3. We refer to it as H_x, since H is its leading coefficient.

As a second example we take the following function of the quartic :—

$$a_0^2 \, \Sigma \, (\beta - \gamma)^2 (a - \delta)^2 = 24 \, (a_0 a_4 - 4 a_1 a_3 + 3 a_2^2) ; \qquad (1)$$

whence, changing the roots into their reciprocals, and a_0, a_1, a_2, a_3, a_4 into a_4, a_3, a_2, a_1, a_0, we have

$$a_0^2 \, \Sigma \, (\gamma - \beta)^2 \, (\delta - a)^2 = 24 \, (a_4 a_0 - 4 a_3 a_1 + 3 a_2^2).$$

These transformations, therefore, do not alter equation (1) : again, since in this case $\psi(a, \beta, \gamma, \delta)$ is a function of the differences of the roots, ψ is unchanged when $a - x$, $\beta - x$, &c. . . . , are substituted for a, β, γ, δ. We infer that $a_0 a_4 - 4 a_1 a_3 + 3 a_2^2$ is an invariant of the quartic U_4.

We observe also, in accordance with what was stated in Art. 146, since

$$\phi \equiv (\beta - \gamma)^2 \, (a - \delta)^2 + (\gamma - a)^2 \, (\beta - \delta)^2 + (a - \beta)^2 \, (\gamma - \delta)^2,$$

that each of the three terms of which ϕ is made up involves all the roots in the degree ϖ which is here equal to 2.

In a similar manner it may be shown that

$$a_0^3 \{ (\gamma - a)(\beta - \delta) - (a - \beta)(\gamma - \delta) \} \{ (a - \beta)(\gamma - \delta)$$
$$- (\beta - \gamma)(a - \delta) \} \{ (\beta - \gamma)(a - \delta) - (\gamma - a)(\beta - \delta) \}$$
$$= - 432 \, (a_0 a_2 a_4 + 2 a_1 a_2 a_3 - a_0 a_3^2 - a_1^2 a_4 - a_2^3)$$

is an invariant of the quartic.

There is no difficulty in determining in any particular case whether ϕ leads to an invariant or covariant, for if ϕ leads to an invariant, $\phi = \pm \psi$, that is ϕ is unchanged (except in sign when its type term is the product of an odd number of differences of the roots, *i.e.* when its weight is odd) when for the roots their reciprocals are substituted, and fractions removed by the simplest multiplier $(a_1 a_2 a_3 \ldots a_n)^\varpi$. From another point of view we may consider an invariant as a covariant reduced to a single term.

148. **Properties of Covariants and Invariants.—** Since ϕ is a homogeneous function of the roots, the covariant derived from it may be written under the form

$$\frac{U^\varpi}{x^\kappa} \phi\left(\frac{x}{a_1 - x}, \frac{x}{a_2 - x}, \ldots \frac{x}{a_n - x}\right),$$

where ϖ is the order, and κ the weight of ϕ.

Also, as ϕ is a function of the differences, we may add 1 to each constituent such as $\dfrac{x}{a_r - x}$, thus obtaining $\dfrac{a_r}{a_r - x}$. Again, multiplying each constituent by x, we have

$$\frac{U^\varpi}{x^{2\kappa}} \phi\left(\frac{a_1 x}{a_1 - x}, \frac{a_2 x}{a_2 - x}, \ldots \frac{a_n x}{a_n - x}\right),$$

which may be reduced to the form

$$(-1)^\kappa U_c^\varpi x^{n\varpi - 2\kappa} \phi\left(\frac{1}{\dfrac{1}{a_1} - \dfrac{1}{x}}, \frac{1}{\dfrac{1}{a_2} - \dfrac{1}{x}}, \ldots \frac{1}{\dfrac{1}{a_n} - \dfrac{1}{x}}\right),$$

where

$$U_c = a_n \left(\frac{1}{x} - \frac{1}{a_1}\right)\left(\frac{1}{x} - \frac{1}{a_2}\right) \ldots \left(\frac{1}{x} - \frac{1}{a_n}\right);$$

whence it is proved that the covariant form

$$U^\varpi \phi\left(\frac{1}{a_1 - x}, \frac{1}{a_2 - x}, \ldots \frac{1}{a_n - x}\right)$$

is unaltered when for $a_1, a_2, \ldots a_n, x$, their reciprocals are substituted; $a_0, a_1, a_2, \ldots a_n$ changed into $a_n, a_{n-1}, \ldots a_0$, respectively, and the result multiplied by $(-1)^\kappa x^{n\varpi - 2\kappa}$.

Now if any covariant whose degree is m be written in the form

$$(B_0, B_1, B_2, \ldots B_m)(x, 1)^m, \tag{1}$$

changing $a_0, a_1, \ldots a_n, x,$ into $a_n, a_{n-1}, \ldots a_0, \dfrac{1}{x},$ we have another form for this covariant, namely,

$$(-1)^\kappa x^{n\varpi - 2\kappa} (C_0, C_1, C_2, \ldots C_m)\left(\frac{1}{x}, 1\right)^m;$$

and as this form is an integral function of x of the same type as (1), we have, by comparing the two forms,

$$m = n\varpi - 2\kappa, \quad \text{and} \quad B_0 = (-1)^\kappa C_m, \ldots B_r = (-1)^\kappa C_{m-r};$$

thus determining the degree of the covariant in terms of the order and weight of the function ϕ, and showing that the conjugate coefficients, *i.e.* those equally removed from the extremes, are related in the following way :—

If $F(a_0, a_1, a_2, \ldots a_n)$ be any coefficient of the covariant, $(-1)^\kappa F(a_n, a_{n-1}, \ldots a_0)$ is its conjugate.

From the expression for the degree of a covariant in terms of ϖ and κ, namely $n\varpi - 2\kappa$, we may draw the following important inferences :—

(1). *If $a_0^\varpi \phi$ is an invariant, $n\varpi = 2\kappa$.*

For, in this case ϕ and ψ are the same function, and consequently their weights κ and $n\varpi - \kappa$ also the same.

(2). *All the invariants of quantics of odd degrees are of even order.*

For if n be odd, it is plain from the equation $n\varpi = 2\kappa$ that ϖ must be even, and κ a multiple of n.

(3). *All covariants of quantics of even degrees are of even degrees.*

For in this case $n\varpi - 2\kappa$ is even.

(4). *The resultant of two covariants is always of an even degree in the coefficients of the original quantic.*

For, the degree of the resultant expressed in terms of the orders and weights of the covariants is

$$\varpi\,(n\varpi' - 2\kappa') + \varpi'(n\varpi - 2\kappa) = 2\,(n\varpi\varpi' - \varpi\kappa' - \varpi'\kappa).$$

We add some examples in illustration of the principles explained in the preceding Articles.

EXAMPLES.

1. Show that the resultant of two equations is an invariant of the system.

2. Show that the discriminant of any quantic is an invariant.

3. Prove directly that any function of the differences of the roots of the covariant

$$U^{\varpi}\phi\left(\frac{1}{a_1 - x},\ \ \frac{1}{a_2 - x},\ \ \frac{1}{a_3 - x},\ \cdots\ \frac{1}{a_n - x}\right)$$

equated to zero is a function of the differences of $a_1, a_2, a_3 \ldots a_n$.

4. If $a,\ \beta,\ \gamma$; and $a',\ \beta'$ be the roots of the equations

$$U \equiv ax^3 + 3bx^2 + 3cx + d = 0,$$

$$U' \equiv a'x^2 + 2b'x + c' = 0\ ;$$

express in terms of the coefficients the function

$$(\beta - \gamma)^2\,(a - a')(a - \beta') + (\gamma - a)^2\,(\beta - a')(\beta - \beta') + (a - \beta)^2\,(\gamma - a')(\gamma - \beta').$$

Denoting this function by ϕ, we easily find

$$-\,a^2 a'\phi = 9\,\{a'\,(bd - c^2) - b'\,(ad - bc) + c'\,(ac - b^2)\}.$$

Attending to the definition at the close of Art. 146 we observe that this function is an invariant of the two equations ; for it involves all the roots of the cubic in the second degree, and all the roots of the quadratic in the first degree. If, in fact, we make the substitutions of Art. 146, and render the function integral by multiplying by $U^2\,U'$, the result will not contain x, and is therefore an invariant of the system.

The geometrical interpretation of the equation $\phi = 0$ is that the quadratic U should form with the Hessian of the cubic U a harmonic system.

5. If $a,\ \beta,\ \gamma$; $a',\ \beta',\ \gamma'$ be the roots of the equations

$$ax^3 + 3bx^2 + 3cx + d = 0,$$

$$a'x^3 + 3b'x^2 + 3c'x + d' = 0\ ;$$

express the following function (when multiplied by aa') in terms of the coefficients, and prove that it is an invariant of the system :—

$$(a - a')\,(\beta - \beta')\,(\gamma - \gamma') + (a - \beta')\,(\beta - \gamma')\,(\gamma - a') + (a - \gamma')\,(\beta - a')\,(\gamma - \beta')\ ;$$

or, differently arranged,

$$(a - a')(\beta - \gamma')(\gamma - \beta') + (a - \beta')(\beta - a')(\gamma - \gamma') + (a - \gamma')(\beta - \beta')(\gamma - a').$$

$$\text{Ans. } 3\{(ad' - a'd) - 3(bc' - b'c)\}.$$

6. If a, β, γ, δ; a', β', γ', δ' be the roots of the biquadratics

$$(a, b, c, d, e)(x, 1)^4 = 0, \quad (a', b', c', d', e')(x, 1)^4 = 0;$$

prove

$$aa' \, \Sigma \, (a - a')(\beta - \beta')(\gamma - \gamma')(\delta - \delta') = 24\{ae' + a'e - 4(bd' + b'd) + 6cc'\},$$

and show that this function is an invariant of the system.

7. Prove that the following function of the roots of a biquadratic and quadratic gives an invariant of the system, and determine its geometrical interpretation :—

$$
\begin{vmatrix} 1 & \beta + \gamma & \beta\gamma \\ 1 & a + \delta & a\delta \\ 1 & a' + \beta' & a'\beta' \end{vmatrix}
\times
\begin{vmatrix} 1 & \gamma + a & \gamma a \\ 1 & \beta + \delta & \beta\delta \\ 1 & a' + \beta' & a'\beta' \end{vmatrix}
\times
\begin{vmatrix} 1 & a + \beta & a\beta \\ 1 & \gamma + \delta & \gamma\delta \\ 1 & a' + \beta' & a'\beta' \end{vmatrix}
= \phi.
$$

The geometrical interpretation of the equation $\phi = 0$ is, that two conjugate foci of the three involutions determined by the biquadratic form along with the quadratic an harmonic system.

8. Prove that the following functions of the roots of a biquadratic and quadratic give invariants of the system, and determine their values in terms of the coefficients :—

$$a_0 b_0{}^2 \, \Sigma \, (a' - a)(a' - \beta)(\beta' - \gamma)(\beta' - \delta),$$

$$a_0{}^2 b_0{}^2 \, \Sigma \, (a - \beta)^2 (\gamma - a')(\delta - \beta')(\gamma - \beta')(\delta - a').$$

9. Find the condition that one pair of roots of a cubic should form an harmonic range with the roots of a given quadratic.

10. Find the condition that the roots of two cubics should determine a system in involution.

The condition is expressed by multiplying together the six determinants of the type

$$
\begin{vmatrix} 1 & a + a' & aa' \\ 1 & \beta + \beta' & \beta\beta' \\ 1 & \gamma + \gamma' & \gamma\gamma' \end{vmatrix},
$$

and equating the result to zero.

149. Formation of Covariants by the Operator D.— From Art. 134 we infer that the expansion of $F(U_n, U_{n-1}, \ldots U_0)$ may be expressed by means of the Differential Calculus in the form

$$F_0 + x D F_0 + \frac{x^2}{1 \cdot 2} D^2 F_0 + \ldots + \frac{x^r}{1 \cdot 2 \cdot 3 \ldots r} D^r F_0 + \ldots,$$

where F_0 is the result of making $x = 0$ in $F(U_n, U_{n-1}, \ldots U_0)$, viz.,

$$F_0 = F(a_n, a_{n-1}, \ldots a_0),$$

and
$$D = a_0 \frac{d}{da_1} + 2a_1 \frac{d}{da_2} + 3a_2 \frac{d}{da_3} + \ldots + na_{n-1} \frac{d}{da_n}.$$

In forming a covariant by this process, the source F_0 with which we set out is altered by the successive operations D till we arrive at the original function $F(a_0, a_1, \ldots a_n)$, from which the source was formed. Since this is a function of the differences, the coefficient derived by the next operation D vanishes, and the covariant is completely formed. The corresponding operations δ on the symmetric function ψ_0 have the effect of reducing the degree in the roots by one each step, the final symmetric function containing the differences only. Thus the successive operations supply between the roots and coefficients a number of relations equal to the number of coefficients in the covariant.

The degree m of the covariant is plainly equal to the number of times δ operates in reducing ψ_0 to ϕ, *i. e.* equal to the difference of the weights of the extreme coefficients. And since

$$\psi_0 = (a_1 a_2 \ldots a_n)^\varpi \, \phi\left(\frac{1}{a_1}, \frac{1}{a_2}, \ldots \frac{1}{a_n}\right),$$

the weight of ψ_0 is $n\varpi - \kappa$, where κ is the weight of $\phi(a_1, a_2, \ldots a_n)$; hence the degree of the covariant whose leading coefficient is $a_0^\varpi \phi$ is $n\varpi - 2\kappa$, the same value as before obtained. We add two simple examples in illustration of this method.

EXAMPLES.

1. Form the Hessian of the cubic

$$a_0 x^3 + 3a_1 x^2 + 3a_2 x + a_3 = 0.$$

Taking the function $H \equiv a_0 a_2 - a_1^2$, we find, as in Art. 147,

$$a_0^2 \, \Sigma \, a^2 (\beta - \gamma)^2 = 18(a_2^2 - a_1 a_3).$$

Operating on the left-hand side by δ, and on the right-hand side by D, we obtain

$$- a_0^2 \, \Sigma \, 2\alpha (\beta - \gamma)^2 = 18(a_1 a_2 - a_0 a_3) \, ;$$

and operating in the same way again,

$$a_0^2 \, \Sigma \, 2(\beta - \gamma)^2 = 36\,(a_1^2 - a_0 a_2).$$

The next operation causes both sides of the equation to vanish. Hence the required covariant is, as in Art. 147,

$$(a_1 a_3 - a_2^2) + (a_0 a_3 - a_1 a_2)\,x + (a_0 a_2 - a_1^2)\,x^2.$$

We find at the same time the corresponding expression in terms of x and the roots.

2. Form the Hessian of the biquadratic

$$a_0 x^4 + 4a_1 x^3 + 6a_2 x^2 + 4a_3 x + a_4 = 0.$$

The covariant whose leading coefficient is $H \equiv a_0 a_2 - a_1^2$ is called the Hessian of the biquadratic. Its degree is 4, since $\varpi = 2$, and $\kappa = 2$; and $\therefore n\varpi - 2\kappa = 4$. Changing the coefficients into their complementaries, the source of the covariant is $a_4 a_2 - a_3^2$, and we easily find

$$H_x \equiv (a_0 a_2 - a_1^2)x^4 + 2(a_0 a_3 - a_1 a_2)x^3 + (a_0 a_4 + 2a_1 a_3 - 3a_2^2)x^2$$

$$+ \, 2(a_1 a_4 - a_2 a_3)x + (a_2 a_4 - a_3^2).$$

150. **Theorem.**[*]—In the discussion of covariants through the medium of the roots, as in the previous Articles, the following proposition, due to Mr. Michael Roberts, is of importance :—

Any function of the differences of the roots of two covariants is a function of the differences of the roots of the original quantic.

Let

$$(B_0,\ B_1,\ B_2,\ \ldots\ B_p)(x,\ y)^p \equiv B_0(x - \beta_1 y)(x - \beta_2 y)\ldots(x - \beta_p y),$$

$$(C_0,\ C_1,\ C_2,\ \ldots\ C_q)(x,\ y)^q \equiv C_0(x - \gamma_1 y)(x - \gamma_2 y)\ldots(x - \gamma_q y)$$

be two covariants of the quantic

$$(a_0,\ a_1,\ a_2,\ \ldots\ a_n)\,(x,\ y)^n.$$

Operating with D or δ on the identical equation

$$B_0\beta_r{}^p + pB_1\beta_r{}^{p-1} + \frac{p \cdot p - 1}{1 \cdot 2}\,B_2\beta_r{}^{p-2} + \ldots + B_p = 0,$$

<hr>

and remembering that, in general, $Df = a_0{}^\varpi \delta\phi$, where

$$f(a_0,\ a_1,\ a_2,\ \ldots\ a_n) = a_0{}^\varpi \phi\,(a_1,\ a_2,\ \ldots\ a_n),$$

we have

$$p\,(B_0\beta_r{}^{p-1} + (p-1)\,B_1\beta_r{}^{p-2} + \ldots + B_{p-1})(1 + \delta\beta_r) = 0\,;$$

and, therefore,

$$\delta\beta_r = -\,1\,;$$

similarly

$$\delta\gamma_s = -\,1,$$

whence

$$\delta\,(\beta_r - \gamma_s) = 0,$$

proving that $\beta_r - \gamma_s$ is a function of the differences of the roots $a_1,\ a_2,\ a_3,\ \ldots\ a_n.$

151. Homographic Transformation applied to the Theory of Covariants.—Hitherto we have discussed the theory of covariants and invariants through the medium of the roots of equations. We now adopt a different mode of treatment which will render the discussion of this subject more complete, and will open the way for the extension of the theory to quantics homogeneous in any number of variables. It is through the medium of such functions that the most important geometrical applications of this theory are carried on. With this object we give in the present Article two important propositions.

Prop. I.—*Let any quantic U_n be transformed by the homographic transformation*

$$x = \frac{\lambda x' + \mu}{\lambda' x' + \mu'}\,;$$

if I and I' be corresponding invariants of the two forms, we have

$$I' = (\lambda\mu' - \lambda'\mu)^\kappa\,I.$$

To prove this, let

$$I = a_0{}^\varpi\,\Sigma\,(a_1 - a_2)^a\,(a_2 - a_3)^b\,\ldots\,(a_1 - a_n)^l,$$

each root entering in the degree $\varpi.$

Now, transforming the similar value of I', since $x' = \dfrac{\mu'x - \mu}{\lambda - \lambda'x}$, we have

$$a'_p - a'_q = \frac{(\lambda\mu' - \lambda'\mu)(a_p - a_q)}{(\lambda - \lambda'a_p)(\lambda - \lambda'a_q)}.$$

Again, transforming U_n, and rendering the result integral, U_n' takes the form

$$a_0'(x' - a_1')(x' - a_2') \ldots (x' - a_n'),$$

where

$$a_0' = a_0(\lambda - \lambda'a_1)(\lambda - \lambda'a_2) \ldots (\lambda - \lambda'a_n) ;$$

making these substitutions for all the differences, and for a_0', the denominators of the fractions which enter by the transformation disappear ; and we have, finally,

$$I' = (\lambda\mu' - \lambda'\mu)^\kappa I.$$

Prop. II.—*If $\phi(x)$ be a covariant of the quantic U_n, the new value of $\phi(x)$, after homographic transformation, is (when rendered integral)*

$$(\lambda\mu' - \lambda'\mu)^\kappa \phi(x).$$

The proof is similar to that of the preceding Proposition. We have

$$\phi(x) = a_0^\varpi \Sigma (a_1 - a_2)^a (a_2 - a_3)^b \ldots (x - a_1)^p (x - a_2)^q \ldots,$$

which expression for $\phi(x)$ is obtained by substituting

$$x - a_1, \quad x - a_2, \ldots x - a_n \text{ for } a_1, a_2, \ldots a_n$$

in the source of the covariant $\phi(x)$ expressed in terms of the roots. Now, transforming, as in the previous Proposition, the value of $\phi(x)$ thus derived; since the factors $\lambda - \lambda'a_1, \lambda - \lambda'a_2, \ldots$ all enter in the same degree ϖ in the denominator (for each root enters the source in the degree ϖ), they will all be removed by the multiplier $a_0'^\varpi$, and the transformed value of $\phi(x)$ is

$$(\lambda\mu' - \lambda'\mu)^\kappa \phi(x).$$

152. Reduction of Homographic Transformation to a double Linear Transformation.—With a view to this reduction let the quantic be written under the homogeneous form

$$U_n = a_0 x^n + n a_1 x^{n-1} y + \frac{n(n-1)}{1 \cdot 2} a_2 x^{n-2} y^2 + \ldots + a_n y^n ;$$

and, in place of putting as before $x = \dfrac{\lambda x' + \mu}{\lambda' x' + \mu'}$, and removing fractions to make U_n integral, let now $\dfrac{x}{y} = \dfrac{\lambda x' + \mu y'}{\lambda' x' + \mu' y'}$, where $\dfrac{x}{y}$ and $\dfrac{x'}{y'}$, are the variables in the ordinary sense. The transformation may therefore be reduced to a linear transformation of both the variables x and y, and can be effected by putting in the original quantic

$$x = \lambda x' + \mu y', \quad y = \lambda' x' + \mu' y',$$

the introduction of fractions being in this way avoided.

Thus we pass from a homographic transformation of functions of a single variable to the linear transformation of homogeneous functions of two variables.

The quantity $\lambda \mu' - \lambda' \mu$ is called the *modulus of transformation.*

We are now enabled to restate Propositions I. and II. of Art. 151, in the following way :—

PROP. I.—*An invariant is a function of the coefficients of a quantic, such that when the quantic is transformed by linear transformation of the variables, the same function of the new coefficients is equal to the original function multiplied by a power of the modulus of transformation.*

PROP. II.—*A covariant is a function of the coefficients of a quantic, and also of the variables, such that when the quantic is transformed by linear transformation, the same function of the new variables and coefficients shall be equal to the original function multiplied by a power of the modulus of transformation.*

EXAMPLES.

1. Performing the linear transformation

$$x = \lambda X + \mu Y, \quad y = \lambda_1 X + \mu_1 Y,$$

if

$$ax^2 + 2b\,xy + cy^2 = AX^2 + 2BXY + CY^2,$$

prove that

$$AC - B^2 = (\lambda\mu_1 - \lambda_1\mu)^2 (ac - b^2).$$

2. Performing the same transformation, if

$$(a,\ b,\ c,\ d,\ e)(x,\ y)^4 = (A,\ B,\ C,\ D,\ E)(X,\ Y)^4,$$

prove that

$$AE - 4BD + 3C^2 = (\lambda\mu_1 - \lambda_1\mu)^4 (ae - 4\,bd + 3c^2).$$

3. Performing the same transformation, if

$$ax^2 + 2bxy + cy^2 = AX^2 + 2BXY + CY^2,$$

and

$$a_1x^2 + 2b_1xy + c_1y^2 = A_1X^2 + 2B_1XY + C_1Y^2,$$

prove that

$$AC_1 + A_1C - 2BB_1 = (\lambda\mu_1 - \lambda_1\mu)^2 (ac_1 + a_1c - 2bb_1).$$

This follows from Ex. 1, applied to the quadratic forms

$$(a + \kappa a_1)x^2 + 2(b + \kappa b_1)xy + (c + \kappa c_1)y^2 = (A + \kappa A_1)X^2 + 2(B + \kappa B_1)XY + (C + \kappa C_1)Y^2,$$

by comparing the coefficients of κ on both sides.

Whence we may infer that, if two quadratics determine a harmonic system, the new quadratics obtained by linear transformation also form an harmonic system. For their roots being α, β and α_1, β_1, we have

$$aa_1\{(\alpha - \alpha_1)(\beta - \beta_1) + (\alpha - \beta_1)(\beta - \alpha_1)\} = 2(ac_1 + a_1c - 2bb_1).$$

153. Properties of Covariants derived from Linear Transformation.—We proceed now to show, taking the second proposition of Art. 152 as the definition of a covariant, that the law of derivation of the coefficients given in Art. 149 immediately follows; that is, *given any one coefficient, all the rest may be determined.*

For this purpose, performing the linear transformation

$$x = X + hY,$$

$$y = 0X + Y,$$

whose modulus is unity, the quantic

$$(a_0, a_1, a_2, \ldots a_n)(x, y)^n \text{ becomes } (A_0, A_1, A_2, \ldots A_n)(X, Y)^n,$$

where

$$A_0 = a_0, \; A_1 = a_1 + a_0 h, \quad A_2 = a_2 + 2a_1 h + a_0 h^2, \; \&c. \quad \text{(See Art. 36.)}$$

Now, if $\phi(a_0, a_1, a_2, \ldots a_n, x, y)$ be any covariant of this quantic, we have by the definition

$$\phi(a_0, a_1, a_2, \ldots a_n, x, y) = \phi(A_0, A_1, A_2, \ldots A_n, X, Y),$$

or

$$\phi(a_0, a_1, a_2, \ldots a_n, x, y) = \phi(A_0, A_1, A_2 \ldots A_n, x - hy, y).$$

Expanding the second member of this equation, and confining our attention to the terms which multiply h; since also $\dfrac{dA_r}{dh} = r a_{r-1}$, when terms are omitted which would be multiplied in the result by h^2, h^3, &c., we have

$$\phi + h\left(- y\frac{d\phi}{dx} + D\phi\right) + h^2\left(\qquad\qquad\right) + \&c. \ldots = \phi,$$

which must hold whatever value h may have; hence

$$y\frac{d\phi}{dx} = a_0 \frac{d\phi}{da_1} + 2a_1 \frac{d\phi}{da_2} + 3a_2 \frac{d\phi}{da_3} + \ldots + na_{n-1} \frac{d\phi}{da_n}, \qquad (1)$$

and, substituting for ϕ the value

$$(B_0, B_1, B_2, \ldots B_m)(x, y)^m,$$

we have

$$m B_0 x^{m-1} y + m(m-1) B_1 x^{m-2} y^2 + \ldots + m B_{m-1} y^m$$

$$= D B_0 x^m \quad + m D B_1 x^{m-1} y \qquad + \ldots + D B_m y^m;$$

whence, comparing coefficients, we have the following equations:

$$D B_0 = 0, \quad D B_1 = B_0, \quad D B_2 = 2B_1, \ldots D B_m = m B_{m-1},$$

which determine the law of derivation of the coefficients from the source B_m; the leading coefficient B_0 being a function of the differences, since $DB_0 = 0$.

The calculation of the coefficients is facilitated by the following theorem which has been proved already on different principles:—

Two coefficients of a covariant equally removed from the extremes become equal (plus or minus) when in either of them, $a_0, a_1, \ldots a_n$ are replaced by $a_n, a_{n-1}, \ldots a_0$, respectively.

To prove this, let the quantic be transformed by the linear substitution

$$x = 0X + Y, \quad y = X + 0Y, \quad \text{whose modulus} = -1.$$

Thus

$$(a_0, a_1, a_2, \ldots a_n)(x, y)^n = (a_n, a_{n-1}, a_{n-2}, \ldots a_0)(X, Y)^n,$$

and, by definition, any covariant

$$\phi(a_n, a_{n-1}, a_{n-2}, \ldots a_0, X, Y) = (-1)^\kappa \phi(a_0, a_1, a_2, \ldots a_n, x, y)$$

$$= (-1)^\kappa \phi(a_0, a_1, a_2, \ldots a_n, Y, X);$$

whence the coefficients of the covariant equally removed from the extremes are similar in form, and become identical (except in sign when κ is odd) when for the suffixes their complementary values are substituted.

We may infer similarly that a covariant satisfies the differential equation

$$x\frac{d\phi}{dy} = a_n \frac{d\phi}{da_{n-1}} + 2a_{n-1}\frac{d\phi}{da_{n-2}} + 3a_{n-2}\frac{d\phi}{da_{n-3}} + \ldots + na_1 \frac{d\phi}{da_0},$$

as well as the equation (1) already given.

Again, if $\phi(a_0, a_1, a_2, \ldots a_n)$ be an invariant of the quantic, the former transformation of the present Article gives, employing the definition of Art. 152,

$$\phi(a_0, a_1, a_2, \ldots a_n) = \phi(A_0, A_1, A_2, \ldots A_n);$$

and proceeding as before in the case of a covariant, we prove that an invariant must satisfy both the differential equations

$$a_0 \frac{d\phi}{da_1} + 2a_1 \frac{d\phi}{da_2} + 3a_2 \frac{d\phi}{da_3} + \ldots + na_{n-1} \frac{d\phi}{da_n} = 0,$$

$$a_n \frac{d\phi}{da_{n-1}} + 2a_{n-1} \frac{d\phi}{da_{n-2}} + 3a_{n-2} \frac{d\phi}{da_{n-3}} + \ldots + na_1 \frac{d\phi}{da_0} = 0,$$

either of which may be regarded as contained in the other, since if we make the linear transformation $x = Y$, $y = X$ (whose modulus $= -1$), we have from the definition of an invariant

$$\phi\left(a_n, a_{n-1}, a_{n-2}, \ldots a_0\right) = (-1)^\kappa \phi\left(a_0, a_1, a_2, \ldots a_n\right);$$

proving that an invariant is a function of the coefficients of a quantic which does not alter (except in sign if the weight be odd) when the coefficients are written in direct or reverse order.

Having now explained the nature of Covariants and Invariants of quantics, and the connexion between the two modes in which these functions may be discussed, we proceed to prove certain propositions which are of wide application in the formation of the Covariants and Invariants of quantics transformed by a linear substitution.   The student who is reading this subject for the first time may pass at once to the next chapter, where the principles already explained are applied to the cases of the quadratic, cubic, and quartic.

154. PROP. I.—*Let any homogeneous quantic of the nth degree $f(x, y)$ become $F(X, Y)$ by the linear transformation*

$$x = \lambda X + \mu Y, \quad y = \lambda' X + \mu' Y;$$

also let any function u of x, y become U by the same transformation; then we have

$$M^n f\left(\frac{du}{dy}, -\frac{du}{dx}\right) = F\left(\frac{dU}{dY}, -\frac{dU}{dX}\right), \tag{1}$$

where M is the modulus of transformation.

To prove this proposition, solving the equations

$$x = \lambda X + \mu Y, \quad y = \lambda' X + \mu' Y,$$

we have

$$MX = \mu'x - \mu y, \quad MY = -\lambda'x + \lambda y;$$

whence

$$M\frac{dX}{dx} = \mu', \quad M\frac{dX}{dy} = -\mu, \quad M\frac{dY}{dx} = -\lambda', \quad M\frac{dY}{dy} = \lambda.$$

Again,

$$\frac{du}{dx} = \frac{dU}{dX}\frac{dX}{dx} + \frac{dU}{dY}\frac{dY}{dx} = \frac{1}{M}\left(\mu'\frac{dU}{dX} - \lambda'\frac{dU}{dY}\right),$$

$$\frac{du}{dy} = \frac{dU}{dX}\frac{dX}{dy} + \frac{dU}{dY}\frac{dY}{dy} = \frac{1}{M}\left(-\mu\frac{dU}{dX} + \lambda\frac{dU}{dY}\right),$$

which equations may be put under the form

$$\frac{du}{dy} = \lambda\left(\frac{1}{M}\frac{dU}{dY}\right) + \mu\left(-\frac{1}{M}\frac{dU}{dX}\right),$$

$$-\frac{du}{dx} = \lambda'\left(\frac{1}{M}\frac{dU}{dY}\right) + \mu'\left(-\frac{1}{M}\frac{dU}{dX}\right);$$

and since

$$f(\lambda X + \mu Y, \quad \lambda'X + \mu'Y) = F(X, Y),$$

changing X and Y into $\dfrac{1}{M}\dfrac{dU}{dY}$, and $-\dfrac{1}{M}\dfrac{dU}{dX}$, respectively, the proposition is proved.

In an exactly similar manner, changing X and Y into

$$\frac{1}{M}\frac{d}{dY}, \quad -\frac{1}{M}\frac{d}{dX},$$

it may be proved that

$$M^n f\left(\frac{d}{dy}, -\frac{d}{dx}\right)u = F\left(\frac{d}{dY}, -\frac{d}{dX}\right)U. \tag{2}$$

The results (1) and (2) may be applied to generate covariants and invariants, as we proceed to show.

Suppose $f(x, y)$ and u to be covariants of any third quantic v, where v may become identical with either as a particular case. Also, denoting by $F_c(X, Y)$ and U_c the same covariants in

X, Y, variables, we have, from Props. I. and II., Art. 152, the identical equations

$$M^p F(X, Y) = F_c(X, Y), \text{ and } M^q U = U_c;$$

whence, substituting from these equations in (1),

$$M^r f\left(\frac{du}{dy}, -\frac{du}{dx}\right) = F_c\left(\frac{dU_c}{dY}, -\frac{dU_c}{dX}\right), \text{ where } r = n + p + q;$$

proving that $f\left(\dfrac{du}{dy}, -\dfrac{du}{dx}\right)$ is a covariant of v.

And in a similar manner it is proved from (2) that

$$f\left(\frac{d}{dy}, -\frac{d}{dx}\right) u$$

leads to an invariant or covariant of v, according as u is of the nth or any higher order.

EXAMPLES.

1. If $\dfrac{d}{dy}, -\dfrac{d}{dx}$ be substituted for x and y in the quartic $(a, b, c, d, e)(x, y)^4 = U$, and the resulting operation performed on the quartic itself, show that the invariant I is obtained.

We find

$$(a, b, c, d, e)\left(\frac{d}{dy}, -\frac{d}{dx}\right)^4 U = 48\,(ae - 4bd + 3c^2).$$

2. Prove, by performing the same operation on H_x, the Hessian of the quartic (see Ex. 2, Art. 149), that the invariant J is obtained.

Here we find

$$(a, b, c, d, e)\left(\frac{d}{dy}, -\frac{d}{dx}\right)^4 H_x = 72\,(ace + 2bcd - ad^2 - eb^2 - c^3).$$

3. Prove that

$$(a, b, c, d)\left(\frac{d}{dy}, -\frac{d}{dx}\right)^3 G_x = -12\,(a^2 d^2 - 6abcd + 4ac^3 + 4b^3 d - 3b^2 c^2),$$

where G_x is the cubic covariant of the cubic $(a, b, c, d)(x, y)^3$.

4. Find the value of

$$(ac - b^2)\left(\frac{du}{dy}\right)^2 - (ad - bc)\frac{du}{dy}\frac{du}{dx} + (bd - c^2)\left(\frac{du}{dx}\right)^2,$$

where $u = (a, b, c, d)(x, y)^3$.

Ans. $-9\,H_x^2$.

155. PROP. II.—*If $\phi(a_0, a_1, a_2, \ldots a_n)$ be an invariant of the quantic $(a_0, a_1, a_2, \ldots a_n)(x, y)^n$, and u a covariant of the nth or any higher degree, including the quantic itself; then*

$$\phi\left(\frac{d^n u}{dx^n}, \frac{d^n u}{dx^{n-1}dy}, \frac{d^n u}{dx^{n-2}dy^2}, \ldots, \frac{d^n u}{dy^n}\right)$$

is an invariant or covariant of the quantic. To prove this, let

$$x = \lambda X + \mu Y, \quad x' = \lambda X' + \mu Y',$$

$$y = \lambda' X + \mu' Y, \quad y' = \lambda' X' + \mu' Y';$$

and, transforming as in the last Proposition,

$$x'\frac{d}{dx} + y'\frac{d}{dy} = X'\frac{d}{dX} + Y'\frac{d}{dY};$$

also, transforming u, we have

$$U = M^p u;$$

whence

$$\left(X'\frac{d}{dX} + Y'\frac{d}{dY}\right)^n U = M^p \left(x'\frac{d}{dx} + y'\frac{d}{dy}\right)^n u;$$

and writing this equation when expanded under the form

$$(D_0, D_1, D_2, \ldots D_n)(X', Y')^n = M^p (d_0, d_1, d_2, \ldots d_n)(x', y')^n,$$

we have, from the definition of an invariant,

$$\phi(D_0, D_1, D_2, \ldots D_n) = M^q \phi(d_0, d_1, d_2 \ldots d_n),$$

showing that $\phi(d_0, d_1, d_2, \ldots d_n)$ is an invariant or covariant. When x, y and x', y' are transformed similarly, as in the present Proposition, they are said to be *cogredient* variables.

EXAMPLES.

1. Let the quadratic

$$a_0 x^2 + 2a_1 xy + a_2 y^2 \quad \text{become} \quad A_0 X^2 + 2A_1 XY + A_2 Y^2.$$

We have then, as in Ex. 1, Art. 152,

$$A_0 A_2 - A_1^2 = M^2(a_0 a_2 - a_1^2).$$

To apply this, let $U = M^p u$, as in the last Article; and since

$$X'^2 \frac{d^2 U}{dX^2} + 2X'Y' \frac{d^2 U}{dX\,dY} + Y'^2 \frac{d^2 U}{dY^2} = M^p \left(x'^2 \frac{d^2 u}{dx^2} + 2x'\,y' \frac{d^2 u}{dx\,dy} + y'^2 \frac{d^2 u}{dy^2} \right),$$

it follows from the last result, considering X', Y' and x', y' as variables, that

$$\frac{d^2 U}{dX^2} \frac{d^2 U}{dY^2} - \left(\frac{d^2 U}{dX\,dY} \right)^2 = M^{2p} \left\{ \frac{d^2 u}{dx^2} \frac{d^2 u}{dy^2} - \left(\frac{d^2 u}{dx\,dy} \right)^2 \right\}.$$

This covariant is called the *Hessian* of U.

2. When u has the values

$$(a,\ b,\ c,\ d)(x,\ y)^3, \quad \text{and} \quad (a,\ b,\ c,\ d,\ e)(x,\ y)^4,$$

what covariants are derived by the process of the last example?

Ans. (1). $(ac - b^2)\,x^2 + (ad - bc)\,xy + (bd - c^2)\,y^2.$

(2). $(ac - b^2)\,x^4 + 2\,(ad - bc)\,x^3 y + (ae + 2bd - 3c^2)\,x^2 y^2$

$$+\ 2\,(be - cd)\,xy^3 + (ce - d^2)\,y^4.$$

156. Prop. III.—*If any invariant of the quantic*

$$U + k\,(xy' - x'y)^n$$

be formed, the coefficients of the different powers of k are covariants of U; x', y' being the variables.

For, transforming U by linear transformation, let

$$(a_0,\ a_1,\ a_2,\ \dots\ a_n)(x,\ y)^n = (A_0,\ A_1,\ A_2,\ \dots\ A_n)(X,\ Y)^n;$$

also if x, y and x', y' be cogredient variables,

$$xy' - x'y = M(XY' - X'Y).$$

Whence

$$(a_0,\ a_1,\ a_2,\ \dots\ a_n)(x,\ y)^n + k\,(xy' - x'y)^n$$

becomes when transformed

$$(A_0,\ A_1,\ A_2,\ \dots\ A_n)(X,\ Y)^n + kM^n(XY' - X'Y)^n;$$

and forming any invariant ϕ of both these forms, we have

$$(\phi,\ \phi_1,\ \phi_2,\ \dots\ \phi_p)(1,\ k)^p = M^\kappa(\Phi,\ \Phi_1,\ \Phi_2,\ \dots\ \Phi_p)(1,\ M^n k)^n,$$

proving that

$$\phi_r = M^q\,\Phi_r,$$

or that ϕ_r is a covariant.

When $(xy' - x'y)^n$ is replaced by $(b_0, b_1, b_2, \ldots b_n)(x, y)^n$, the preceding Proposition becomes by extension the following :—

If $\phi(a_0, a_1, a_2, \ldots a_n)$ *be an invariant of* $(a_0, a_1, a_2, \ldots a_n)(x, y)^n$, *all the coefficients of* k *in*

$$\phi(a_0 + kb_0, a_1 + kb_1, \ldots a_n + kb_n)$$

are invariants of the system of quantics

$$(a_0, a_1, a_2, \ldots a_n)(x, y)^n, \ (b_0, b_1, b_2, \ldots b_n)(x, y)^n.$$

Or, which is the same thing,

$$\left(b_0\frac{d}{da_0} + b_1\frac{d}{da_1} + \ldots b_n\frac{d}{da_n}\right)^r \phi, \ \&c., \ \&c.,$$

are invariants of the system.

If, further, ϕ be replaced by a covariant, we may in like manner generate new covariants, a similar proof applying in this case.

157. Prop. IV.—*If* $\phi(x, y)$ *and* $\psi(x, y)$ *are homogeneous quantics, the determinant*

$$\begin{vmatrix} \dfrac{d\phi}{dx} & \dfrac{d\phi}{dy} \\[2ex] \dfrac{d\psi}{dx} & \dfrac{d\psi}{dy} \end{vmatrix}$$

is a covariant of these quantics.

For, transforming ϕ and ψ by the linear substitution

$$x = \lambda X + \mu Y, \quad y = \lambda' X + \mu' Y,$$

we have

$$\Phi(X, Y) = \phi(x, y), \quad \Psi(X, Y) = \psi(x, y),$$

giving

$$\frac{d\Phi}{dX} = \lambda\frac{d\phi}{dx} + \lambda'\frac{d\phi}{dy}, \quad \frac{d\Psi}{dX} = \lambda\frac{d\psi}{dx} + \lambda'\frac{d\psi}{dy},$$

$$\frac{d\Phi}{dY} = \mu\frac{d\phi}{dx} + \mu'\frac{d\phi}{dy}, \quad \frac{d\Psi}{dY} = \mu\frac{d\psi}{dx} + \mu'\frac{d\psi}{dy}.$$

Whence

$$\begin{vmatrix} \dfrac{d\Phi}{dX} & \dfrac{d\Phi}{dY} \\[2ex] \dfrac{d\Psi}{dX} & \dfrac{d\Psi}{dY} \end{vmatrix} = \begin{vmatrix} \lambda\dfrac{d\phi}{dx} + \lambda'\dfrac{d\phi}{dy}, & \mu\dfrac{d\phi}{dx} + \mu'\dfrac{d\phi}{dy} \\[2ex] \lambda\dfrac{d\psi}{dx} + \lambda'\dfrac{d\psi}{dy}, & \mu\dfrac{d\psi}{dx} + \mu'\dfrac{d\psi}{dy} \end{vmatrix},$$

which reduces to

$$M\left(\frac{d\phi}{dx}\,\frac{d\psi}{dy} - \frac{d\phi}{dy}\,\frac{d\psi}{dx}\right);$$

and the proposition is proved.

This covariant is called the *Jacobian* of ϕ and ψ, and is often written under the form $J(\phi, \psi)$.

We now conclude this Chapter with some examples selected to illustrate the foregoing theory. The student is referred for further information on this subject to Salmon's *Lessons Introductory to the Modern Higher Algebra*, and to Clebsch's *Theorie Der Binären Algebraischen Formen*.

Miscellaneous Examples.

1. From the definitions, Art. 146, prove that all the invariants of the quantic $U(xy' - x'y)$ are covariants of U, the variable being $x' : y'$.

If U is a cubic, what covariants are thus derived?

2. If I_1, I_2, I_3, $\ldots I_n$ be the same invariant for each of the quantics
$$\frac{\phi(x)}{x - a_1}, \quad \frac{\phi(x)}{x - a_2}, \quad \frac{\phi(x)}{x - a_3}, \quad \ldots \frac{\phi(x)}{x - a_n}$$
of the order ϖ, where a_1, a_2, $\ldots a_n$ are the roots of $\phi(x) = 0$, prove that

$$\sum_{r=1}^{r=n} I_r (x - a_r)^{\varpi}$$

is a covariant of $\phi(x)$.

3. If a_1, a_2, a_3, $\ldots a_n$ be the roots of the equation
$$(a_0,\ a_1,\ a_2,\ \ldots a_n)(x,\ 1)^n = 0;$$
and if
$$a_0{}^{\varpi}\phi_1\phi_2 \ldots \phi_m = F(a_0,\ a_1,\ a_2,\ \ldots a_n),$$

where ϕ_1, ϕ_2, $\ldots \phi_m$ are all the values of a rational and integral function of some or all the roots obtained by substitution, find the equation whose roots are the m values of $-\dfrac{\phi}{\delta\phi}$, given $\delta^2\phi = 0$.

$$\textit{Ans. } F(U_0,\ U_1,\ U_2,\ \ldots U_n).$$

4. Let α, β, γ ; and α', β', γ' be the roots of the cubic equations

$$(a,\ b,\ c,\ d)\,(x,\ 1)^3 = 0,\quad (a',\ b',\ c',\ d)\,(x,\ 1)^3 = 0,$$

and prove that the following covariant of the system

$$aa'\ \Sigma\ \{(\beta - \gamma)\,(\beta' - \gamma') + (\beta - \gamma')\,(\beta' - \gamma)\}\,(x - \alpha)\,(x - \alpha')$$

expressed in terms of the coefficients is

$$-18\{(ac' + a'c - 2bb')\,x^2 + (ad' + a'd - bc' - b'c)\,x + (bd' + b'd - 2cc')\}.$$

5. Express the identical relation connecting three quadratics in terms of their invariants.

Let
$$U = a_1 x^2 + 2b_1 xy + c_1 y^2,$$
$$V = a_2 x^2 + 2b_2 xy + c_2 y^2,$$
$$W = a_3 x^2 + 2b_3 xy + c_3 y^2\ ;$$

multiplying together the two determinants

$$\begin{vmatrix} a_1 & b_1 & c_1 & 0 \\ a_2 & b_2 & c_2 & 0 \\ a_3 & b_3 & c_3 & 0 \\ y^2 & -xy & x^2 & 0 \end{vmatrix}\begin{vmatrix} c_1 & -2b_1 & a_1 & 0 \\ c_2 & -2b_2 & a_2 & 0 \\ c_3 & -2b_3 & a_3 & 0 \\ x^2 & 2xy & y^2 & 0 \end{vmatrix},$$

we have

$$4\begin{vmatrix} I_{11} & I_{12} & I_{13} & U \\ I_{12} & I_{22} & I_{23} & V \\ I_{13} & I_{23} & I_{33} & W \\ U & V & W & 0 \end{vmatrix} \equiv 0, \quad \text{where} \quad 2I_{pq} = a_p c_q + a_q c_p - 2b_p b_q.$$

Expanding this determinant we have

$$(I_{22} I_{33} - I_{23}{}^2)\,U^2 + (I_{33} I_{11} - I_{31}{}^2)\,V^2 + (I_{11} I_{22} - I_{12}{}^2)\,W^2 + 2(I_{31} I_{12} - I_{11} I_{23})\,VW$$
$$+ 2(I_{23} I_{12} - I_{22} I_{31})\,WU + 2(I_{23} I_{31} - I_{33} I_{12})\,UV \equiv 0. \tag{1}$$

There are two particular cases worth noticing :—

(1). *When the three quadratics are mutually harmonic.* — In this case $I_{23} = 0$, $I_{31} = 0$, $I_{12} = 0$; and making these quantities vanish in equation (1), we have

$$\frac{U^2}{I_{11}} + \frac{V^2}{I_{22}} + \frac{W^2}{I_{33}} \equiv 0.$$

(2). *When one of the quadratics $W = 0$ determines the foci of the involution of the points given by the other two, $U = 0$, and $V = 0$.* In this case $I_{13} = 0$, and $I_{23} = 0$; and making this reduction in the general equation (1), we have

$$(I_{12}{}^2 - I_{11} I_{22})\,W^2 = I_{33}(I_{22}\,U^2 - 2I_{12}\,UV + I_{11}\,V^2)\ ;$$

but from the equations $I_{13} = 0$, and $I_{23} = 0$, we find

$$a_3 = \kappa\,(a_1 b_2), \quad -2b_3 = \kappa\,(c_1 a_2), \quad c_3 = \kappa\,(b_1 c_2)\,;$$

whence

$$4\,(a_3 c_3 - b_3{}^2) = \kappa^2\,\{(a_1 b_2)(b_1 c_2) - (c_1 a_2)^2\},$$

or

$$2I_{33} = \kappa^2\,\{I_{11} I_{22} - I_{12}{}^2\},$$

and reducing, when $\kappa = 1$

$$-2W^2 = I_{22}\,U^2 - 2I_{12}\,UV + I_{11}\,V^2.$$

6. Prove that the quartics

$$(a_0 x^2 + 2b_0 xy + c_0 y^2)(a_2 x^2 + 2b_2 xy + c_2 y^2) - (a_1 x^2 + 2b_1 xy + c_1 y^2)^2 = 0,$$

$$(a_0 x^2 + 2a_1 xy + a_2 y^2)(c_0 x^2 + 2c_1 xy + c_2 y^2) - (b_0 x^2 + 2b_1 xy + b_2 y^2)^2 = 0,$$

have the same invariants.

7. Prove that the condition that four roots of an equation of the nth degree should determine on a right line a harmonic system of points may be expressed by equating to zero an invariant of the degree $\dfrac{(n-1)(n-2)(n-3)}{2}$.

8. If $\phi\,(a_0,\ a_1,\ a_2,\ \ldots\ a_n)$ be any rational, integral, and homogeneous function which depends on the differences of the roots of the quantic $(a_0,\ a_1,\ a_2, \ldots a_n)\,(x, 1)^n$; prove that $\dfrac{d\phi}{da_n}$ depends on the differences of the roots also.

9. Prove that the functions

$$a_0 a_2 - a_1{}^2, \quad a_0 a_4 - 4a_1 a_3 + 3a_2{}^2, \quad a_0{}^2 a_3 - 3a_0 a_1 a_2 + 2a_1{}^3,$$

which depend on the differences of the roots of the equation

$$(a_0,\ a_1,\ a_2,\ \ldots\ a_n)\,(x,\ 1)^n = 0,$$

give rise to covariants of the degrees

$$2n - 4, \qquad 2n - 8, \qquad 3n - 6.$$

10. Prove that the coefficient of the penultimate term in the equation of the squares of the differences of any quantic leads to a covariant of that quantic of the fourth degree in the variables.

11. Prove that the product of two covariants whose sources are ϕ and ψ may be written under the form

$$\phi\psi + xD\,(\phi\psi) + \frac{x^2}{1\,.\,2}\,D^2\,(\phi\psi) + \&\text{c.} \ldots$$

Mr. M. Roberts.

12. Prove that the mth power of the quantic

$$(a_0,\ a_1,\ a_2,\ \ldots\ a_n)\,(x,\ 1)^n$$

may be represented by

$$a_n{}^m + xD\,(a_n{}^m) + \frac{x^2}{1\,.\,2}\,D^2\,(a_n{}^m) + \frac{x^3}{1\,.\,2\,.\,3}\,D^3\,(a_n{}^m) + \&\text{c.}$$

Mr. M. Roberts.

13. Prove from both definitions of a covariant that any covariant of a covariant is a covariant of the original quantic or quantics.

14. If $a_1, a_2, a_3, \ldots a_m,$ and $\beta_1, \beta_2, \beta_3, \ldots \beta_n$ be the roots of the equations

$$U \equiv (a_0, a_1, a_2, \ldots a_m)(x, 1)^m = 0, \quad \text{and} \quad V \equiv (b_0, b_1, b_2, \ldots b_n)(x, 1)^n = 0\ ;$$

from the simplest function of the differences of their roots, viz., $\Sigma(a_p - \beta_q),$ it is required to derive a covariant of the system U and V.

This question will be solved if we express

$$UV \sum \frac{a_p - \beta_q}{(x - a_p)(x - \beta_q)}$$

in terms of the coefficients of U and V.

For this purpose we have

$$\sum \frac{a_p - \beta_q}{(x - a_p)(x - \beta_q)} \equiv \sum \frac{a}{x - a} \sum \frac{1}{x - \beta} - \sum \frac{\beta}{x - \beta} \sum \frac{1}{x - a},$$

and if U and V be written as homogeneous functions of x and y,

$$\sum \frac{1}{x - ay} = \frac{d \log U}{dx}, \quad \sum \frac{a}{x - ay} = -\frac{d \log U}{dy}, \ \&c.$$

Whence, substituting these values in the last equation, we have

$$UV \sum \frac{a_p - \beta_q}{(x - ay)(x - \beta y)} = \frac{dU}{dx}\frac{dV}{dy} - \frac{dU}{dy}\frac{dV}{dx}\ ;$$

which is the Jacobian of U and V. It should be noticed also that the leading coefficient of $J(U, V)$ is $a_0 b_1 - a_1 b_0$.

15. To reduce, by the linear transformation,

$$x = \lambda X + \mu Y, \quad y = \lambda' X + \mu' Y,$$

the two cubics

$$U \equiv (a, b, c, d)(x, y)^3, \quad V \equiv (a', b', c', d')(x, y)^3$$

to the forms

$$U = \frac{1}{4}\frac{dF}{dX}, \quad V = \frac{1}{4}\frac{dF}{dY}.$$

Let $\qquad F = (A, B, C, D, E)(X, Y)^4\ ;$

then $\qquad U \equiv (a, b, c, d)(x, y)^3 = (A, B, C, D)(X, Y)^3,$

$$V \equiv (a', b', c', d')(x, y)^3 = (B, C, D, E)(X, Y)^3.$$

Now, substituting the differential symbols $D_y, -D_x$ for $x, y,$ and $\dfrac{1}{M}D_Y, -\dfrac{1}{M}D_X$ for X and Y in the Hessian of both forms of U, we find the operational equation

$$\begin{vmatrix} D_{x^2}^2 & D_{xy}^2 & D_{y^2}^2 \\ a & b & c \\ b & c & d \end{vmatrix} = \frac{1}{M^4} \begin{vmatrix} D_{X^2}^2 & D_{XY}^2 & D_{Y^2}^2 \\ A & B & C \\ B & C & D \end{vmatrix}\ ;$$

whence, operating on both forms of V, we have

$$\psi(x, y) \equiv \begin{vmatrix} a' & b' & c' \\ a & b & c \\ b & c & d \end{vmatrix} x + \begin{vmatrix} b' & c' & d' \\ a & b & c \\ b & c & d \end{vmatrix} y = \frac{JY}{M^4}.$$

Similarly,

$$\phi(x, y) \equiv \begin{vmatrix} a & b & c \\ a' & b' & c' \\ b' & c' & d' \end{vmatrix} x + \begin{vmatrix} b & c & d \\ a' & b' & c' \\ b' & c' & d' \end{vmatrix} y = \frac{JX}{M^4},$$

where J is the ternary invariant of F.

Again, since

$$\phi(D_y, -D_x) = \frac{J}{M^5} D_Y, \quad \text{and} \quad -\psi(D_y, -D_x) = \frac{J}{M^5} D_X,$$

performing the operation

$$\phi(D_y, -D_x)\,\psi(x, y), \quad \text{or} \quad \psi(D_y, -D_x)\,\phi(x, y),$$

on equivalent forms, we have

$$Q \equiv \begin{vmatrix} a & b & c \\ a' & b' & c' \\ b' & c' & d' \end{vmatrix} \begin{vmatrix} b' & c' & d' \\ a & b & c \\ b & c & d \end{vmatrix} - \begin{vmatrix} a' & b' & c' \\ a & b & c \\ b & c & d \end{vmatrix} \begin{vmatrix} b & c & d \\ a' & b' & c' \\ b' & c' & d' \end{vmatrix} = \frac{J^2}{M^9}.$$

We are now in a position to determine the coefficients of F in terms of the coefficients of U and V.

For we have from former equations

$$Q x = \begin{vmatrix} b' & c' & d' \\ a & b & c \\ b & c & d \end{vmatrix} \phi - \begin{vmatrix} b & c & d \\ a' & b' & c' \\ b' & c' & d' \end{vmatrix} \psi,$$

$$Q y = -\begin{vmatrix} a' & b' & c' \\ a & b & c \\ b & c & d \end{vmatrix} \phi + \begin{vmatrix} a & b & c \\ a' & b' & c' \\ b' & c' & d' \end{vmatrix} \psi;$$

whence, substituting these values of x and y in U and V, we find

$$Q^3 . U = (A_0, B_0, C_0, D_0)(\phi, \psi)^3,$$

$$Q^3 . V = (B_0, C_0, D_0, E_0)(\phi, \psi)^3,$$

and, therefore,

$$Q^3 U = \frac{1}{4}\frac{dF_0}{dX}, \quad Q^3 V = \frac{1}{4}\frac{dF_0}{dY}, \quad \text{where } F_0 = (A_0,\, B_0,\, C_0,\, D_0,\, E_0)\,(\phi,\, \psi)^4 ;$$

also

$$\frac{A}{A_0} = \frac{B}{B_0} = \frac{C}{C_0} = \frac{D}{D_0} = \frac{E}{E_0} = \frac{M^{15}}{J^3}.$$

16. Determine the invariants of F_0 in the preceding example.

We have from the equations of Ex. 15,

$$J^{10} = M^{45} J_0, \quad \text{and} \quad J^6 I = M^{30} I_0 ;$$

and, substituting differential symbols for x, y and X, Y in both forms of V, and operating on U, we find

$$P \equiv ad_1 - a_1 d - 3\,(bc_1 - b_1 c) = \frac{I}{M^3},$$

which equation, along with the equation $Q = \dfrac{J^2}{M^9}$, enables us by previous results to express I_0 and J_0 in terms of P and Q in the following way :

$$I_0 = PQ^3, \quad \text{and} \quad J_0 = Q^5.$$

We add certain observations. In the first place, since

$$\frac{I_0^3}{J_0^2} = \frac{I^3}{J^2} = \frac{P^3}{Q},$$

when $I^3 = 27 J^2$, we have necessarily $P^3 = 27 Q$; but the first relation holds when F has a square factor, which necessitates U and V having a common factor ; whence we infer that $P^3 - 27 Q$ is the resultant of the cubics U and V.

Again, if $Q = 0$, this transformation of U and V fails, for the values of X and Y become identical. It is, moreover, easy to show that if κ can be determined so that $U + \kappa V$ be a perfect cube, $Q = 0$, for in this case the derived functions with regard to x and y become identical ; whence we have the equations

$$\frac{a + \kappa a'}{b + \kappa b'} = \frac{b + \kappa b'}{c + \kappa c'} = \frac{c + \kappa c'}{d + \kappa d'}.$$

Equating these fractions separately to $-\kappa'$, we find the equations

$$a + \kappa a' + \kappa' b + \kappa\kappa' b' = 0,$$
$$b + \kappa b' + \kappa' c + \kappa\kappa' c' = 0,$$
$$c + \kappa c' + \kappa' d + \kappa\kappa' d' = 0 ;$$

and solving for κ, κ', $\kappa\kappa'$, we may eliminate them, and thus find the condition that $U + \kappa V$ be a perfect cube in the form

$$\begin{vmatrix} a & b & c \\ a' & b' & c' \\ b' & c' & d' \end{vmatrix} \begin{vmatrix} b' & c' & d' \\ a & b & c \\ b & c & d \end{vmatrix} - \begin{vmatrix} a' & b' & c' \\ a & b & c \\ b & c & d \end{vmatrix} \begin{vmatrix} b & c & d \\ a' & b' & c' \\ b' & c' & d' \end{vmatrix} = 0.$$

Or, eliminating κ and κ^2 without introducing κ', we have from the above equations another form for Q, i. e.,

$$\begin{vmatrix} ac - b^2 & ac' + a'c - 2bb' & a'c' - b'^2 \\ ad - bc & ad' + a'd - bd' - b'd & a'd' - b'c' \\ bd - c^2 & bd' + b'd - 2cc' & b'd' - c'^2 \end{vmatrix}.$$

17. Prove that the quartic

$$f(x, y) \equiv (a, b, c, d, e)(x, y)^4$$

may be reduced by the linear transformation

$$x = \lambda X + \mu Y, \quad y = \lambda' X + \mu' Y$$

to the form

$$f(\lambda, \lambda') X^4 + f(\mu, \mu') Y^4 + 6\rho M^2 X^2 Y^2,$$

where

$$4\rho^3 - I\rho + J = 0, \quad M \equiv \lambda\mu' - \lambda'\mu.$$

Dr. Sylvester.

18. Prove that the common factors of two quantics are double factors of their Jacobian $J(U, V)$, when the quantics are of the same degree.

19. Prove that the $2(n-1)$ double factors of $lU + mV$, obtained by varying l and m, are the factors of $J(U, V)$, where U and V are both of the nth degree.

20. Find the resultant of two cubics U and V by eliminating between

$$U = 0, \quad V = 0, \quad \frac{dJ(U, V)}{dx} = 0, \quad \frac{dJ(U, V)}{dy} = 0.$$

21. Prove that every double factor of U is a double factor of its Hessian

$$\frac{d^2 U}{dx^2} \frac{d^2 U}{dy^2} - \left(\frac{d^2 U}{dx\,dy}\right)^2.$$

CHAPTER XV.

COVARIANTS AND INVARIANTS OF THE QUADRATIC, CUBIC,

AND QUARTIC.

158. The Quadratic.—*The quadratic has only one inva-riant, and no covariant other than the quadratic itself.*

For, if a and β be the roots of the quadratic

$$U = ax^2 + 2bx + c = 0,$$

the only functions of their difference which can lead to an inva-riant or covariant are powers of $a - \beta$ of the type $(a - \beta)^{2p}$; the odd powers of $a - \beta$ not being expressible by the coefficients in a rational form.   Whence, expressing

$$U^{2p}\left(\frac{1}{a - x} - \frac{1}{\beta - x}\right)^{2p}$$

by the coefficients, we conclude that the quadratic has only the one distinct invariant $ac - b^2$, and no covariant distinct from U itself.

159. The Cubic and its Covariants.—In the present Article the covariants of the cubic will be discussed as examples of the principles already explained, and in the following Article the definite number of covariants and invariants will be deter-mined.

In the case of the cubic, a covariant is obtained from a function of the differences of the roots most simply by sub-stituting

$$\beta\gamma + ax, \quad \gamma a + \beta x, \quad a\beta + \gamma x \text{ for } -a, \ -\beta, \ -\gamma,$$

and thus avoiding fractions; for, transforming $a - \beta$, we have

$$\frac{1}{a - x} - \frac{1}{\beta - x} = \frac{-(\beta\gamma + ax) + (\gamma a + \beta x)}{(x - a)(x - \beta)(x - \gamma)};$$

z 2

and when fractions are removed we arrive at the above transformation (the order being equal to the weight in the case of either function of the differences H or G). This mode of transforming functions of the differences will now be applied to the covariants of the cubic.

(1). *The Quadratic Covariant, or Hessian, H_x.*

Transforming both sides of the equation

$$a_0^2 \left(a + \omega\beta + \omega^2\gamma\right) \left(a + \omega^2\beta + \omega\gamma\right) = 9 \left(a_1^2 - a_0 a_2\right),$$

we have

$$a_0^2 \left\{\left(a + \omega\beta + \omega^2\gamma\right) x + \beta\gamma + \omega\gamma a + \omega^2 a\beta\right\}$$

$$\times \left\{\left(a + \omega^2\beta + \omega\gamma\right) x + \beta\gamma + \omega^2\gamma a + \omega a\beta\right\} = 9 \left(U_2^2 - U_3 U_1\right);$$

thus showing that

$$Lx + L_1 \text{ and } Mx + M_1 \qquad \text{(see Art. 59)}$$

are the factors of

$$H_x \equiv \left(a_0 a_2 - a_1^2\right) x^2 + \left(a_0 a_3 - a_1 a_2\right) x + \left(a_1 a_3 - a_2^2\right),$$

where

$$L_1 = \beta\gamma + \omega\gamma a + \omega^2 a\beta, \quad M_1 = \beta\gamma + \omega^2\gamma a + \omega a\beta.$$

From the form of the Hessian in terms of the roots in Art. 147, or from the relations of Art. 42, we conclude that *when a cubic is a perfect cube, each of the coefficients of the Hessian vanishes identically.*

(2). *The Cubic Covariant, G_x.*

We have, by Ex. 4, p. 111,

$$a_0^3 \left\{\left(a + \omega\beta + \omega^2\gamma\right)^3 + \left(a + \omega^2\beta + \omega\gamma\right)^3\right\} = -27 \left(a_0^2 a_3 + 2 a_1^3 - 3 a_0 a_1 a_2\right).$$

Transforming both sides of this equation as before, we find

$$a_0^3 \left\{\left(Lx + L_1\right)^3 + \left(Mx + M_1\right)^3\right\} = -27 \left(U^2 U_0 + 2 U_2^3 - 3 U_1 U_2 U\right)$$
$$= -27 G_x,$$

where G_x denotes the covariant formed from the function of differences G; and by the method of Art. 149, by means of the source $a_3^2 a_0 - 3 a_3 a_2 a_1 + 2 a_2^3$, we easily obtain

$$G_x = \left(a_0^2 a_3 - 3 a_0 a_1 a_2 + 2 a_1^3\right) x^3 + 3 \left(a_0 a_1 a_3 + a_1^2 a_2 - 2 a_0 a_2^2\right) x^2$$
$$- \left(a_3^2 a_0 - 3 a_3 a_2 a_1 + 2 a_2^3\right) \quad - 3 \left(a_3 a_2 a_0 + a_2^2 a_1 - 2 a_3 a_1^2\right) x.$$

Resolving $(Lx + L_1)^3 + (Mx + M_1)^3$, we may obtain the factors of G_x; or, more simply, since the factors of G are $\beta + \gamma - 2a$, $\gamma + a - 2\beta$, $a + \beta - 2\gamma$, the factors of G_x are

$$\frac{1}{\beta - x} + \frac{1}{\gamma - x} - \frac{2}{a - x}, \quad \frac{1}{\gamma - x} + \frac{1}{a - x} - \frac{2}{\beta - x}, \quad \frac{1}{a - x} + \frac{1}{\beta - x} - \frac{2}{\gamma - x},$$

when fractions are removed.

We have obviously the following geometrical interpretation of the equation $G_x = 0$:—If we take three points on a line A, B, C determined by $U = 0$; and three points A', B', C', such that A' is the harmonic conjugate of A with regard to B and C; B' of B with regard to C and A; and C' of C with regard to A and B; the points A', B', C' are determined by $G_x = 0$. (Compare Ex. 13, p. 86, and Ex. Art. 65.)

(3). *Expression of the Cubic as the difference of two cubes.*

This can be effected, by means of the factors of the Hessian, as follows:—

$$(Lx + L_1)^3 - (Mx + M_1)^3 = 27\, U \frac{\sqrt{\Delta}}{a_0^{\,3}}.$$

For, from Ex. 2, p. 111, we have

$$L^3 - M^3 = \sqrt{-27}\,(\beta - \gamma)\,(\gamma - a)\,(a - \beta).$$

Transforming this equation as before, the first side becomes

$$(Lx + L_1)^3 - (Mx + M_1)^3,$$

and the second side

$$\sqrt{-27}\,(\beta - \gamma)\,(\gamma - a)\,(a - \beta)\,(x - a)\,(x - \beta)\,(x - \gamma).$$

Substituting from previous equations, we have

$$(Lx + L_1)^3 - (Mx + M_1)^3 = 27\,\frac{U}{a_0^{\,4}}\sqrt{G^2 + 4H^3} = 27\,\frac{U\sqrt{\Delta}}{a_0^{\,3}}.$$

(Compare Art. 58.)

(4). *Relation between the Cubic and its Covariants.*

The following relation exists:—

$$G_x^{\,2} + 4H_x^{\,3} = \Delta\,U^2.$$

For, from Ex. 2, p. 111,

$$a_0^6 (\beta - \gamma)^2 (\gamma - a)^2 (a - \beta)^2 = - 27 (G^2 + 4H^3) = - 27 a_0^2 \Delta,$$

and transforming this equation as before,

$$a_0^6 (\beta - \gamma)^2 (\gamma - a)^2 (a - \beta)^2 (x - a)^2 (x - \beta)^2 (x - \gamma)^2 = - 27 (G_x^2 + 4 H_x^3);$$

whence $\qquad\qquad\qquad \Delta U^2 = G_x^2 + 4 H_x^3.$

(5). *Solution of the Cubic.*

The expression

$$(U \sqrt{\Delta} + G_x)^{\frac{1}{3}} + (U \sqrt{\Delta} - G_x)^{\frac{1}{3}}$$

is a linear factor of U.

For, from Ex. 2 and Ex. 3, we have

$$a_0^3 (Lx + L_1)^3 = 27 (U \sqrt{\Delta} - G_x),$$

$$- a_0^3 (Mx + M_1)^3 = 27 (U \sqrt{\Delta} + G_x);$$

and since

$$(Lx + L_1) - (Mx + M_1)$$

is a factor of U, the proposition follows.

This form of solution of the cubic is due to Prof. Cayley.

160. **Number of Covariants and Invariants of the Cubic.**—The following method of determining the number of covariants and invariants of the cubic is virtually equivalent to that proposed by Professor Cayley :—

The cubic has only two covariants, their leading terms being H *and* G ; *and only one invariant, viz., its discriminant* Δ, *where*

$$a^2 \Delta = G^2 + 4H^3, \text{ or } \Delta = a^2 d^2 + 4 ac^3 - 6 abcd + 4 d b^3 - 3 b^2 c^2.$$

To prove this, let $\phi (a, \beta, \gamma)$ be any integral function of the *differences* of the roots (of order ϖ), expressible by the coefficients in a rational form.

We have then (see *Remark*, Art. 37),

$$a^r \phi (a, \beta, \gamma) = F (a, H, G) \qquad\qquad (1)$$

(where r remains to be determined) ; and, in the first place, if ϕ be an even function of the roots, G can enter this equation in even powers only, since H is an even function of the roots.

Now, eliminating the even powers of G by means of the relation

$$G^2 + 4H^3 = a^2\Delta,$$

equation (1) takes the form

$$a^r\phi(a,\ \beta,\ \gamma) = F(a,\ H,\ \Delta)\ ;$$

and, therefore,

$$a^\varpi\phi(a,\ \beta,\ \gamma) = F_0(a,\ H,\ \Delta) + \Sigma\frac{F_p(H,\ \Delta)}{a^p}, \qquad (2)$$

where ϖ is the order of $\phi(a,\ \beta,\ \gamma)$, and F_0 an integral function.

It is now necessary to prove the following Lemma :—

No function of H and Δ exists that is divisible by a.

For, suppose $F_p(H,\ \Delta)$ to be divisible by a ; then making a vanish, we have

$$F_p(H',\ \Delta') = 0,$$

where $H' = -b^2$, $\Delta' = 4db^3 - 3b^2c^2$, the values of H and Δ when a vanishes.   This equation is plainly impossible ; for, eliminating b by means of the equation $H' = -b^2$, c and d remain in the equation connecting H' and Δ'.

Wherefore equation (2) must assume the form

$$a^\varpi\phi(a,\ \beta,\ \gamma) = F_0(a,\ H,\ \Delta)\ ;$$

for the first side of the equation is expressible as an integral function of the coefficients; therefore so must the second side also, and consequently the fractional part disappears.

Now, to extend this result to odd functions of the roots, we have only to multiply the first side of the equation by

$$a^3(2a - \beta - \gamma)(2\beta - \gamma - a)(2\gamma - a - \beta),$$

and the second side by $27G$, for G must be a factor of every odd function, since H is even.

We are now in a position to prove the original proposition as to the number of invariants and covariants.   For since $a^\varpi\phi$ is of the form

$$GF(a,\ H,\ \Delta),\ \text{ or }\ F(a,\ H,\ \Delta),$$

according as ϕ is an odd or even function of the roots, it follows in the first place that there cannot be an invariant of an odd degree in the roots, since $GF(a, H, \Delta)$ does not remain the same function when a, b, c, d are changed into d, c, b, a, respectively; and the only invariant of an even degree must be a power of Δ, since if $F(a, H, \Delta)$ contained a or H besides Δ, it could not remain the same function when the coefficients are similarly interchanged.

Again, the cubic has only two distinct covariants; for it has been proved that every function of the differences $a^x\phi$ is of one of the forms

$$F(a, H, \Delta), \quad \text{or} \quad GF(a, H, \Delta).$$

Now, considering these forms as the leading terms of covariants, every covariant must be expressible as

$$F(U, H_x, \Delta), \quad \text{or} \quad G_x F(U, H_x, \Delta);$$

that is, every covariant is expressible in a rational and integral form in terms of H_x and G_x, along with U and Δ; or in other words, there are only two distinct covariants.

161. The Quartic. Its Covariants and Invariants.— We have shown already that the quartic has two invariants, I and J (see Art. 147). From the functions H and G of the differences of the roots we can derive two covariants H_x and G_x, whose leading coefficients are H and G; for from the relation

$$a_0^2 \, \Sigma (a - \beta)^2 = 48 \, (a_0 a_2 - a_1^2)$$

we derive, by the process of Art. 147,

$$a_0^2 \, \Sigma (a - \beta)^2 (x - \gamma)^2 (x - \delta)^2 = 48 \, (U_0 U_2 - U_1^2);$$

and, expanding $U_0 U_2 - U_3^2$, we have

$$H_x = (a_0 a_2 - a_1^2) x^4 + 2 \, (a_0 a_3 - a_1 a_2) x^3 + (a_0 a_4 + 2a_1 a_3 - 3a_2^2) x^2$$
$$+ 2 \, (a_1 a_4 - a_2 a_3) x + (a_2 a_4 - a_3^2).$$

In a similar manner, since

$$G = a_0^2 \, a_3 + 2a_1^3 - 3a_0 a_1 a_2,$$

we obtain the covariant

$$G_x \equiv U^2 U_1 + 2 U_3{}^3 - 3 U U_3 U_2,$$

which reduces to the sixth degree; and if it be written as follows :—

$$G_x \equiv A_0 x^6 + A_1 x^5 + A_2 x^4 + A_3 x^3 + A_4 x^2 + A_5 x + A_6,$$

we find, by expanding the above, or more simply, by forming the source A_6, and performing the successive operations of Art. 149, the following values of the coefficients :—

$$A_6 = a_4{}^2 a_1 - 3a_4 a_3 a_2 + 2a_3{}^3, \quad A_5 = a_4{}^2 a_0 + 2a_4 a_3 a_1 + 6a_3{}^2 a_2 - 9a_4 a_2{}^2,$$

$$A_4 = 5a_4 a_3 a_0 + 10 a_3{}^2 a_1 - 15 a_4 a_2 a_1, \quad A_3 = 10 a_0 a_3{}^2 - 10 a_1{}^2 a_4,$$

$$A_2 = -5a_0 a_1 a_4 - 10a_1{}^2 a_3 + 15a_0 a_2 a_3, \quad A_1 = -a_0{}^2 a_4 - 2a_0 a_1 a_3 - 6a_1{}^2 a_2 + 9a_0 a_2{}^2,$$

$$A_0 = -a_0{}^2 a_3 + 3a_0 a_1 a_2 - 2a_1{}^3.$$

Here it will be observed that, after A_3 is determined, A_2, A_1, and A_0 may be obtained from A_4, A_5, and A_6 by changing the suffixes into their complementary values, and altering the sign of the whole, in accordance with what was proved in Art. 148.

We proceed in the following Articles to discuss the leading properties of these two covariants of the quartic.

162. **Quadratic Factors of the Sextic Covariant**,[*] G_x.—As the quadratic factors of G_x enter prominently into the following discussion, we proceed in the first place to find those factors expressed in terms of the roots of the quartic, and to deduce their principal properties.

Since the factors of G, expressed in terms of a, β, γ, δ, are

$$\beta + \gamma - a - \delta, \quad \gamma + a - \beta - \delta, \quad a + \beta - \gamma - \delta,$$

the factors of G_x are obtained from these by substituting $\dfrac{1}{x-a}$, $\dfrac{1}{x-\beta}$, $\dfrac{1}{x-\gamma}$, $\dfrac{1}{x-\delta}$, for a, β, γ, δ, respectively, and multiplying each factor by $\dfrac{U}{a}$ to remove fractions.

---

[*] See a Paper by Prof. Ball, *Quarterly Journal of Mathematics*, vol. vii. p. 368.

Whence, denoting these factors by u, v, w, we have

$$au = U\left(\frac{1}{x-\beta} + \frac{1}{x-\gamma} - \frac{1}{x-a} - \frac{1}{x-\delta}\right),$$

$$av = U\left(\frac{1}{x-\gamma} + \frac{1}{x-a} - \frac{1}{x-\beta} - \frac{1}{x-\delta}\right), \qquad (1)$$

$$aw = U\left(\frac{1}{x-a} + \frac{1}{x-\beta} - \frac{1}{x-\gamma} - \frac{1}{x-\delta}\right),$$

which values of u, v, w, arranged in powers of x, are

$$u = (\beta + \gamma - a - \delta)\, x^2 - 2(\beta\gamma - a\delta)\, x + \beta\gamma\, (a + \delta) - a\delta\, (\beta + \gamma),$$

$$v = (\gamma + a - \beta - \delta)\, x^2 - 2\, (\gamma a - \beta\delta)\, x + \gamma a\, (\beta + \delta) - \beta\delta\, (\gamma + a), \qquad (2)$$

$$w = (a + \beta - \gamma - \delta)\, x^2 - 2\, (a\beta - \gamma\delta)\, x + a\beta\, (\gamma + \delta) - \gamma\delta\, (a + \beta)\,;$$

and, consequently, $32\, G_x = a^3\, uvw.$

From formulas (1) we easily find

$$v = (a - \delta)(x - \beta)(x - \gamma) - (\beta - \gamma)(x - a)(x - \delta),$$

$$w = (a - \delta)(x - \beta)(x - \gamma) + (\beta - \gamma)(x - a)(x - \delta)\,;$$

and from these and similar equations we have

$$\frac{v^2 - w^2}{\mu - \nu} = \frac{w^2 - u^2}{\nu - \lambda} = \frac{u^2 - v^2}{\lambda - \mu} = 4\frac{U}{a}, \qquad (3)$$

and, consequently,

$$(\mu - \nu)\, u^2 + (\nu - \lambda)\, v^2 + (\lambda - \mu)\, w^2 = 0\,;$$

whence

$$-(\mu - \nu)\, u^2 = (w\sqrt{\lambda - \mu} + v\sqrt{\lambda - \nu})(w\sqrt{\lambda - \mu} - v\sqrt{\lambda - \nu}).$$

Since, as this equation shows, the factors on the second side are both perfect squares, we may assume

$$w\sqrt{\lambda - \mu} + v\sqrt{\lambda - \nu} = 2\, u_1{}^2,$$

$$w\sqrt{\lambda - \mu} - v\sqrt{\lambda - \nu} = 2\, u_2{}^2\,;$$

we have, therefore,

$$w \sqrt{\lambda - \mu} = u_1^2 + u_2^2,$$

$$v \sqrt{\lambda - \nu} = u_1^2 - u_2^2,$$

$$u \sqrt{\nu - \mu} = 2 u_1 u_2 ;$$

from which values we conclude that u, v, w, *the quadratic factors of G_x, are mutually harmonic.*

163. **Expression of the Hessian by the Quadratic Factors of G_x.**—Since

$$- 48 \frac{H_x}{a^2} = \Sigma (a - \beta)^2 (x - \gamma)^2 (x - \delta)^2 ;$$

combining the terms in pairs, and noticing that

$$\Sigma (\beta - \gamma) (a - \delta) U = 0,$$

$$\Sigma (a - \beta)^2 (x - \gamma)^2 (x - \delta)^2$$
$$= \Sigma \{ (\beta - \gamma) (x - a) (x - \delta) + (a - \delta) (x - \beta) (x - \gamma) \}^2,$$

the quantities between brackets being u, v, w, we have

$$- 48 \frac{H_x}{a^2} = u^2 + v^2 + w^2,$$

which is the required expression for H_x.

164. **Expression of the Quartic itself by the Quadratic Factors of G_x.**—From equations (3) a symmetrical value may be obtained for U; for, substituting in those equations in place of λ, μ, ν their values in terms of the roots ρ_1, ρ_2, ρ_3 of the equation $4\rho^3 - I\rho + J = 0$, we find

$$a^2 (v^2 - w^2) = 16 (\rho_2 - \rho_3) U, \quad a^2 (w^2 - u^2) = 16 (\rho_3 - \rho_1) U,$$
$$a^2 (u^2 - v^2) = 16 (\rho_1 - \rho_2) U,$$

from which equations, by means of the value of H_x in the preceding Article, we obtain

$$(au)^2 = 16 (\rho_1 U - H_x), \quad (av)^2 = 16 (\rho_2 U - H_x), \qquad (4)$$
$$(aw)^2 = 16 (\rho_3 U - H_x).$$

We now make the assumption

$$u^2 = \Delta_1 X^2, \quad v^2 = \Delta_2 Y^2, \quad w^2 = \Delta_3 Z^2,$$

where

$$\Delta_1 = -(\lambda - \mu)(\lambda - \nu), \quad \Delta_2 = -(\mu - \nu)(\mu - \lambda), \quad \Delta_3 = -(\nu - \lambda)(\nu - \mu).$$

The object of this transformation is to replace u, v, w by three quadratics whose discriminants are each equal to unity. The quadratics X, Y, Z determined by the above equations are of this nature, for if Δ_1 be the discriminant of u, we have

$$\Delta_1 = (\beta + \gamma - a - \delta)\{\beta\gamma(a + \delta) - \gamma a(\beta + \delta)\} - (\beta\gamma - a\delta)^2,$$

which may be put under the form

$$(\beta + \gamma)(a + \delta)(\beta\gamma + a\delta) - (\gamma a + \beta\delta)(a\beta + \gamma\delta) - (\beta\gamma + a\delta)^2,$$

from which we easily find, as above,

$$\Delta_1 = -(\lambda - \mu)(\lambda - \nu), \text{ with similar values for } \Delta_2 \text{ and } \Delta_3.$$

Making these substitutions, the preceding equations become

$$(\rho_1 - \rho_2)(\rho_1 - \rho_3) X^2 = H_x - \rho_1 U, \qquad (5)$$

$$(\rho_2 - \rho_3)(\rho_2 - \rho_1) Y^2 = H_x - \rho_2 U,$$

$$(\rho_3 - \rho_1)(\rho_3 - \rho_2) Z^2 = H_x - \rho_3 U;$$

from which are easily deduced the following values of U and H_x, and the identical equation connecting X, Y, Z :—

$$H_x = \rho_1^2 X^2 + \rho_2^2 Y^2 + \rho_3^2 Z^2,$$

$$- U = \rho_1 X^2 + \rho_2 Y^2 + \rho_3 Z^2,$$

$$0 = X^2 + Y^2 + Z^2;$$

where, as has been proved, X, Y, Z are three mutually harmonic quadratics whose discriminants are reduced to unity in each case. The value of G_x may be expressed in terms of X, Y, Z as follows. Since

$$32\, G_x = a^3\, u\, v\, w,$$

and

$$u^2 v^2 w^2 = (\mu - \nu)^2 (\nu - \lambda)^2 (\lambda - \mu)^2 X^2 Y^2 Z^2 = \frac{256}{a^6}(I^3 - 27 J^2) X^2 Y^2 Z^2,$$

we have

$$G_x = \tfrac{1}{2} \sqrt{I^3 - 27 J^2} \,.\, XYZ.$$

165. Resolution of the Quartic.—From the equations

$$-U = \rho_1 X^2 + \rho_2 Y^2 + \rho_3 Z^2,$$
$$0 = X^2 + Y^2 + Z^2,$$

we find

$$U = (\rho_1 - \rho_2) Y^2 + (\rho_1 - \rho_3) Z^2, \quad U = (\rho_2 - \rho_3) Z^2 + (\rho_2 - \rho_1) X^2,$$
$$U = (\rho_3 - \rho_1) X^2 + (\rho_3 - \rho_2) Y^2.$$

Now, substituting for X^2, Y^2, Z^2 from equations (5), and breaking up these values of U into their factors, we have three ways of resolving U depending on the solution of the equation

$$4\rho^3 - I\rho + J = 0.$$

The resolution of the quartic has been presented differently by Professor Cayley, in a symmetrical form, which may be easily derived from the expressions already given for U and H_x. For, since in general

$$l(a_1 x^2 + 2 b_1 xy + c_1 y^2) + m(a_2 x^2 + 2 b_2 xy + c_2 y^2) + n(a_3 x^2 + 2 b_3 xy + c_3 y^2)$$

is a perfect square, when

$$\Sigma\, l^2 (a_1 c_1 - b_1^2) + \Sigma\, mn (a_2 c_3 + a_3 c_2 - 2 b_2 b_3) = 0,$$

$lX + mY + nZ$ is a perfect square when $l^2 + m^2 + n^2 = 0$,

X, Y, Z being mutually harmonic, and the discriminants of each reduced to unity.

The resolution of U is therefore reduced to finding values of l, m, n such that $lX + mY + nZ$, or

$$l \sqrt{\rho_2 - \rho_3}\, \sqrt{H_x - \rho_1 U} + m \sqrt{\rho_3 - \rho_1}\, \sqrt{H_x - \rho_2 U}$$
$$+ n \sqrt{\rho_1 - \rho_2}\, \sqrt{H_x - \rho_3 U},$$

being a perfect square, may vanish when U vanishes; or in fact to satisfy the two equations

$$l\sqrt{\rho_2 - \rho_3} + m\sqrt{\rho_3 - \rho_1} + n\sqrt{\rho_1 - \rho_2} = 0, \quad l^2 + m^2 + n^2 = 0.$$

These equations are plainly satisfied if

$$\frac{l}{\sqrt{\rho_2-\rho_3}} = \frac{m}{\sqrt{\rho_3-\rho_1}} = \frac{n}{\sqrt{\rho_1-\rho_2}} \, ;$$

whence, finally,

$$(\rho_2-\rho_3)\sqrt{H_x-\rho_1 U} + (\rho_3-\rho_1)\sqrt{H_x-\rho_2 U} + (\rho_1-\rho_2)\sqrt{H_x-\rho_3 U}$$

is the square of a linear factor of the quartic U.

If it be required to resolve the quartic $\kappa U - \lambda H_x$, it appears in a similar manner that

$$l\sqrt{\rho_2-\rho_3}\sqrt{H_x-\rho_1 U} + m\sqrt{\rho_3-\rho_1}\sqrt{H_x-\rho_2 U}$$
$$+ n\sqrt{\rho_1-\rho_2}\sqrt{H_x-\rho_3 U},$$

being a perfect square, must vanish when $\kappa U - \lambda H_x$ vanishes; or, values of l, m, n must be determined so as to satisfy the equations

$$l^2 + m^2 + n^2 = 0,$$

$$l\sqrt{(\rho_2-\rho_3)(\kappa-\rho_1\lambda)} + m\sqrt{(\rho_3-\rho_1)(\kappa-\rho_2\lambda)} + n\sqrt{(\rho_1-\rho_2)(\kappa-\rho_3\lambda)} = 0.$$

These equations are plainly satisfied if

$$\frac{l}{\sqrt{(\rho_1-\rho_3)(\kappa-\rho_1\lambda)}} = \frac{m}{\sqrt{(\rho_3-\rho_1)(\kappa-\rho_2\lambda)}} = \frac{n}{\sqrt{(\rho_1-\rho_2)(\kappa-\rho_3\lambda)}} \, ;$$

whence

$$(\rho_2-\rho_3)\sqrt{\kappa-\rho_1\lambda}\sqrt{H_x-\rho_1 U} + (\rho_3-\rho_1)\sqrt{\kappa-\rho_2\lambda}\sqrt{H_x-\rho_2 U}$$
$$+ (\rho_1-\rho_2)\sqrt{\kappa-\rho_2\lambda}\sqrt{H_x-\rho_3 U}$$

is the square of a linear factor of $\kappa U - \lambda H_x$.

166. **The Invariants and Covariants of** $\kappa U - \lambda H_x$.—From the equations of Art. 164, viz.,

$$\rho_1^2 X^2 + \rho_2^2 Y^2 + \rho_3^2 Z^2 = H_x,$$
$$\rho_1 X^2 + \rho_2 Y^2 + \rho_3 Z^2 = -U,$$
$$X^2 + Y^2 + Z^2 = V = 0,$$

we may, by adding $-\dfrac{\lambda I}{6} V$ to $\lambda H_x - \kappa U$, reduce it to the form

$R_1 X^2 + R_2 Y^2 + R_3 Z^2$, where $R_1 + R_2 + R_3 = 0$. When this is done, we have the following reduced values of R_1, R_2, R_3 :—

$$3R_1 = \kappa\,(2\rho_1 - \rho_2 - \rho_3) + \lambda\,(2\rho_2\rho_3 - \rho_3\rho_1 - \rho_1\rho_2),$$

$$3R_2 = \kappa\,(2\rho_2 - \rho_3 - \rho_1) + \lambda\,(2\rho_3\rho_1 - \rho_1\rho_2 - \rho_2\rho_3),$$

$$3R_3 = \kappa\,(2\rho_3 - \rho_1 - \rho_2) + \lambda\,(2\rho_1\rho_2 - \rho_2\rho_3 - \rho_3\rho_1).$$

On account of the similarity of the forms

$$\rho_1 X^2 + \rho_2 Y^2 + \rho_3 Z^2 \quad\text{and}\quad R_1 X^2 + R_2 Y^2 + R_3 Z^2,$$

which are of a fixed type, we calculate the invariants and covariants of $\kappa U - \lambda H_x$ by simply changing ρ_1, ρ_2, ρ_3 into R_1, R_2, R_3 in the expressions for the invariants and covariants of U.

Therefore, since

$$I = \tfrac{2}{3}\{(\rho_2 - \rho_3)^2 + (\rho_3 - \rho_1)^2 + (\rho_1 - \rho_2)^2\}, \quad J = -\,4\rho_1\rho_2\rho_3,$$

and

$$R_2 - R_3 = (\rho_2 - \rho_3)(\kappa - \lambda\rho_1), \quad R_3 - R_1 = (\rho_3 - \rho_1)(\kappa - \lambda\rho_2),$$

$$R_1 - R_2 = (\rho_1 - \rho_2)(\kappa - \lambda\rho_3),$$

we find the following values for the invariants of $\kappa U - \lambda H_x$:—

$$I_{(\kappa,\,\lambda)} = I\kappa^2 - 3J\kappa\lambda + \frac{I^2}{12}\lambda^2,$$

$$J_{(\kappa,\,\lambda)} = J\kappa^3 - \frac{I^2}{6}\kappa^2\lambda + \frac{IJ}{4}\kappa\lambda^2 - \frac{54J^2 - I^3}{216}.$$

If we form the covariants $H_{(\kappa,\,\lambda)}$, and $G_{(\kappa,\,\lambda)}$, of

$$4\,\Omega = 4\kappa^3 - I\kappa\lambda^2 + J\lambda^3$$

(the reducing cubic rendered homogeneous in κ, λ), we find, as M. Hermite has remarked,

$$I_{(\kappa,\,\lambda)} = -\tfrac{3}{4}H_{(\kappa,\,\lambda)}, \quad J_{(\kappa,\,\lambda)} = \tfrac{1}{16}G_{(\kappa,\,\lambda)}.$$

Again, to calculate the Hessian of $\kappa U - \lambda H_x$, we reduce

$$R_1^2 X^2 + R_2^2 Y^2 + R_3^2 Z^2$$

by the substitutions

$$\rho_1^3 X^2 + \rho_2^3 Y^2 + \rho_3^3 Z^2 \equiv -\tfrac{1}{4}(IU + JV) \equiv -\tfrac{1}{4}IU,$$

$$\rho_1^4 X^2 + \rho_2^4 Y^2 + \rho_3^4 Z^2 \equiv \tfrac{1}{4}(IH_x + JU),$$

the first of which follows from the equations

$$\rho_1^2 = \rho_2\rho_3 + \tfrac{1}{4}I, \quad \rho_2^2 = \rho_3\rho_1 + \tfrac{1}{4}I, \quad \rho_3^2 = \rho_1\rho_2 + \tfrac{1}{4}I,$$

multiplying by $\rho_1 X^2$, $\rho_2 Y^2$, $\rho_3 Z^2$, respectively; and the second from the first by changing X^2, Y^2, Z^2 into $\rho_1 X^2$, $\rho_2 Y^2$, $\rho_3 Z^2$.

In this way we find the following form for the Hessian of $\kappa U - \lambda H_x$:—

$$H_x\left(4\kappa^2 - \frac{I}{3}\lambda^2\right) + U\left(\frac{2}{3}I\kappa\lambda - J\lambda^2\right);$$

which may be expressed in the form

$$\frac{4}{3}\left(H_x\frac{d\Omega}{d\kappa} - U\frac{d\Omega}{d\lambda}\right).$$

Again, since

$$I^3 - 27J^2 = 16(\rho_2 - \rho_3)^2(\rho_3 - \rho_1)^2(\rho_1 - \rho_2)^2,$$

and

$$G_\omega = \tfrac{1}{2}\sqrt{I^3 - 27J^2}\,.\,XYZ.$$

Transforming ρ_1, ρ_2, ρ_3 into R_1, R_2, R_3, we find

$$I^3_{(\kappa,\lambda)} - 27J^2_{(\kappa,\lambda)} = \Omega^2(I^3 - 27J^2),$$

$$G_{(\kappa,\lambda)x} = \Omega G_x.$$

We have therefore expressed the invariants and covariants of $\kappa U - \lambda H_x$ in terms of the invariants and covariants of U.

167. **Number of Covariants and Invariants of the Quartic.**—We proceed to prove the following proposition, which determines the number of these functions :—

The quartic has only the two distinct invariants I and J, and two distinct covariants whose leading coefficients are H and G. This proposition asserts that every invariant is a *rational* and *integral* function of I and J, and every covariant a *rational* and *integral* function of U, H_x, G_x, I, J. The following discussion, like that before given in the case of the cubic, is the same in principle as that proposed by Prof. Cayley.

Attending to the observations in Arts. 37, 38, it is plain that if $\phi(a, \beta, \gamma, \delta)$ be any integral function of the differences of the roots expressible by the coefficients in a rational form, we have, in general, considering the equation with the second term removed,

$$a^r \phi(a, \beta, \gamma, \delta) = F(a, H, I, G),$$

where F is a rational and integral function, and r remains to be determined.

And if, in the first place, ϕ be an odd function of the roots; changing their signs, and subtracting the two values of ϕ, we find

$$2a^r \phi(a, \beta, \gamma, \delta) = F(a, H, I, G) - F(a, H, I, -G).$$

This value of ϕ plainly vanishes with G; whence, eliminating the powers of G beyond the first by the identical equation

$$- G^2 = 4H^3 - a^2 HI + a^3 J,$$

we have

$$a^r \phi(a, \beta, \gamma, \delta) = GF_1(a, H, I, J).$$

It follows that every odd function ϕ of the differences of the roots is divisible by

$$(\beta + \gamma - a - \delta)(\gamma + a - \beta - \delta)(a + \beta - \gamma - \delta);$$

and removing this factor on the first side of the equation, and $32\dfrac{G}{a^3}$ on the second side, we have

$$a^{r-3} \phi_1(a, \beta, \gamma, \delta) = F_1(a, H, I, J),$$

where ϕ_1 is an even function of the roots, and F_1 a rational and integral function.

We proceed to prove, in the second place, if $\phi(a, \beta, \gamma, \delta)$ be any even integral function of the differences of the roots, of the order ϖ, expressible by the coefficients in a rational form, that $a^\varpi \phi(a, \beta, \gamma, \delta)$ can be expressed as a rational and *integral* function of a, H, I, J.

2 A

To prove this, the following lemma is necessary :—

There exists no function of H, I, J which is divisible by a. For, suppose $F(H, I, J)$ to be divisible by a. Making a vanish, we have $F(H', I', J') = 0$, where $H' = -b^2$, $I' = -4bd + 3c^2$, $J' = 2bcd - eb^2 - c^3$ (the values of H, I, J, when $a = 0$); and as it is impossible to eliminate b, c, d, e, so as to obtain a relation between H', I', J', we conclude that no relation such as $F(H', I', J') = 0$ exists ; and therefore there is no function of the form $F(H, I, J)$ which is divisible by a.

We now proceed with the proof of the proposition ; and since, as has been already proved in the case of an even function of the roots,

$$a^\varpi \phi (a, \beta, \gamma, \delta) = F(a, H, I, J),$$

we have, dividing by $a^{\nu - \varpi}$,

$$a^\varpi \phi (a, \beta, \gamma, \delta) = F_0 (a, H, I, J) + \Sigma \frac{F_p (H, I, J)}{a^p}.$$

Now, since the first side of this equation is expressible as a rational and integral function of the coefficients not divisible by a, the second side must be a similar function of the coefficients ; and this, by the lemma just established, is impossible unless such terms as $\Sigma \dfrac{F_p (H, I, J)}{a^p}$ disappear.

Wherefore

$$a^\varpi \phi (a, \beta, \gamma, \delta) = F_0 (a, H, I, J) ;$$

and, finally, we have proved that $a^\varpi \phi (a, \beta, \gamma, \delta)$ may be expressed by the forms

$$GF (a, H, I, J), \quad \text{or} \quad F(a, H, I, J),$$

according as ϕ is odd or even.

We are now in a position to prove the original proposition as to the number of invariants and covariants. For, if $F(a, H, I, J)$ be an invariant, a and H must disappear, since if they were present this function could not remain the same when the coefficients are written in direct or reverse order. Similarly, no odd function such as $GF(a, H, I, J)$ can give an invariant. It follows that every invariant is a function of I and J.

Again, the quartic has only two distinct covariants; for we have proved that every function of the differences $a^w\phi$ is of one of the forms

$$F(a, H, I, J) \quad \text{or} \quad GF(a, H, I, J).$$

Now, considering these forms as the leading terms of covariants, it has been proved that every covariant is expressible as

$$F(U, H_x, I, J) \quad \text{or} \quad G_xF(U, H_x, I, J);$$

that is, every covariant is expressible in terms of H_x and G_x, along with U, I, and J; and this is the proposition which was required to be proved.

MISCELLANEOUS EXAMPLES.

1. If U be any cubic, and G_x its cubic covariant, prove that the Hessian of $\lambda U + \mu G_x$ has the same roots as the Hessian of U, λ and μ being constants.

2. If $\alpha_1, \beta_1, \gamma_1$ be the roots of $G_x = 0$, prove that

$$\left(\frac{d}{d\alpha_1} + \frac{d}{d\beta_1} + \frac{d}{d\gamma_1}\right)\phi_1(\alpha_1, \beta_1, \gamma_1) = \left(\frac{d}{d\alpha} + \frac{d}{d\beta} + \frac{d}{d\gamma}\right)\phi(\alpha, \beta, \gamma),$$

where

$$\phi(\alpha, \beta, \gamma) = \phi_1(\alpha_1, \beta_1, \gamma_1);$$

and also that

$$\delta\alpha_1 = \delta\beta_1 = \delta\gamma_1 = -1.$$

3. Given

$$U \equiv (a, b, c, d)(x, y)^3, \quad \text{and} \quad V \equiv (a', b', c', d')(x, y)^3,$$

find the relation which connects the coefficients of these cubics when it is possible to determine the ratio $\lambda : \mu$, so that

$$\lambda \dot{U} + \mu V$$

should be a perfect cube.

In this case the Hessian of $\lambda U + \mu V$ must vanish identically; and writing it under the two forms

$$\lambda^2 H_x + \lambda\mu K_x + \mu^2 H_x' \equiv Lx^2 + Mxy + Ny^2,$$

where

$$K_x \equiv (ac' + a'c - 2bb')x^2 + (ad' + a'd - bc' - b'c)xy + (bd' + b'd - 2cc')y^2,$$

we have

$$L = 0, \quad M = 0, \quad N = 0;$$

and eliminating λ^2, $\lambda\mu$, μ^2 from these equations, the condition is obtained in the following form:—

$$\begin{vmatrix} ac - b^2 & ad - bc & bd - c^2 \\ ac' + a'c - 2bb' & ad' + a'd - bc' - b'c & bd' + b'd - 2cc' \\ a'c' - b'^2 & a'd' - b'c' & b'd' - c'^2 \end{vmatrix} = 0.$$

2 A 2

4. If a quartic have a square factor, prove that the same square factor enters its Hessian; and determine the relations between the coefficients when the quartic and its Hessian are identical.

5. Prove that the sextic covariant G_x of the quartic $\phi(x)$ may be written under the form

$$\Sigma \, \phi'(a) \, \frac{\{\phi(x)\}^2}{(x-a)^2} \, ;$$

and that, when the quartic has a *double* factor, the covariant G_x has that factor as a *quadruple* factor.

6. Find the value of the determinant

$$\Delta \equiv \begin{vmatrix} \beta+\gamma-\alpha-\delta & \beta\gamma-\alpha\delta & \beta\gamma\,(\alpha+\delta)-\alpha\delta\,(\beta+\gamma) \\ \gamma+\alpha-\beta-\delta & \gamma\alpha-\beta\delta & \gamma\alpha\,(\beta+\delta)-\beta\delta\,(\gamma+\alpha) \\ \alpha+\beta-\gamma-\delta & \alpha\beta-\gamma\delta & \alpha\beta\,(\gamma+\delta)-\gamma\delta\,(\alpha+\beta) \end{vmatrix},$$

whose constituents are the coefficients in the quadratic factors of the sextic covariant of the quartic (see Art. 162).

If these constituents be represented by a_1, b_1, c_1, a_2, b_2, c_2, &c., and the inverse constituents by A_1, B_1, C_1, &c., we find

$$\begin{vmatrix} A_1 & B_1 & C_1 \\ A_2 & B_2 & C_2 \\ A_3 & B_3 & C_3 \end{vmatrix} = \begin{vmatrix} -Ac_1 & 2Ab_1 & -Aa_1 \\ -Bc_2 & 2Bb_2 & -Ba_2 \\ -Cc_3 & 2Cb_3 & -Ca_3 \end{vmatrix},$$

where

$$A = (\beta-\gamma)\,(\alpha-\delta), \quad B = (\gamma-\alpha)\,(\beta-\delta), \quad C = (\alpha-\beta)\,(\gamma-\delta)\,;$$

hence

$$\Delta^2 = -2\,ABC\Delta,$$

or

$$\Delta = -2ABC = -2\,(\beta-\gamma)\,(\gamma-\alpha)\,(\alpha-\beta)\,(\alpha-\delta)\,(\beta-\delta)\,(\gamma-\delta).$$

CHAPTER XVI.

168. Tschirnhausen's Transformation.—Theorem.—
The most general rational algebraic transformation of a root of an equation of the nth degree can be reduced to an integral transformation of the degree $n - 1$ at most.

For every rational function of a root a_r of the equation $f(x) = 0$ is of the form

$$\frac{\chi(a_r)}{\psi(a_r)},$$

where χ and ψ are integral functions; also

$$\frac{\chi(a_r)}{\psi(a_r)} = \chi(a_r) \frac{\psi(a_1) \ldots \ldots \psi(a_{r-1}) \psi(a_{r+1}) \ldots \ldots \psi(a_n)}{\psi(a_1) \psi(a_2) \ldots \ldots \ldots \ldots \psi(a_{n-1}) \psi(a_n)},$$

and the denominator $\psi(a_1) \psi(a_2) \ldots \psi(a_n)$, being a symmetric function of the roots of $f(x) = 0$, can be expressed as a rational function of the coefficients. Whence $\frac{\chi(a_r)}{\psi(a_r)}$ is reduced to an integral form.

Moreover, the numerator of the former fraction is a symmetric function of the roots of the equation $\dfrac{f(x)}{x - a_r} = 0$, and may consequently be expressed as a rational function of the coefficients of that equation; that is, in terms of a_r and the coefficients of $f(x)$.

Now, denoting by $F(a_r)$ this integral form of $\dfrac{\chi(a_r)}{\psi(a_r)}$, we have by division

$$F(a_r) = Q f(a_r) + \phi(a_r) = \phi(a_r),$$

where $\phi(a_r)$ does not exceed the degree $n - 1$; which proves the proposition.

This transformation, in which, from the equation $f(x) = 0$, a new equation in y is formed by means of the substitution $y = \phi(x)$ (where $\phi(x)$ is a rational integral function of degree inferior to that of $f(x)$), was first employed by *Tschirnhausen*.

169. Formation of the Transformed Equation.—We proceed to explain the method of forming the equation whose roots are

$$\phi(a_1), \ \phi(a_2), \ \phi(a_3), \ \dots \ \phi(a_n),$$

where $\phi(x)$ is a rational and integral function of x of the degree $n - 1$.

Let
$$\phi(x) \equiv a_0 + a_1 x + a_2 x^2 + \dots + a_{n-1} x^{n-1}.$$

Raising successively $\phi(x)$ to the different powers 2, 3, ... n, and reducing the exponents of x in each case below n (by dividing by $f(x)$ and retaining the remainder), we have

$$\phi^2 = b_0 + b_1 x + b_2 x^2 + \dots + b_{n-1} x^{n-1},$$

$$\phi^3 = c_0 + c_1 x + c_2 x^2 + \dots + c_{n-1} x^{n-1},$$

$$\cdot \quad \cdot \quad \cdot \quad \cdot \quad \cdot \quad \cdot \quad \cdot \quad \cdot \quad \cdot$$

$$\phi^n = l_0 + l_1 x + l_2 x^2 + \dots + l_{n-1} x^{n-1}.$$

Substituting for x in these equations each of the roots of the equation $f(x) = 0$, and adding, we find, if S_1, S_2, S_3, &c., denote the sums of the powers of the roots of the required equation,

$$S_1 = na_0 + a_1 s_1 + a_2 s_2 + \dots + a_{n-1} s_{n-1},$$

$$S_2 = nb_0 + b_1 s_1 + b_2 s_2 + \dots + b_{n-1} s_{n-1},$$

$$\cdot \quad \cdot \quad \cdot \quad \cdot \quad \cdot \quad \cdot \quad \cdot \quad \cdot \quad \cdot$$

$$S_n = nl_0 + l_1 s_1 + l_2 s_2 + \dots + l_{n-1} s_{n-1}.$$

Now, expressing s_1, s_2, ... s_{n-1} in terms of the coefficients of $f(x)$, we have S_1, S_2, ... S_n determined in terms of the coefficients of $\phi(x)$ and $f(x)$; we are also enabled by Art. 129 to express the coefficients of the equation whose roots are $\phi(a_1)$, $\phi(a_2)$, ... $\phi(a_n)$ in terms of S_1, S_2, ... S_n, and therefore finally in terms of the coefficients of $\phi(x)$ and $f(x)$; thus the transformation is completed.

170. Second Method of forming the Transformed Equation.—There is another way of finding the final equation in ϕ by elimination, which we now give. Since

$$a_0 - \phi + a_1 x + a_2 x^2 + \ldots + a_{n-1} x^{n-1} = 0,$$

if this equation be multiplied by x, x^2, $\ldots x^{n-1}$, and the exponents of x reduced below n by means of the equation $f(x) = 0$, we have in all n equations to eliminate dialytically the $n - 1$ quantities x, x^2, $\ldots x^{n-1}$. We thus obtain the transformed equation in the form of a determinant of the nth order, ϕ entering into the diagonal constituents only. For example, if $f(x) = x^n - 1$, we obtain the transformed equation in the following form:—

$$\begin{vmatrix} a_0 - \phi & a_1 & a_2 & . & . & a_{n-1} \\ a_{n-1} & a_0 - \phi & a_1 & . & . & a_{n-2} \\ . & . & . & . & . & . \\ . & . & . & . & . & . \\ a_1 & a_2 & a_3 & . & . & a_0 - \phi \end{vmatrix} = 0.$$

Although these methods of performing Tschirnhausen's transformation appear simple, yet if they be applied to particular cases the result usually appears in a complicated form. Professor Cayley, by choosing a form of the transformation suggested by M. Hermite, was enabled to take advantage of the theory of covariants, and thus to complete the transformation for the cubic, quartic, and quintic. We shall content ourselves with showing in an elementary way how Professor Cayley's results for the cubic and quartic may be obtained.

171. Tschirnhausen's Transformation applied to the Cubic.—Let the cubic equation

$$a x^3 + 3 b x^2 + 3 c x + d = 0$$

be written under the form

$$z^3 + 3 H z + G = 0;$$

and let it be transformed by the substitution

$$y = \lambda + \kappa z + z^2.$$

If z_1, z_2, z_3 be the roots of the cubic, and y_1, y_2, y_3 the corresponding values of y, we have

$$y_2 - y_3 = (z_2 - z_3)\,(\kappa - z_1),$$
$$y_3 - y_1 = (z_3 - z_1)\,(\kappa - z_2), \qquad (1)$$
$$y_1 - y_2 = (z_1 - z_2)\,(\kappa - z_3),$$

and, consequently,

$$2y_1 - y_2 - y_3 = (2z_1 - z_2 - z_3)\,\kappa + (2z_2 z_3 - z_3 z_1 - z_1 z_2),$$
$$2y_2 - y_3 - y_1 = (2z_2 - z_3 - z_1)\,\kappa + (2z_3 z_1 - z_1 z_2 - z_2 z_3), \qquad (2)$$
$$2y_3 - y_1 - y_2 = (2z_3 - z_1 - z_2)\,\kappa + (2z_1 z_2 - z_2 z_3 - z_3 z_1).$$

Wherefore, if the equation in y with the second term removed be

$$Y^3 + 3H'Y + G' = 0,$$

we have from equations (1) and (2)

$$H' = H_\kappa, \quad G' = G_\kappa,$$

where H_κ and G_κ are the Hessian and cubic covariant of

$$\kappa^3 + 3H\kappa + G\,;$$

and the transformation is therefore completed, since $y_1 + y_2 + y_3$ can be easily determined.

172. **Tschirnhausen's Transformation applied to the Quartic.**—In this case we do not attempt to form directly the transformed quartic, but prove the following theorem, which shows how this transformation may be resolved into two others.

Theorem.—*Tschirnhausen's transformation changes a quartic U into one having the same invariants as $lU + mH_x$, and therefore in general reducible to the latter form by linear transformation.*

To prove this, let the quartic

$$x^4 + p_1 x^3 + p_2 x^2 + p_3 x + p_4 = 0$$

be transformed by the substitution

$$y = a_0 + a_1 x + a_2 x^2 + a_3 x^3.$$

If x_1, x_2, x_3, x_4 be the roots of the quartic, and y_1, y_2, y_3, y_4 the corresponding values of y, we have

$$\frac{y_2 - y_3}{x_2 - x_3} = a_1 + a_2 (x_2 + x_3) + a_3 (x_2{}^2 + x_2 x_3 + x_3{}^2),$$

$$\frac{y_1 - y_4}{x_1 - x_4} = a_1 + a_2 (x_1 + x_4) + a_3 (x_1{}^2 + x_1 x_4 + x_4{}^2).$$

From these equations we proceed to show that

$$\frac{(y_2 - y_3)(y_1 - y_4)}{(x_2 - x_3)(x_1 - x_4)} = P_0 + Q_0 (x_2 x_3 + x_1 x_4),$$

where P_0 and Q_0 involve the roots of the quartic symmetrically.

In the first place, we find

$$(x_2{}^2 + x_2 x_3 + x_3{}^2)(x_1{}^2 + x_1 x_4 + x_4{}^2) = p_2{}^2 - p_4 p_3 + p_4 - p_2 \lambda,$$

where λ has its usual value, i.e. $x_2 x_3 + x_1 x_4$; and secondly, since

$$x_2{}^2 + x_2 x_3 + x_3{}^2 = (x_2 + x_3)^2 - x_2 x_3, \ \&c.,$$

we find again

$$(x_2 + x_3)(x_1{}^2 + x_1 x_4 + x_4{}^2) + (x_1 + x_4)(x_2{}^2 + x_2 x_3 + x_3{}^2) = p_3 - p_1 p_2 + p_1 \lambda.$$

Finally, since the other terms in the product are obviously of the same form as $P_0 + Q_0 \lambda$, we have proved that

$$\frac{(y_2 - y_3)(y_1 - y_4)}{(x_2 - x_3)(x_1 - x_4)} = P_0 + Q_0 (x_2 x_3 + x_1 x_4) \ ;$$

whence

$$(y_2 - y_3)(y_1 - y_4) = (\nu - \mu)(P_0 + Q_0 \lambda).$$

Now, introducing ρ_1, ρ_2, ρ_3, in place of λ, μ, ν, this and the similar equations preserve their forms; whence, altering P_0 and Q_0 into similar quantities, we obtain the equations

$$(y_2 - y_3)(y_1 - y_4) = 4 (\rho_3 - \rho_2)(P - Q\rho_1),$$

$$(y_3 - y_1)(y_2 - y_4) = 4 (\rho_1 - \rho_3)(P - Q\rho_2),$$

$$(y_1 - y_2)(y_3 - y_4) = 4 (\rho_2 - \rho_1)(P - Q\rho_3),$$

which lead at once to the invariants of the transformed quartic; and comparing their values with the invariants of $\kappa U - \lambda H_x$ given in Art. 166, the theorem follows at once.

173. Reduction of the Cubic to a Binomial form by Tschirnhausen's Transformation.—Let the cubic

$$ax^3 + 3bx^2 + 3cx + d$$

be reduced to the form $\;y^3 - V\;$ by the transformation

$$y = q + px + x^2.$$

If x_1, x_2, x_3 be the roots of the given cubic, and y_1 a root of the transformed cubic, we have the following equations to determine p and q :—

$$x_1{}^2 + px_1 + q = y_1,$$

$$x_2{}^2 + px_2 + q = \omega y_1,$$

$$x_3{}^2 + px_3 + q = \omega^2 y_1;$$

from which we find

$$p = -\,\frac{x_1{}^2 + \omega x_2{}^2 + \omega^2 x_3{}^2}{x_1 + \omega x_2 + \omega^2 x_3},\quad q = -\tfrac{1}{3}(s_2 + ps_1).$$

Adding $x_1 + x_2 + x_3$ to this value of p, we have

$$p + x_1 + x_2 + x_3 = -\,\frac{x_2 x_3 + \omega x_3 x_1 + \omega^2 x_1 x_2}{x_1 + \omega x_2 + \omega^2 x_3};$$

it follows (see Ex. 23, p. 57), that there are only two ways of completing this transformation, as the values of p, q ultimately depend on the solution of the Hessian of the cubic.

174. Tschirnhausen's Transformation applied to Reduce the Quartic to a Trinomial Form, in which the Second and Fourth Terms are absent.—Let the quartic

$$ax^4 + 4bx^3 + 6cx^2 + 4dx + e$$

be reduced to the form $\;y^4 + Py^2 + Q\;$ by the transformation

$$y = q + px + x^2.$$

If $\;x_1$, x_2, x_3, $x_4\;$ be the roots of the quartic; also y_1, y_2 *two distinct* roots of the transformed quartic, we have the following equations to determine p and q :—

$$x_1^2 + px_1 + q = y_1, \qquad x_3^2 + px_3 + q = y_2,$$

$$x_2^2 + px_2 + q = -y_1, \qquad x_4^2 + px_4 + q = -y_2 ;$$

from which we find

$$p = -\frac{x_1^2 + x_2^2 - x_3^2 - x_4^2}{x_1 + x_2 - x_3 - x_4}, \qquad q = -\tfrac{1}{4}\left(s_2 + p\,s_1\right).$$

And, adding $x_1 + x_2 + x_3 + x_4$ to this value of p, we have

$$p + x_1 + x_2 + x_3 + x_4 = \frac{2\left(x_1 x_2 - x_3 x_4\right)}{x_1 + x_2 - x_3 - x_4} ;$$

hence, by Ex. 7, p. 126, it follows that there are three ways of reducing the quartic to the proposed form, the determination of which ultimately depends on the solution of the reducing cubic of the quartic.

175. Removal of the Second, Third, and Fourth Terms from an Equation of the nth Degree.—We commence by proving the following proposition, which we shall subsequently apply :—

A homogeneous function V of the second degree in r quantities $x_1, x_2, x_3, \dots x_n$ can be expressed in general as the sum of n squares.

To prove this, let V, arranged in powers of x_1, take the following form :—

$$V = P_1 x_1^2 + 2Q_1 x_1 + R_1,$$

where P_1 does not contain $x_1, x_2, \dots x_n$; also Q_1 and R_1 are linear and quadratic functions, respectively, of $x_2, x_3, \dots x_n$.

Again, $$V = \left(\sqrt{P_1}\,x_1 + \frac{Q}{\sqrt{P_1}}\right)^2 + R_1 - \frac{Q^2}{P_1} ;$$

also $$V_1 = R_1 - \frac{Q^2}{P_1} = P_2 x_2^2 + 2Q_2 x_2 + R_2,$$

where P_2 is a constant, and Q_2 and R_2 do not contain x_1 and x_2; and similarly

$$V_1 = \left(\sqrt{P_2}\, x_2 + \frac{Q_2}{\sqrt{P_2}}\right)^2 + R_2 - \frac{Q_2^2}{P_2},$$

so that

$$V = \left(\sqrt{P_1}\, x_1 + \frac{Q_1}{\sqrt{P_1}}\right)^2 + \left(\sqrt{P_2}\, x_2 + \frac{Q_2}{\sqrt{P_2}}\right)^2 + R_2 - \frac{Q_2^2}{P_2}.$$

Proceeding in this way we arrive ultimately at $R_{n-1} - \dfrac{Q_{n-1}^2}{P_{n-1}}$, which is equal to $P_n x_n^2$; and the proposition is proved.

It may happen that P_n vanishes, in which case V is reduced to the sum of $n - 1$ squares. (Compare Ex. 25, p. 143.)

Now, returning to the original problem, let the equation be

$$x^n + p_1 x^{n-1} + p_2 x^{n-2} + \ldots + p_n = 0;$$

and, putting

$$y = ax^4 + \beta x^3 + \gamma x^2 + \delta x + \varepsilon,$$

let the transformed equation be

$$y^n + Q_1 y^{n-1} + Q_2 y^{n-2} + \ldots + Q_n = 0,$$

where, by Art. 169, $Q_1, Q_2, \ldots Q_r, \ldots$ are homogeneous functions of the first, second, $\ldots r^{th}$ degrees in $a, \beta, \gamma, \delta, \varepsilon$.

Now, if $a, \beta, \gamma, \delta, \varepsilon$ can be determined so that

$$Q_1 = 0, \quad Q_2 = 0, \quad Q_3 = 0,$$

the problem will be solved. For this purpose, eliminating ε from Q_2 and Q_3, by substituting its value derived from $Q_1 = 0$, we obtain two homogeneous equations,

$$R_2 = 0, \quad R_3 = 0,$$

of the second and third degrees in a, β, γ, δ; and by the proposition proved above we may write R_2 under the form

$$u^2 - v^2 + w^2 - t^2,$$

which is satisfied by putting $u = v$ and $w = t$. From these simple equations we find $\gamma = la + m\beta$, and $\delta = l_1 a + m_1 \beta$; and substituting these values in $Q_3 = 0$, we have a cubic equation to determine the ratio $\beta : \alpha$. Whence, giving any one of the quantities α, β, γ, δ, ε a definite value, the rest are determined, and the equation is reduced to the form

$$y^n + Q_4 y^{n-4} + Q_5 y^{n-5} + \ldots + Q_n = 0.$$

In a similar way we may remove the coefficients Q_1, Q_2, Q_4, by solving an equation of the fourth degree.

Applying this method to the quintic, we may reduce it to either of the trinomial forms*

$$x^5 + Px + Q,$$
$$x^5 + Px^2 + Q;$$

or again, changing x into $\dfrac{1}{x}$, to either of the forms

$$x^5 + Px^3 + Q,$$
$$x^5 + Px^4 + Q.$$

In this investigation we have followed M. Serret (see his *Cours d'Algèbre Supérieure*, vol. i., Art. 192).

176. Reduction of the Quintic to the Sum of Three Fifth Powers.—This reduction can be effected by the solution of an equation of the third degree, as we proceed to show. Let

$$(a_0, a_1, a_2, a_3, a_4, a_5)(x, y)^5 \equiv b_1(x + \beta_1 y)^5 + b_2(x + \beta_2 y)^5 + b_3(x + \beta_3 y)^5,$$

where β_1, β_2, β_3 are the roots of the equation

$$p_3 x^3 + p_2 x^2 + p_1 x + p_0 = 0.$$

Now, comparing coefficients in the two forms of the quintic,

$$a_0 = b_1 + b_2 + b_3 , \quad a_1 = b_1\beta_1 + b_2\beta_2 + b_3\beta_3,$$

$$a_2 = b_1\beta_1^2 + b_2\beta_2^2 + b_3\beta_3^2, \quad a_3 = b_1\beta_1^3 + b_2\beta_2^3 + b_3\beta_3^3,$$

$$a_4 = b_1\beta_1^4 + b_2\beta_2^4 + b_3\beta_3^4, \quad a_5 = b_1\beta_1^5 + b_2\beta_2^5 + b_3\beta_3^5;$$

---

* See Note A.

whence

$$p_0 a_0 + p_1 a_1 + p_2 a_2 + p_3 a_3 = 0,$$

$$p_0 a_1 + p_1 a_2 + p_2 a_3 + p_3 a_4 = 0,$$

$$p_0 a_2 + p_1 a_3 + p_2 a_4 + p_3 a_5 = 0.$$

When these equations are taken in conjunction with the equation

$$p_0 + p_1 x + p_2 x^2 + p_3 x^3 = 0,$$

we have the following equation to determine β_1, β_2, β_3 :—

$$\begin{vmatrix} 1 & x & x^2 & x^3 \\ a_0 & a_1 & a_2 & a_3 \\ a_1 & a_2 & a_3 & a_4 \\ a_2 & a_3 & a_4 & a_5 \end{vmatrix} = 0.$$

Also, b_1, b_2, b_3 are determined by the equations

$$b_1 \quad + b_2 \quad + b_3 \quad = a_0,$$

$$b_1 \beta_1 + b_2 \beta_2 + b_3 \beta_3 = a_1,$$

$$b_1 \beta_1^2 + b_2 \beta_2^2 + b_3 \beta_3^2 = a_2 \,;$$

whence the question is completely solved when β_1, β_2, β_3 are known.

This important transformation of the quintic is a particular case of the following general theorem due to Dr. Sylvester :—

Any homogeneous function of x, y, of the degree $2n - 1$, can be reduced to the form

$$b_1 (x + \beta_1 y)^{2n-1} + b_2 (x + \beta_2 y)^{2n-1} + \ldots + b_n (x + \beta_n y)^{2n-1}$$

by the solution of an equation of the nth degree.

The proof of the general theorem is exactly similar to that above given for the case of the quintic.

177. Quartics Transformable into each other.—We proceed to determine under what conditions two quartics can be transformed, the one into the other, by linear transformation.

Let the quartics be

$$U = (a,\ b,\ c,\ d,\ e)(x,\ y)^4 = a\,(x - \alpha y)(x - \beta y)(x - \gamma y)(x - \delta y),$$

$$V = (a',\ b',\ c',\ d',\ e')(x',\ y')^4 = a'\,(x' - \alpha'y')(x' - \beta'y')(x' - \gamma'y')(x' - \delta'y');$$

and if they become identical by the transformation

$$x' = \lambda x + \mu y, \qquad y' = \lambda'x + \mu'y,$$

we have, by Art. 39,

$$\frac{(\beta' - \gamma')(\alpha' - \delta')}{(\beta - \gamma)(\alpha - \delta)} = \frac{(\gamma' - \alpha')(\beta' - \delta')}{(\gamma - \alpha)(\beta - \delta)} = \frac{(\alpha' - \beta')(\gamma' - \delta')}{(\alpha - \beta)(\gamma - \delta)},$$

showing that the six anharmonic ratios determined by the roots must be the same for both equations.

From these equations we have also the following relations between the invariants of the two forms:—

$$I' = r^4 I, \qquad J' = r^6 J; \tag{1}$$

whence

$$\frac{I'^3}{J'^2} = \frac{I^3}{J^2}. \tag{2}$$

Or, what is termed the *absolute invariant* of the quartics is the same for both.

The conditions expressed by the equations (1), (2), are always *necessary*; but not always *sufficient*, as we proceed to illustrate by two exceptional cases.

Suppose, in the first place,

$$U = u^2 vw, \qquad V = u'^2 v'^2,$$

where $u, v, w, u', v',$ are of the linear form $lx + my$.

Although the condition $\dfrac{I^3}{J^2} = \dfrac{I'^3}{J'^2}$ is satisfied in this case, the common value of these fractions being 27, it is impossible to transform U into V, since it is impossible to make vw a perfect square by linear transformation.

Secondly, if $\qquad U = u^3 v, \qquad V = u'^4;$

although the equations $I' = r^4 I,\ J' = r^6 J$ are satisfied, since $I' = 0,\ I = 0,\ J' = 0,\ J = 0,$ it is, nevertheless, impossible to transform U into V.

In both these cases it would be impossible to identify the six anharmonic ratios depending on the roots of the quartics. And, in general, it is impossible to transform one quantic into another by linear transformation, when any relation exists between the invariants of one of them which does not exist between the invariants of the other (see Clebsch's *Theorie der Binären Algebraischen Formen*, Art. 92).

Miscellaneous Examples.

1. If the coefficients of three quadratics

$$a_1 x^2 + 2b_1 xy + c_1 y^2, \quad a_2 x^2 + 2b_2 xy + c_2 y^2, \quad a_3 x^2 + 2b_3 xy + c_3 y^2$$

be connected by the relation

$$\begin{vmatrix} a_1 & b_1 & c_1 \\ a_2 & b_2 & c_2 \\ a_3 & b_3 & c_3 \end{vmatrix} = 0;$$

prove that they may be reduced by linear transformation to the forms

$$A_1 X^2 + C_1 Y^2, \quad A_2 X^2 + C_2 Y^2, \quad A_3 X^2 + C_3 Y^2.$$

2. Prove that the most general rational transformation of a quartic $f(x)$ may be reduced to the transformation

$$y = \frac{P}{p-x} + \frac{Q}{q-x}.$$

When $P = R f(p) f'(q)$, and $Q = - R f(q) f'(p)$, show that the second term of the transformed quartic is absent

3. Prove that the transformation

$$y = \frac{\alpha x^2 + 2\beta x + \gamma}{\alpha_1 x^2 + 2\beta_1 x + \gamma_1}$$

may be resolved into the three successive transformations—(1) a homographic transformation; (2) a transformation of the roots into their squares; (3) a homographic transformation.

4. If p be any integer, prove that

$$\frac{(x_1{}^p - x_2{}^p)(x_3{}^p - x_4{}^p)}{(x_1 - x_2)(x_3 - x_4)} = \Sigma_0 + (x_1 x_2 + x_3 x_4) \Sigma_1,$$

where Σ_0 and Σ_1 are symmetric functions of x_1, x_2, x_3, x_4.

5. Reduce $(a, b, c, d)(x, y)^3$ to the sum of two cubes by the method of Art. 176.

6. Show how to transform two quadratics to the forms

$$au^2 + bv^2, \quad a'u^2 + b'v^2,$$

where u and v are linear functions of x and y.

7. Prove that two cubics can in general be transformed, the one into the other, by linear transformation, when their discriminants do not vanish; and determine the transformation.

8. Prove that the three roots of a cubic may be expressed as

$$x, \quad \theta(x), \quad \theta^2(x),$$

where

$$\theta(x) = \frac{lx + m}{l'x + m'}, \quad lm' - l'm = 1, \quad l + m' = 1,$$

and $\theta^3(x) = 1$. (Compare Art. 60.)

The meaning of the notation here employed is that $\theta^2(x)$ is derived from $\theta(x)$, and $\theta^3(x)$ from $\theta^2(x)$, in the same way as $\theta(x)$ is from x.

9. If $(a, b, c, d, e)(x, 1)^4$ become $(A, B, C, D, E)(y, 1)^4$ by the transformation

$$y = \frac{a + 2\beta x + \gamma x^2}{a_1 + 2\beta_1 x + \gamma_1 x^2},$$

find the invariants of the latter form.

10. Show that the quartic $U \equiv (a, b, c, d, e)(x, y)^4$ may be reduced to the form

$$U \equiv (\rho_2 - \rho_3)(\xi^4 + \eta^4) + 6\rho_1 \xi^2 \eta^2,$$

where the modulus of transformation is equal to unity.

By Article 164 we have

$$-U \equiv \rho_1 X^2 + \rho_2 Y^2 + \rho_3 Z^2,$$

and making the substitutions

$$X = 2i\xi\eta, \quad Y = i(\xi^2 - \eta^2), \quad Z = \xi^2 + \eta^2, \quad \text{where } i^2 = -1,$$

we obtain the required form. These substitutions are allowable, for ξ and η are proportional to u_1 and u_2 (Art. 162); and, putting for ξ, η the values

$$\xi = m_2 x - m_1 y, \quad \eta = -l_2 x + l_1 y,$$

we find that X, Y, Z have the same discriminant $(l_1 m_2 - l_2 m_1)^2$. Since this value is unity, it is proved that the modulus of the transformation

$$x = l_1 \xi + m_1 \eta, \quad y = l_2 \xi + m_2 \eta,$$

viz., $l_1 m_2 - l_2 m_1$, is equal to unity.

2 B

CHAPTER XVII.

178. Graphic Representation of Imaginary Quantities.—The imaginary expression $a + b\sqrt{-1}$ may be written in the form

$$\mu\left(\cos a + \sin a \sqrt{-1}\right),$$

where

$$\mu = \sqrt{a^2 + b^2}, \quad \text{and} \quad \tan a = \frac{b}{a}.$$

It may be regarded, therefore, as determined by the linear magnitude μ, and the angle a; μ being called the *modulus*, and a the *argument* of the imaginary quantity.

Let rectangular axes OX, OY (fig. 7) be taken; and a point A such that $XOA = a$, and $OA = \mu$. We have then $OM = \mu \cos a = a$, and $AM = \mu \sin a = b$. The expression $a + b\sqrt{-1}$ may therefore be represented graphically by the right line drawn from O to a

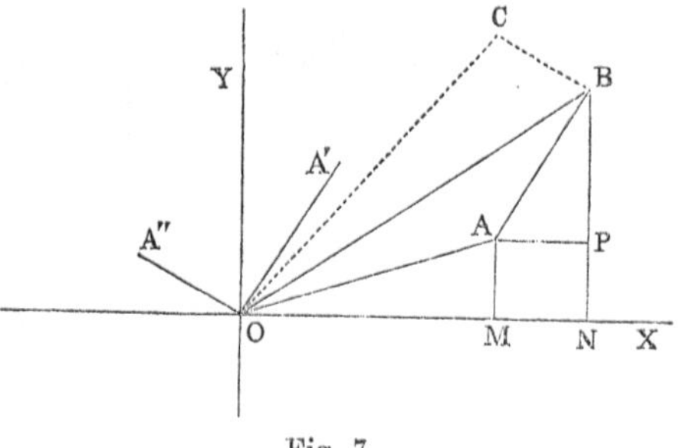

Fig. 7.

point in a plane whose co-ordinates referred to the fixed axes are a, b; the distance OA of this point from the origin being equal to the modulus, and the angle XOA equal to the argument of the imaginary quantity.

The magnitude of an imaginary quantity is estimated by the magnitude of its modulus. When the imaginary quantity vanishes (that is, when a and b separately vanish) its modulus vanishes; and, conversely, when the modulus vanishes, since then $a^2 + b^2 = 0$, a and b must separately vanish, and therefore the imaginary quantity itself. Two imaginary quantities $a + ib$ and $a' + ib'$, are equal when $a = a'$ and $b = b'$, i.e. when the moduli are equal and when the arguments either are equal or differ by a multiple of 2π.

In what follows we shall for brevity represent the modulus and argument of $a + b\sqrt{-1}$ by the notation

$$mod.\ (a + ib), \qquad arg.\ (a + ib),$$

where i as usual represents $\sqrt{-1}$.

179. Addition and Subtraction of Imaginaries.— Let a second imaginary quantity $a' + ib'$ be represented by the right line OA', so that

$$OA' = mod.\ (a' + ib'), \quad XOA' = arg.\ (a' + ib').$$

We proceed to determine the mode of representing the sum

$$a + ib + a' + ib'.$$

Writing this sum in the form $a + a' + i(b + b')$, we observe, in accordance with the convention of Art. 178, that it will be represented by the line drawn from the origin to the point whose co-ordinates are $a + a'$, $b + b'$. To find this point, draw AB parallel and equal to OA'; since AP, BP, are equal to a', b', B is the required point, and we have

$$OB = mod.\ \{a + a' + i(b + b')\}, \quad XOB = arg.\ \{a + a' + i(b + b')\}.$$

To add two imaginary quantities, therefore, we draw OA to represent one of them; and, at its extremity, AB to represent the second (that is, so that its length is equal to the modulus, and the angle it makes with OX equal to the argument, of the second); then OB represents the sum of the two imaginary quantities. Since OB is less than $OA + AB$, it follows that *the modulus of the sum of two imaginary quantities is less than the sum of their moduli.*

This mode of representation may be extended to the addition of any number of imaginary quantities. Thus, to add a third $a'' + ib''$, represented by OA'', we draw BC parallel and equal to OA'', and join OC. Then OC represents the sum of the three imaginary quantities OA, OA', OA''. It is evident also that we may conclude in general that *the modulus of the sum of any number of imaginary quantities is less than the sum of their moduli.*

Subtraction of imaginaries can be represented in a similar way. Since OB represents the sum of OA and OA', OA will represent the difference of OB and OA'. To subtract two imaginary quantities, therefore, we draw at the extremity of the line representing the first a line parallel and equal to the second, but in an opposite direction (*i.e.* a direction which makes with OX an angle greater by π than the argument of the first). We join O to the extremity of this line to find the right line which represents the difference of the two given imaginaries.

180. **Multiplication and Division of Imaginaries.—** To multiply the two imaginary quantities $a + ib$, $a' + ib'$, we write them in the form

$$a + ib = \mu\,(\cos a + i \sin a), \quad a' + ib' = \mu'\,(\cos a' + i \sin a').$$

We have then, by De Moivre's theorem,

$$(a + ib)(a' + ib') = \mu\mu'\{\cos (a + a') + i \sin (a + a')\},$$

which proves that *the product of two imaginary quantities is an imaginary quantity of the same form, whose modulus is the product of the two moduli, and whose argument is the sum of the two arguments.*

In the same way it appears that the product of any number of imaginary factors is an imaginary quantity, whose modulus is the product of all the moduli, and whose argument is the sum of all the arguments.

To divide $a + ib$ by $a' + ib'$, we have similarly

$$\frac{a + ib}{a' + ib'} = \frac{\mu}{\mu'} \{\cos (a - a') + i \sin (a - a')\},$$

which proves that *the quotient of two imaginary quantities is an imaginary quantity of the same form, whose modulus is the quotient of the two moduli, and whose argument is the difference of the two arguments.*

It is evident from the foregoing propositions that any power of an imaginary quantity, e.g., $(a + ib)^m$, can be expressed in the form $A + iB$, where A and B are real quantities. And, more generally, if in any polynomial

$$a_0 z^n + a_1 z^{n-1} + a_2 z^{n-2} + \ldots + a_{n-1} z + a_n,$$

whose coefficients are either real or imaginary quantities, an imaginary quantity $a + ib$ be substituted for the variable z, the result can be expressed in the standard form of imaginary quantities, i. e., $A + iB$.

It was assumed in the proof of the theorem of Art. 16 that when a product of any number of factors (real or imaginary) vanishes, one of the factors must vanish. This is evident when the factors are all real. From what is above proved the same conclusion holds when the factors are imaginary ; for, in order that the modulus of the product may vanish, one of its factors must vanish, and therefore the imaginary quantity of which that factor is the modulus.

181. **The Complex Variable.**—In the earlier Chapters of the present work the variation of a polynomial was studied corresponding to the passage of the variable through real values from $-\infty$ to $+\infty$; and the mode of representing by a figure the form of the polynomial was explained. Such a mode of treatment is only a particular case of a more general inquiry. Given a polynomial

$$f(z) = a_0 z^n + a_1 z^{n-1} + a_2 z^{n-2} + \ldots + a_{n-1} z + a_n,$$

we may study its variations corresponding to the different values of z, where z has the imaginary form $x + iy$, and where x and y both take all possible real values. This form $x + iy$ is called the *complex variable.* All possible *real* values of the variable are of course included in the values of $x + iy$, being those values which arise by varying x and putting $y = 0$. In accordance with the principles of Art. 178 we may represent the imaginary quantity $x + iy$ by the line OP (fig. 8) drawn from a fixed origin O to the point whose co-ordinates are x, y. Or we may say, $x + iy$ is represented by the point P. Thus all possible values of $x + iy$ will be represented by all the points in a plane. Since for any particular value of z, $f(z)$ takes the form $A + iB$ (Art. 180), the values of $f(z)$ may be represented in a similar manner by points in a plane. We confine ourselves in the present Article to the representation of the variable $x + iy$ itself. We conceive the variation of $x + iy$ to take place in a continuous manner; for example, by the motion of the point x, y, along a curve. If OP and OP' represent two consecutive values of the variable, we write the corresponding values $x + iy$, $x' + iy'$, as follows :—

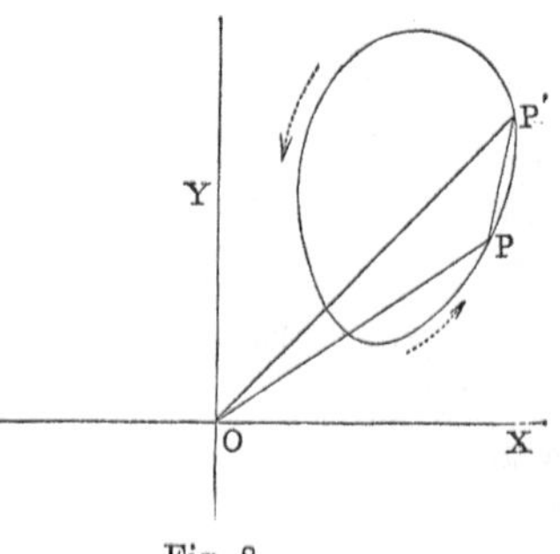

Fig. 8.

$$z = x + iy = r\,(\cos\theta + i\sin\theta), \quad z' = x' + iy' = r'\,(\cos\theta' + i\sin\theta').$$

Since OP' represents the sum of OP and PP' (Art. 179), it follows that PP' represents the imaginary increment of z; and if $z' = z + h$, h may be written in the form

$$h = \rho\,(\cos\phi + i\sin\phi),$$

where $\rho = PP'$, and ϕ is the angle PP' makes with OX.

The variation of the modulus of z is $OP' - OP$ or $r' - r$; the variation of the argument of z is $P'OP$ or $\theta' - \theta$; the variation of z itself is h or $\rho\,(\cos\phi + i\sin\phi)$, as just explained.

Let the point be supposed to describe a closed curve. When it returns to P the modulus takes again its original value; and the argument takes its original value if the point O is exterior to the curve, or is increased by 2π if O is interior to the curve.

If the complex variable describes the same line in two opposite directions, the variations of its argument are equal and of opposite signs, *i.e.* the total variation is nothing. From this we can derive a property of the variation of the argument of the complex variable, which will be found of importance in our succeeding investigations.

Let a plane area be divided into any number of parts by lines *BD, AF, EC,* &c. (fig. 9); then *the variation of the argument relatively to the perimeter of the whole area is equal to the sum of its variations relatively to the perimeters of the partial areas:*

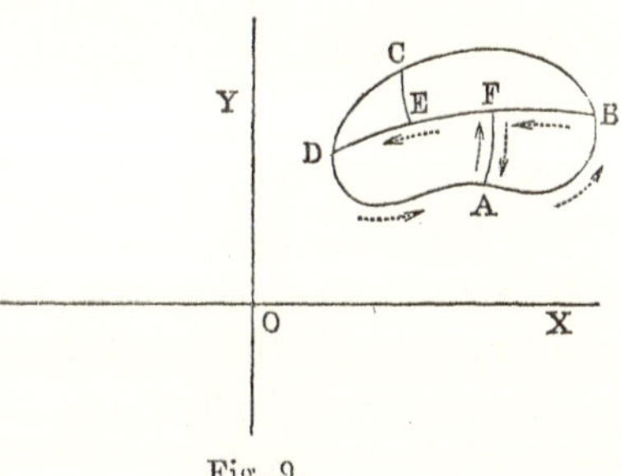

Fig. 9.

all the areas being supposed to be described by the variable moving in the same sense. This is evident; for when the point is made to describe all the partial areas in the same sense, each of the internal dividing lines will be described twice, the two descriptions being in opposite directions; and the external perimeter will be described once; hence the total variation of the argument relatively to the dividing lines vanishes, and the variation relatively to the external perimeter alone remains. In the figure, for example, when the point describes the areas *ABF, AFD* in the sense indicated by the arrows, the total variation relatively to the line *AF* vanishes.

182. Continuity of a Function of the Complex Variable.—Suppose the complex variable z, starting from a fixed value z_0, to receive a small imaginary increment $h = \rho\,(\cos\phi + i\,\sin\phi)$; we have then, if $f(z)$ be the given function,

$$f(z) = f(z_0 + h) = f(z_0) + f'(z_0)\,h + \frac{f''(z_0)}{1 \cdot 2}\,h^2 + \&c.,$$

and the increment of $f(z)$, being equal to $f(z_0 + h) - f(z_0)$, is

$$f'(z_0)\, h + \frac{f''(z_0)}{1 \cdot 2}\, h^2 + \frac{f'''(z_0)}{1 \cdot 2 \cdot 3}\, h^3 + \&\text{c.} \ldots$$

In this expression the coefficients of the powers of h are all imaginary expressions of the usual form; and if their moduli be a, b, c, &c., the moduli of the successive terms are $a\rho$, $b\rho^2$, $c\rho^3$, &c.; and since, by Art. 179, the modulus of a sum is less than the sum of the moduli, it follows that the modulus of the increment of $f(z)$ is less than

$$a\rho + b\rho^2 + c\rho^3 + \&\text{c.}$$

Now a value may be assigned to ρ (Art. 4), such that for it, or any less value of ρ, the value of this expression will be less than any assigned quantity. It follows that to an infinitely small variation of the complex variable corresponds an infinitely small variation of the function ; in other words, *the function varies continuously at the same time as the complex variable itself.*

183. **Variation of the Argument of** $f(z)$ **corresponding to the Description of a small Closed Curve by the Complex Variable.**—Corresponding to a continuous series of values of z we have a continuous series of values of $f(z)$, which can be represented, like the values of z itself, by points in a plane. We represent these series of points by two figures (fig. 10) side

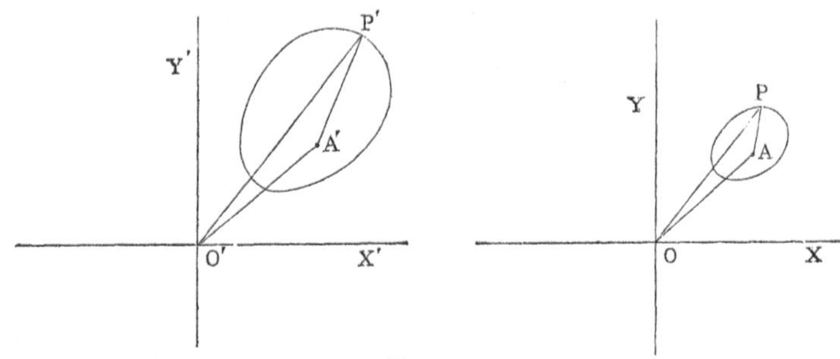

Fig. 10.

by side, which, to avoid confusion, may be supposed to be drawn on different planes. To each point P, representing $x + iy$, corresponds one determinate point P' representing $f(z)$. When P describes a continuous curve, P' describes also a continuous curve; and when P returns to its original position after describing a closed curve, P' returns also to its original position.

Our present object is to discuss the variation of the argument of $f(z)$ corresponding to the description of a small closed curve by P. Let A be any determinate point whose co-ordinates are x_0, y_0, i.e. $z_0 = x_0 + iy_0$. We divide the discussion into two cases :—

(1). When $x_0 + iy_0$ is not a root of $f(z) = 0$, *i.e.* when $f(z_0)$ is different from zero.

(2). When $x_0 + iy_0$ is a root of $f(z) = 0$, or $f(z_0) = 0$.

(1). In the first case, to the point A corresponds a point A' representing the value of $f(z_0)$, and $O'A'$ is different from zero. Let $z = z_0 + h$, where $h = \rho\,(\cos\phi + i\sin\phi)$; and suppose P, which represents z, to describe a small closed curve round A. Let P' represent $f(z)$; then $A'P'$ represents the increment of $f(z)$ corresponding to the increment AP of z. By the previous Article it appears that values so small may be assigned to ρ, that the modulus of the increment of $f(z)$, namely $A'P'$, may be always less than the assigned quantity $O'A'$; hence P may be supposed to describe round A a closed curve so small that the corresponding closed curve described by P' will be exterior to O'. It follows, by Art. 181, that *corresponding to the description by P of a small closed curve, which does not contain a point satisfying the equation $f(z) = 0$, the total variation of the argument of $f(z)$ is nothing.*

(2). In the second case, suppose $x_0 + iy_0$ is a root of the equation $f(z) = 0$ repeated m times, and let

$$f(z) \equiv (z - z_0)^m\,\psi(z)\,;$$

then

$$f(z) = h^m\psi(z) = \rho^m\,(\cos m\phi + i\sin m\phi)\,\psi(z).$$

In this case $O'A' = 0$; and when P describes a closed curve round A, P' describes a closed curve round O', and the argument of $f(z)$ will be increased by a multiple of 2π. To determine this increment, we have from the above equation

$$arg. \; f(z) = m\phi + arg. \; \psi(z) \; ;$$

and the increment of $arg. \; f(z)$ will be obtained by adding the increment of $m\phi$ to the increment of $arg. \; \psi(z)$. Now the latter increment is nothing by (1), since the curve described by P may be supposed to contain no root of $\psi(z) = 0$; and since the increment of ϕ is 2π in one revolution of P, the increment of $m\phi$ is $2m\pi$. It follows *that when P describes a small closed curve containing a root of the equation $f(z) = 0$, repeated m times, the argument of $f(z)$ is increased by $2\,m\pi$.*

184. **Cauchy's Theorem.**—When z describes the same line in a plane in two opposite directions, $f(z)$ describes the corresponding line in its plane in two opposite directions, and the $arg. \; f(z)$ undergoes equal and opposite variations. It follows that if any plane area be divided into parts, as in Art. 181, the variation of the $arg. \; f(z)$, corresponding to the description in the same sense by z of all the partial areas, is equal to the variation of $arg. \; f(z)$ corresponding to the description by z of the external perimeter only. Now let any closed perimeter in the plane XY be described ; and suppose, in the first place, that it contains no point which satisfies the equation $f(z) = 0$. It can be broken up into a number of small areas, with respect to each of which the conclusions of (1) Art. 183 hold ; and by what has been just proved it follows that the variation of $arg. \; f(z)$ corresponding to the description by z of the closed perimeter is nothing. Suppose, in the second place, that the closed perimeter contains a point which is a root of the equation $f(z) = 0$ repeated m times. Let a small closed curve $PQRS$ be described round this point.

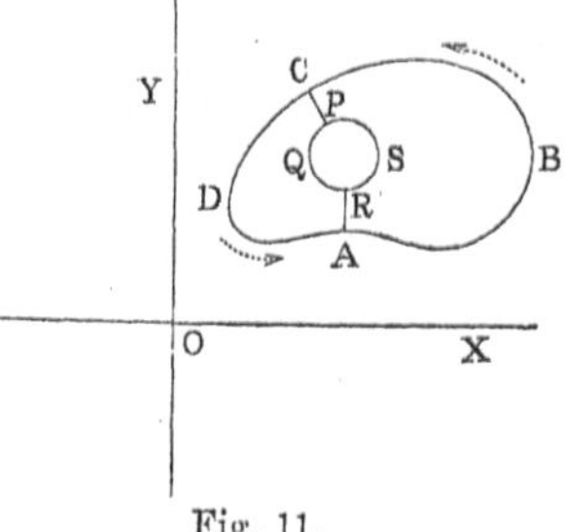

Fig. 11.

The variation of $arg. \; f(z)$ corresponding to the description by z of the whole perimeter, is equal to the sum of its

variations corresponding to the description of the areas $ABCPSR$, $CDARQP$, $PQRS$. The two former variations vanish by what is above proved; and the latter is, by (2), Art. 183, equal to $2\,m\pi$. The total variation, therefore, of $f(z)$ is $2\,m\pi$. Similarly, if the area includes a second, third, &c., points which represent roots repeated m', m'', &c., times, the total variation $= 2\,(m + m' + m'' + \&\text{c.})\,\pi$. Hence we derive the following theorem due to Cauchy :—

The number of roots of any polynomial, comprised within a given plane area, is obtained by dividing by 2π the total variation of the argument of this polynomial, corresponding to the complete description by the complex variable of the perimeter of the area.

185. Number of Roots of the General Equation.— We are enabled by means of the principles established in the preceding Articles to prove the theorem contained in Arts. 15 and 16 ; namely, *every rational and integral equation of the nth degree has n roots real or imaginary.*

In the former Articles the reasons were given why the proof of this theorem, which may be regarded as the fundamental theorem of the Theory of Equations, was deferred.

Let

$$f(z) \equiv a_0 z^n + a_1 z^{n-1} + a_2 z^{n-2} + \ldots + a_{n-1} z + a_n$$

be a rational and integral function of z. Without making any supposition as to the existence of roots of $f(z) = 0$ further than that $f(z)$ cannot vanish for any infinite values of the variable, we can suppose z to describe in its plane a circle so large that no root exists outside of it. If, then,

$$f(z) = z^n \{a_0 + a_1 z' + a_2 z'^2 + \ldots + a_n z'^n\}$$

$$= z^n \phi(z'), \text{ where } z' = \frac{1}{z},$$

z', whose modulus is the reciprocal of the modulus of z, will describe a small circle containing a portion of the plane corresponding to the part outside of the circle described by z; and no root of $\phi(z') = 0$ will be included within this small circle.

Hence, corresponding to the description of the whole circle by z, the variation of *arg.* $\phi(z') = 0$, and, therefore,

$$\text{variation of } arg. \ f(z) = \text{variation of } arg. \ z^n;$$

and if

$$z = r\left(\cos\theta + i\sin\theta\right), \quad \text{or} \quad z^n = r^n\left(\cos n\theta + i\sin n\theta\right),$$

θ is increased by 2π, and, therefore, *arg.* z^n is increased by $2\,n\pi$.

It follows from Cauchy's Theorem, Art. 184, that the number of roots comprised within the circle described by z, i.e. the total number of roots of the equation $f(z) = 0$ is n; and the theorem is proved.*

----

* Up to the period of Lagrange it appears to have been taken for granted that every equation must have a root. In his *Traité de la Résolution des Équations Numériques,* note X., Lagrange has an investigation, the object of which is to prove *à priori* the possibility of decomposing any polynomial into real factors of the first or second degree. Gauss also gave a proof of the proposition. Cauchy occupied himself with the problem, and gave two forms of demonstration, one of which we have followed in the text. A simple investigation of this problem will be found in a Paper by Mr. John C. Malet, M.A., "On a Proof that every Algebraic Equation has a Root," *Trans. of the Royal Irish Academy,* vol. xxvi. p. 453.

NOTES.

———◆———

NOTE A.

ALGEBRAIC SOLUTION OF EQUATIONS.

THE solution of the quadratic equation was known to the Arabians, and is found in
the works of Mohammed Ben Musa and other writers published in the ninth century.
In a treatise on Algebra by Omar Alkhayyami, which belongs probably to the middle
of the eleventh century, is found a classification of cubic equations, with methods
of geometrical construction; but no attempt at a general solution. The study of
Algebra was introduced into Italy from the Arabian writers by Leonardo of Pisa
early in the thirteenth century; and for a long period the Italians were the chief
cultivators of the science. A work, styled *L'Arte Maggiore*, by Lucas Paciolus
(known as Lucas de Burgo) was published in 1494. This writer adopts the Arabic
classification of cubic equations, and pronounces their solution to be as impossible
in the existing state of the science as the quadrature of the circle. At the same
time he signalizes this solution as the problem to which the attention of mathematicians should be next directed in the development of the science. The solution of
the equation $x^3 + mx = n$ was effected by Scipio Ferreo; but nothing more is
known of his discovery than that he imparted it to his pupil Florido in the year
1505. The attention of Tartaglia was directed to the problem in the year 1530, in
consequence of a question proposed to him by Colla, whose solution depended on
that of a cubic of the form $x^3 + px^2 = q$. Florido, learning that Tartaglia had obtained a solution of this equation, proclaimed his own knowledge of the solution of
the form $x^3 + mx = n$. Tartaglia, doubting the truth of his statement, challenged
him to a disputation in the year 1535; and in the mean time himself discovered the
solution of Ferreo's form $x^3 + mx = n$. This solution depends on assuming for x
an expression $\sqrt[3]{t} - \sqrt[3]{u}$ consisting of the difference of two radicals; and, in fact,
constitutes the solution usually known as Cardan's. Tartaglia continued his labours,
and discovered rules for the solution of the various forms of cubics included under the
classification of the Arabic writers. Cardan, anxious to obtain a knowledge of these
rules, applied to Tartaglia in the year 1539; but without success. After many
solicitations Tartaglia imparted to him a knowledge of these rules; receiving from

him, however, the most solemn and sacred promises of secrecy. Regardless of his promises, Cardan published in 1545 Tartaglia's rules in his great work styled *Ars Magna*. It had been the intention of Tartaglia to publish his rules in a work of his own. He commenced the publication of this work in 1556; but died in 1559, before he had reached the consideration of cubic equations. As his work, therefore, contained no mention of his own rules, these rules came in process of time to be regarded as the discovery of Cardan, and to be called by his name.

The solution of equations of the fourth degree was the next problem to engage the attention of algebraists; and here, as well as in the case of the cubic, the impulse was given by Colla, who proposed to the learned the solution of the equation $x^4 + 6x^2 + 36 = 60x$. Cardan appears to have made attempts to obtain a formula for equations of this kind; but the discovery was reserved for his pupil Ferrari. The method employed by Ferrari was the introduction of a new variable, in such a way as to make both sides of the equation perfect squares; this variable itself being determined by an equation of the third degree. It is, in fact, virtually the method of Art. 63. This solution is sometimes ascribed to Bombelli, who published it in his treatise on Algebra, in 1579. The solution known as Simpson's, which was published much later (about 1740), is in no respect essentially different from that of Ferrari. In the year 1637 appeared Descartes' treatise, in which are found many improvements in algebraical science, the chief of which are his recognition of the negative and imaginary roots of equations, and his "Rule of Signs." His expression of the biquadratic as the product of two quadratic factors, although deducible immediately from Ferrari's form, was an important contribution to the study of this quantic. Euler's algebra was published in 1770. His solution of the biquadratic (see Art. 61) is important, inasmuch as it brings the treatment of this form into harmony with that of the cubic by means of the assumed irrational form of the root. The methods of Descartes and Euler were the result of attempts made to obtain a general algebraic solution of equations. Throughout the eighteenth century many mathematicians occupied themselves with this problem; but their labours were unsuccessful in the case of equations of a degree higher than the fourth.

In the solutions of the cubic and biquadratic obtained by the older analysts we observe two distinct methods in operation : the first, illustrated by the assumptions of Tartaglia and Euler, proceeding from an assumed explicit irrational form of the root; the other, seeking by the aid of a transformation of the given function, to change its factorial character, so as to reduce it to a form readily resolvible. In Art. 55 these two methods are illustrated ; together with a third, the conception of which is to be traced to Vandermonde and Lagrange, who published their researches about the same time, in the years 1770 and 1771. The former of these writers was the first to indicate clearly the necessary character of an algebraical solution of any equation, viz., that it must, by the combination of radical signs involved in it, represent any root indifferently when the symmetric functions of the roots are substituted for the functions of the coefficients involved in the formula (see Art. 94). His attempts to construct formulas of this character were successful in the cases of the cubic and biquadratic ; but failed in the case of the quintic. Lagrange undertook a review of the labours of his predecessors in the direction of the general solution of equations, and traced all their results to one uniform principle. This principle consists in reducing the solution of the given equation to that of an equation of lower degree, whose roots are linear functions of the roots of the given equation and of the roots of unity. He shows also that the reduction of a quintic cannot be effected in this way, the equation on which its solution depends being of the sixth degree.

All attempts at the solution of equations of the fifth degree having failed, it was natural that mathematicians should inquire whether any such solution was possible at all. Demonstrations have been given by Abel and Wantzel (see Serret's *Cours d'Algèbre Supérieure*, Art. 516) of the impossibility of resolving algebraically equations unrestricted in form, of a degree higher than the fourth. A transcendental solution, however, of the quintic has been given by M. Hermite, in a form involving elliptic integrals. Among other contributions to the discussion of the quintic since the researches of Lagrange, one of leading importance is its expression in a trinomial form by means of the Tschirnhausen transformation (see Art. 175). Tschirnhausen himself had succeeded in the year 1683, by means of the assumption $y = P + Qx + x^2$, in the reduction of the cubic and quartic, and had imagined that a similar process might be applied to the general equation. The reduction of the quintic to the trinomial form was published by Mr. Jerrard in his *Mathematical Researches*, 1832–1835; and has been pronounced by M. Hermite to be the most important advance in the discussion of this quantic since Abel's demonstration of the impossibility of its solution by radicals. In a Paper published by the Rev. Robert Harley in the *Quarterly Journal of Mathematics*, vol. vi. p. 38, it is shown that this reduction had been previously effected, in 1786, by a Swedish mathematician named Bring. Of equal importance with Bring's reduction is Dr. Sylvester's transformation (Art. 176), by means of which the quintic is expressed as the sum of three fifth powers, a form which gives great facility to the treatment of this quantic. Other contributions which have been made in recent years towards the discussion of quantics of the fifth and higher degrees have reference chiefly to the invariants and covariants of these forms. For an account of these researches the student is referred to Clebsch's *Theorie der Binären Algebraischen Formen*, and to Salmon's *Lessons Introductory to the Modern Higher Algebra*.

There has also grown up in recent years a very wide field of investigation relative to the algebraic solution of equations, known as the "Theory of Substitutions." This theory arose out of the researches of Lagrange before referred to, and has received large additions from the labours of Cauchy, Abel, Galois, and other writers. Although many important results have been arrived at by these investigators, the subject is of such vast extent and difficulty that it must be considered as only in its infancy as yet. The reader desirous of information on this subject is referred to Serret's *Cours d'Algèbre Supérieure*, and to the *Traité des Substitutions et des Equations Algebriques*, by M. Camille Jordan.

NOTE B.

SOLUTION OF NUMERICAL EQUATIONS.

The first attempt at a general solution by approximation of numerical equations was published in the year 1600, by Vieta. Cardan had previously applied the rule of "false position" (called by him "regula aurea") to the cubic; but the results obtained by this method were of little value. It occurred to Vieta that a particular numerical root of a given equation might be obtained by a process analogous to the ordinary processes of extraction of square and cube roots; and he inquired in what way these known processes should be modified in order to afford a root of an equation whose coefficients are given numbers. Taking the equation $f(x) = Q$, where Q is a given number, and $f(x)$ a polynomial containing different powers of x, with numerical coefficients, Vieta showed that, by substituting in $f(x)$ a known approximate value of the root, another figure of the root (expressed as a decimal) might be obtained by division. When this value was obtained, a repetition of the process furnished the next figure of the root ; and so on. It will be observed that the principle of this method is identical with the main principle involved in the methods of approximation of Newton and Horner (Arts. 100, 101). All that has been added since Vieta's time to this mode of solution of numerical equations is the arrangement of the calculation so as to afford facility and security in the process of evolution of the root. How great has been the improvement in this respect may be judged of by an observation in Montucla's *Histoire des Mathématiques*, vol. i. p. 603, where, speaking of Vieta's mode of approximation, the author regards the calculation (performed by Wallis) of the root of a biquadratic to eleven decimal places as a work of the most extravagant labour. The same calculation can now be conducted with great ease by anyone who has mastered Horner's process explained in the text.

Newton's method of approximation was published in 1669 ; but before this period the method of Vieta had been employed and simplified by Harriot, Oughtred, Pell, and others. After the period of Newton, Simpson and the Bernouillis occupied themselves with the same problem. Daniel Bernouilli expressed a root of an equation in the form of a recurring series, and a similar expression was given by Euler ; but both these methods of solution have been shown by Lagrange to be in no respect essentially different from Newton's solution (*Traité de la Résolution des Equations Numériques*). Up to the period of Lagrange, therefore, there was in existence only one distinct method of approximation to the root of a numerical equation ; and this method, as finally perfected by Horner, in 1819, remains at the present time the best practical method yet discovered for this purpose.

Lagrange, in the work above referred to, pointed out the defects in the methods of Vieta and Newton. With reference to the former he observed that it required too many trials ; and that it could not be depended on, except when all the terms on the left-hand side of the equation $f(x) = Q$ were positive. As defects in Newton's method he signalized—first, its failure to give a commensurable root in finite terms ;

secondly, the insecurity of the process which leaves doubtful the exactness of each fresh correction ; and lastly, the failure of the method in the case of an equation with roots nearly equal. The problem Lagrange proposed to himself was the following :—" Etant donnée une équation numérique sans aucune notion préalable de la grandeur ni de l'espèce de ses racines, trouver la valeur numérique exacte, s'il est possible, ou aussi approchée qu'on voudra de chacune de ses racines."

Before giving an account of his attempted solution of this problem, it is necessary to review what had been already done in this direction, in addition to the methods of approximation above described. Harriot discovered in 1631 the composition of an equation as a product of factors, and the relations between the roots and coefficients. Vieta had already observed this relation in the case of a cubic ; but he failed to draw the conclusion in its generality, as Harriot did. This discovery was important, for it led to the observation that any integral root must be a factor of the absolute term of an equation, and Newton's Method of Divisors for the determination of such roots was a natural result. Attention was next directed towards finding limits of the roots, in order to diminish the labour necessary in applying the method of divisors as well as the methods of approximation previously in existence. Descartes, as already remarked, was the first to recognise the negative and imaginary roots of equations; and the inquiry commenced by him as to the determination of the number of real and of imaginary roots of any given equation was continued by Newton, Stirling, De Gua, and others.

Lagrange observed that, in order to arrive at a solution of the problem above stated, it was first necessary to determine the number of the real roots of the given equation, and to separate them one from another. For this purpose he proposed to employ the equation whose roots are the squares of the differences of the roots of the given equation. Waring had previously, in 1762, indicated this method of separating the roots; but Lagrange observes (*Equations Numériques*, Note iii.), that he was not aware of Waring's researches when he composed his own memoir on this subject. It is evident that when the equation of differences is formed, it is possible, by finding an inferior limit to its positive roots, to obtain a number less than the least difference of the real roots of the given equation. By substituting in succession numbers differing by this quantity, the real roots of the given equation will be separated. When the roots are separated in this way Lagrange proposed to determine each of them by the method of continued fractions, explained in the text (Art. 105). This mode of obtaining the roots escapes the objections above stated to Newton's method, inasmuch as the amount of error in each successive approximation is known ; and when the root is commensurable the process ceases of itself, and the root is given in a finite form. Lagrange gave methods also of obtaining the imaginary roots of equations, and observed that if the equation had equal roots they could be obtained in the first instance by methods already in existence (see Art. 74).

Theoretically, therefore, Lagrange's solution of the problem which he proposed to himself is perfect. As a practical method, however, it is almost useless. The formation of the equation of differences for equations of even the fourth degree is very laborious, and for equations of higher degrees becomes well nigh impracticable. Even if the more convenient modes of separating the roots discovered since Lagrange's time be taken in conjunction with the rest of his process, still this process is open to the objection that it gives the root in the form of a continued fraction, and that the labour of obtaining it in this form is greater than the corresponding labour of obtaining it by Horner's process in the form of a decimal. It will be observed also that the latter process, in the perfected form to which Horner has brought it, is free from all the objections to Newton's method above stated.

Since the period of Lagrange, the most important contributions to the analysis of numerical equations, in addition to Horner's improvement of the method of approximation of Vieta and Newton, are those of Fourier, Budan, and Sturm. The researches of Budan were published in 1807 ; and those of Fourier in 1831, after his death. There is no doubt, however, that Fourier had discovered before the publication of Budan's work the theorem which is ascribed to them conjointly in the text. The researches of Sturm were published in 1835. The methods of separation of the roots proposed by these writers are fully explained in Chapter IX. By a combination of these methods with that of Horner we have now a solution of Lagrange's problem far simpler than that proposed by Lagrange himself. And it appears impossible to reach mnch greater simplicity in this direction. In extracting a root of an equation, just as in extracting an ordinary square or cube root, labour cannot be avoided ; and Horner's process appears to reduce this labour to a minimum. The separation of the roots also, especially when two or more are nearly equal, must remain a work of more or less labour. This labour may admit of some reduction by the consideration of the functions of the coefficients which play so important a part in the theory of the different quantics. If, for example, the functions H, I, and J, are calculated for a given quartic, it will be possible at once to tell the character of the roots (see Art. 93). Mathematicians may also invent in process of time some mode of calculation applicable to numerical equations analogous to the logarithmic calculation of simple roots. But at the present time the most perfect solution of Lagrange's problem is to be sought in a combination of the methods of Sturm and Horner.

NOTE C.

DETERMINANTS.

The name "determinant" was introduced by Cauchy, as well as the notation in ordinary use (Art. 107) to represent these functions. Although Leibnitz in 1693 had observed the peculiarity of the functions which arise from the solution of linear equations, no further advance in this direction took place till Cramer in 1750 was led to the study of such functions in connexion with the analysis of curves. During the latter period of the eighteenth century the subject was further enlarged by the labours of Bezout, Laplace, Vandermonde, and Lagrange. These labours were continued in the present century by Gauss and Cauchy; to the former of whom is due the proposition that the product of two determinants is itself a determinant. A great impulse was given to the study of these expressions by the writings of Jacobi in Crelle's *Journal*, and by his memoirs published in 1841. Among more recent mathematicians who have advanced this subject may be mentioned Hermite, Hesse, Joachimstal, Cayley, Sylvester, and Salmon. There is now no department of mathematics, pure or applied, in which the employment of this calculus is not of great assistance, not only furnishing brevity and elegance in the demonstration of known properties, but even leading to new discoveries in mathematical science. Among recent works which have rendered this subject accessible to students may be mentioned Spottiswoode's *Elementary Theorems relating to Determinants*, London, 1851; Brioschi's *La teorica dei determinanti*, Pavia, 1854; Baltzer's *Theorie und Anwendung der determinanten*, Leipzig, 1864; Dostor's *Éléments de la théorie des Déterminants*, Paris, 1877; Scott's *Theory of Determinants*, Cambridge, 1880; and the chapters in Salmon's *Lessons introductory to the Modern Higher Algebra*, Dublin, 1876. For further information on the history of this subject, as well as on that of Eliminants, Invariants, Covariants, and Linear Transformations, the reader is referred to the notes at the end of the work last mentioned.

THE END.

DUBLIN UNIVERSITY PRESS SERIES.

THE Provost and Senior Fellows of Trinity College have undertaken the publication of a Series of Works, chiefly Educational, to be entitled the DUBLIN UNIVERSITY PRESS SERIES.

The following volumes of the Series are now ready, viz. :—

Six Lectures on Physical Geography. By the REV. S. HAUGHTON, M.D., Dubl., D.C.L., Oxon., F.R.S., *Fellow of Trinity College, and Professor of Geology in the University of Dublin.*

An Introduction to the Systematic Zoology and Morphology of Vertebrate Animals. By ALEXANDER MACALISTER, M.D., Dubl., *Professor of Comparative Anatomy in the University of Dublin.*

The Codex Rescriptus Dublinensis of St. Matthew's Gospel (Z). First Published by Dr. Barrett in 1801. A New Edition, Revised and Augmented. Also, Fragments of the Book of Isaiah, in the LXX. Version, from an Ancient Palimpsest, now first Published. Together with a newly discovered Fragment of the Codex Palatinus. By T. K. ABBOTT, B.D., *Fellow of Trinity College, and Professor of Biblical Greek in the University of Dublin.* With two Plates of Facsimiles.

The Parabola, Ellipse, and Hyperbola, treated Geometrically. By ROBERT WILLIAM GRIFFIN, A.M., LL.D., *Ex-Scholar, Trinity College, Dublin.*

An Introduction to Logic. By WILLIAM HENRY STANLEY MONCK, M.A. *Professor of Moral Philosophy in the University of Dublin.*

Essays in Political and Moral Philosophy. By T. E. CLIFFE LESLIE, Hon. LL.D., Dubl., *of Lincoln's Inn, Barrister-at-Law, late Examiner in Political Economy in the University of London, Professor of Jurisprudence and Political Economy in the Queen's University.*

The Correspondence of Cicero: a revised Text, with Notes and Prolegomena.—Vol. I., The Letters to the end of Cicero's Exile. By ROBERT Y. TYRRELL, M.A., *Fellow of Trinity College, and Professor of Latin in the University of Dublin.*

Faust from the German of Goethe. By THOMAS E. WEBB, LL.D., Q.C., *Regius Professor of Laws, and Public Orator in the University of Dublin.*

[Over.

The Correspondence of Robert Southey with Caroline Bowles: to which are added—Correspondence with Shelley, and Southey's Dreams. Edited, with an Introduction, by EDWARD DOWDEN, LL.D., *Professor of English Literature in the University of Dublin.*

The Mathematical and other Tracts of the late James M'Cullagh, F.T.C.D., *Professor of Natural Philosophy in the University of Dublin.* Now first collected, and edited by REV. J. H. JELLETT, B.D., and REV. SAMUEL HAUGHTON, M.D., *Fellows of Trinity College, Dublin.*

A Sequel to the First Six Books of the Elements of Euclid, containing an Easy Introduction to Modern Geometry. With numerous Examples. By JOHN CASEY, LL.D., F.R.S., *Vice-President, Royal Irish Academy ; Member of the London Mathematical Society ; and Professor of the Higher Mathematics and Mathematical Physics in the Catholic University of Ireland.*

Theory of Equations : with an Introduction to the Theory of Binary Algebraic Forms. By WILLIAM SNOW BURNSIDE, M.A., *Erasmus Smith's Professor of Mathematics in the University of Dublin :* and ARTHUR WILLIAM PANTON, M.A., *Fellow and Tutor, Trinity College, Dublin.*

In the Press :—

Evangelia Antehieronymiana ex Codice vetusto Dublinensi. Ed. T. K. ABBOTT, B.D.

The Veil of Isis ; or, Idealism. By THOMAS E. WEBB, LL.D., Q.C., *Regius Professor of Laws, and Public Orator in the University of Dublin.*

DUBLIN : HODGES, FIGGIS, AND CO.

LONDON : LONGMANS, GREEN, AND CO.